Synthesis of β-Lactam antibiotics

Coördination and production
DSM Communication Support Center

Layout
Buro Chris Peeters, Oirsbeek

Corrections
Gerry Ariaans

Lithography and printing
PlantijnCasparie, Eindhoven

ISBN 978-94-010-3851-5 ISBN 978-94-010-0850-1 (eBook)
DOI 10.1007/978-94-010-0850-1

Originally published by Kluwer Academic Publishers in 2001
Softcover reprint of the hardcover 1st edition 2001

No part of the material protected by this copyrighted notice may be reproduced or utilized in any form or by any means, electronic or mechanical,

Synthesis of β-Lactam antibiotics

Chemistry, Biocatalysis & Process Integration

Edited by
Alle Bruggink
DSM Research, Life Sciences R&D
Geleen, The Netherlands
Nijmegen University, Department of Organic Chemistry
Nijmegen, The Netherlands

Springer-Science+Business Media, B.V.

List of Contributors

W.B.L. Alkema[8]
H.H. Beeftink[7]
A. Bruggink[1,3]
L. Cao[5]
B.W. Dijkstra[9]
F.J. Dommerholt[3]
R. de Gelder[4]
W.M. van Gulik[6]
J.J. Heijnen[6]
C.M.H. Hensgens[9]
A.E.M. Janssen[7]
D.B. Janssen[8]
M.H.A. Janssen[5]
R. Keltjens[3]
G.J. Kemperman[3]
A.J.H. Klunder[3]
L.M. van Langen[5]
L.P. Ooijkaas[7]
M.Ottens[6]
J.J. Polderman[8]
F. van Rantwijk[5]
J.L. van Roon[7]
P.D. Roy[2]
C.G.P.H. Schroën[7]
R.A. Sheldon[5]
A.J.J. Straathof[6]
M. Strubel[7]
G.T.M. Titulaer[3]
J. Tramper[7]
E.J. de Vries[8]
M.A. Wegman[5]
L.A.M. van der Wielen[6]
W.A. van Winden[6]
J. Zhu[3]
B. Zwanenburg[3]

1 DSM Research, Life Sciences R&D, P.O Box 18, 6160 MD Geleen, The Netherlands.
2 DSM Fine Chemicals, P.O. Box 81, 5900 AB Venlo, The Netherlands.
3 University of Nijmegen, NSR Institute for Molecular Structure, Design and Synthesis, Department of Organic Chemistry, Toernooiveld 1, 6525 ED Nijmegen, The Netherlands.
4 University of Nijmegen, NSR Institute for Molecular Structure, Design and Synthesis, Department of Inorganic Chemistry, Toernooiveld 1, 6525 ED Nijmegen, The Netherlands.
5 Delft University of Technology, Biocatalysis and Organic Chemistry, Julianalaan 136, 2628 BL Delft, The Netherlands.
6 Delft University of Technology, Kluyver Laboratory for Biotechnology, Julianalaan 67, 2628 BC Delft, The Netherlands.
7 Wageningen University, Food and Bioprocess Engineering Group, P.O. Box 8129, 6700 EV Wageningen, The Netherlands.
8 Department of Biochemistry, BIOSON Research Institute, Groningen Biomolecular Sciences and Biotechnology Institute, University of Groningen, Nijenborgh 4, 9747 AG Groningen, The Netherlands.
9 Laboratory of Biophysical Chemistry, BIOSON Research Institute, Groningen Biomolecular Sciences and Biotechnology Institute, University of Groningen, Nijenborgh 4, 9747 AG Groningen, The Netherlands.

Contents

List of Contributors ***IV***

Foreword ***VII***

Preface ***IX***

Acknowledgements ***XI***

Liability ***XI***

Chapter I **Industrial Synthesis of Semisynthetic Antibiotics.** ***12***
A. Bruggink and P.D. Roy.

Chapter II **Molecular Precision in the Chemistry of Cephalosporin type Antibiotics.** ***56***
B. Zwanenburg, G.J. Kemperman, J. Zhu, R. de Gelder, F.J. Dommerholt, G.T.M. Titulaer, R. Keltjens and A.J.H. Klunder.

Chapter III **Biocatalysts and Biocatalysis in the Synthesis of β–Lactam Antibiotics.** ***102***
R.A. Sheldon, F. van Rantwijk, L.M. van Langen, M.A. Wegman, L. Cao and M.H.A. Janssen.

Chapter IV **Process Technology and Process Integration in the Preparation of Penicillins.** ***150***
L.A.M. van der Wielen, M.Ottens and A.J.J. Straathof.

Chapter V **Biocatalytic Production of Semi-Synthetic Cephalosporins: Process Technology and Integration.** ***206***
J. Tramper, H.H. Beeftink, A.E.M. Janssen, L.P. Ooijkaas, J.L. van Roon, M. Strubel and C.G.P.H. Schroën.

Chapter VI **Engineering Enzymes for the Synthesis of Semi-Synthetic Antibiotics.** ***250***
W.B.L. Alkema, E.J. de Vries, C.M.H. Hensgens, J.J. Polderman, B.W. Dijkstra and D.B. Janssen.

Chapter VII **Modeling the Metabolism of Penicillin-G Formation.** ***280***
W.M. van Gulik, W.A. van Winden and J.J. Heijnen.

Foreword

Antibiotics, particularly β-lactam antibiotics such as penicillins and cephalosporins have revolutionized the treatment of infections and are with us for more than 50 years. Despite the fact that most of the antibiotics that are used today are older than 25 years the processes that are used in their production are still subject to very significant improvement.

New developments in technology offer room for process improvements that often combine sustainable technology and economic benefits.

In the present book a summary is presented of the results that have been obtained in the so-called 'cluster project'. It started as a collaboration between three companies, DSM, Gist-brocades and their joint-venture Chemferm, and six academic groups at four Dutch Universities. During the project the number of companies involved reduced to one by the merger of DSM and Gist-brocades. The project was co-financed by the Dutch Ministry of Economic Affairs.

From our perspective the project has developed into a prime example of fruitful collaboration between academia and industry. In the book many examples are highlighted of scientific challenge in concert with industrial relevance. By the nature of the research that was performed the economic relevance of the results obtained is not made completely clear yet, as this has to await their progressive development into industrial practice.

The book describes a truly multidisciplinary effort and covers a range of technologies from process technology (engineering at a macroscopic scale) to protein engineering (engineering at a molecular scale).

From the perspective of our position in the research management of DSM we want to congratulate the participants in the project (almost 100 scientists have been involved, part-time or full-time) with their achievements, and thank them for their contribution to DSM's competitiveness in the world of antibiotics.

Our thanks also extend to the Ministry of Economic Affairs for its contribution to the financing of our efforts.

We have experienced the project as exciting both from our industrial and our scientific perspective. We invite you, as a reader, to share our excitement.

Emmo Meijer
Joop Roels

Preface

This book is the result of five years of intensive cooperation between DSM and six academic research groups in the Netherlands. Funded by government subsidies and by participating industries and academia, a multi-disciplinary team of experts has studied the scope and limitations of biocatalytic routes to the industrially most relevant penicillins and cephalosporins. These studies represent an excellent example of how traditional (stoïchiometric organic) synthesis and classical fermentations have evolved into modern (bio-) catalysis and biosynthesis based on insights into metabolic pathways and enzyme actions. The need for a multi-disciplinary approach including integration with process technology and reactor design is also clearly demonstrated.

The results of this teamwork, summarized in this book and published in over 100 scientific papers and several patents and patent applications, represent guidelines for the technological future of DSM Anti Infectives within the context of its business development. The present and future research themes of the participating academic groups are also greatly influenced by the results of this cooperation. Further fruitful collaborations are to be expected.

May, 2001,
Alle Bruggink

Acknowledgements

We would like to express our thanks to all of the contributors of this book. Not all the students, PhD's, postdocs and technical staff can be mentioned, though without their enthusiastic contributions to these studies this publication would not have been possible. Also, the numerous people in DSM Research, DSM Anti Infectives and DSM Fine Chemicals have been indispensable in achieving the results reported in this book, and their translation into useful industrial applications. We remember with great pleasure our frequent progress meetings, in particular our Vaalsbroek conferences. The lessons that we have learned on how industry and university can collaborate fruitfully in a multi-disciplinary team will remain with us throughout our professional careers.
We greatly acknowledge the motivating trust of the DSM management in developing and effecting this challenging project. In addition, the Dutch Ministry of Economic Affairs and the Senter department are greatly acknowledged for their enthusiastic efforts towards starting this unique collaboration and for their ensuing financial support.

Liability

Chapter I Industrial Synthesis of Semisynthetic Antibiotics *13*

§1 The Early Days (1950 - 1970) ***13***
1.1 Development of semi-synthetic antibiotics *13*
1.2 The discovery of 6-APA, 7-ACA and 7-ADCA *14*
1.3 Pre-industrial Processes *16*

§2 First Generation of Industrial Processes ***17***
2.1 The largest products *17*
2.2 Life cycle of Ampicillin processes *21*
2.3 Development of a Phenylglycine product line *26*
2.4 Industrial Processes for Amoxicillin *30*
2.5 Development of p-Hydroxyphenylglycine *31*
2.6 Processes for Cephalosporins *34*
2.7 Market versus Process Development *39*

§3 Second Generation of Industrial Processes ***45***
3.1 Introduction of Biocatalysis *45*
3.2 Enzymatic couplings with 6-APA and 7-ADCA *46*
3.3 First steps towards (Process) Integrations *48*

§4 Third Generation Processes? ***50***
4.1 Marketing prospects *50*
4.2 Process Prospects *52*

§5 Further Reading ***54***
5.1 Books *54*
5.2 Reviews *54*

Chapter I Industrial Synthesis of Semisynthetic Antibiotics

ALLE BRUGGINK (DSM RESEARCH, NIJMEGEN UNIVERSITY) AND PETER D. ROY (DSM FINE CHEMICALS)

§1 The Early Days (1950 - 1970)

1.1 Development of semi-synthetic antibiotics

Even in the early days of investigations on penicillin synthesis by direct fermentation, the importance of incorporating different side chains for antibiotic activity was evident. Addition of various mono-substituted acetic acids to the Penicillium fermentation medium gave the corresponding new penicillins with varying activity profiles. Notably, such studies led to phenoxymethyl penicillin (Pen V), which is acid stable and orally active. Up to the late 50's, only two of the many penicillins produced by fermentation, namely Pen G and Pen V, proved to be clinically useful, though indeed these have stood the test of time. Further progress here was limited, as a result of the selective biochemical incorporation of side-chains by the Penicillium fungus (restricted mainly to aliphatic or aryl aliphatic carboxylic acids).

In the mid 1950's these fermentative penicillin's were seen as excellent antibiotics, though broad clinical use exposed their inactivation by bacterial penicillinase and the lack of a wider spectrum of activity, particularly against gram negative bacteria. There were many groups working on this so-called 'penicillin problem'. Essentially what was needed here were more significant structural modifications in the N-acyl side chains.

One of the Beecham lines of research was to modify the aromatic ring of the phenylacetic acid side chain. When 6-APA became available, this line of work led to Methicillin (2,6-dimethoxy benzoyl side chain) which was the first of the semi-synthetic penicillins to be clinically effective against staphylococcal β-lactamase.

At the same time, Abraham's investigations on Cephalosporins carried out at Oxford, noted the presence of an α-amino aliphatic group in the side-chain of penicillin N (D-α-amino adipyl side chain), which showed a slightly enhanced gram-negative activity. This prompted the incorporation of an α-amino-group as part of the side chain of semi-synthetic penicillins. From structure-activity relationship studies, D(-) phenylglycine coupled to 6-APA (Ampicillin) emerged as a potent oral broad spectrum antibiotic, launched by Beecham in 1961 under the patriotic name 'Penbritin'.

The fact that the D (-) phenylglycine derivative was the predominantly active optical isomer was an important finding, yet this posed the eventual challenge of efficient resolution of the racemic amino acid on production scale.

1.2 The discovery of 6-APA, 7-ACA and 7-ADCA

6-APA

The historical events which led to the serendipitous discovery of 6-APA started with the observed discrepancy between the microbiological and chemical assays for penicillin in fermentation broths. The chemical assay based on β-lactam reaction with hydroxylamine (colorimetric test) was consistently some 10 - 15 % higher than the bio-assay. This was not investigated but accepted as being due to an unknown factor in the complex fermentation mixture reacting with hydroxylamine. Beecham meanwhile happened to be working on the para-aminophenyl analogue of Pen G, which proved difficult to isolate. They attempted to acylate the para-amino group to facilitate isolation but were surprised to find that the microbiological assay went up 10 - 15 % on acylation (of 6-APA!).

Further investigation led to the isolation of 6-APA by Beecham in England from a side chain depleted fermentation of *P.chrysogenum* (Batchelor et.al., 1959). This together with the re-acylation of synthetic 6-APA (Sheehan, 1959), marked the dawn of a fruitful era in semi-synthetic penicillins admirably exemplified by Ampicillin (Beecham, 1962) and Amoxycillin (Beecham, 1972).

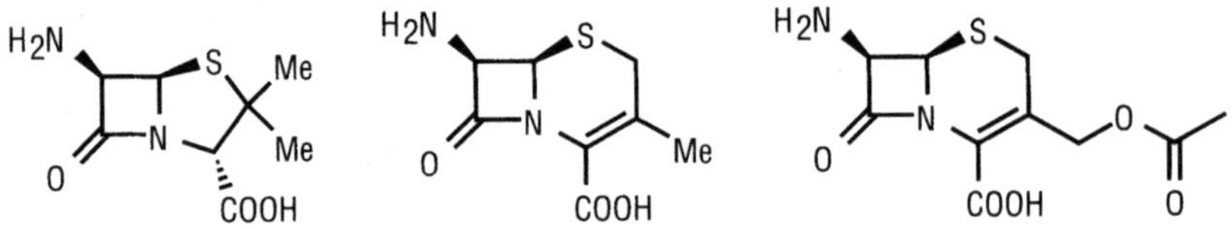

Fig. I.1 The major nuclei for semisynthetic antibiotics

For an excellent narrative and historical account of the fascinating penicillin story, including the Anglo-American effort during and following the second world war, as well as the controversy surrounding the discovery of 6-APA, the reader is directed to a book by J.H. Sheehan entitled 'The enchanted Ring' (referring to the β-Lactam ring).

6-APA also became readily available in industrial quantities by chemical removal of the side chain (so-called Delft cleavage), and more significantly by the eventual industrial route of enzymatic hydrolysis of penicillin using immobilized penicillin acylase as biocatalyst. The 6-APA process of today has become technologically simple, producing excellent yields of high quality product.

7-ADCA

As in the case of 6-APA, the finding of 7-ADCA was an opportunist event. The driving force here was the need to prepare the cephalosporin C nucleus 7-ACA. Cephalosporin C had been isolated earlier from *C.acremonium* and the structure elucidated in 1961. Since cephalosporin C was not sufficiently active clinically, the aim was to duplicate the success seen in the SSP's by replacing the α-aminoadipyl side chain with other acyl groups. However, early attempts at enzymatic hydrolysis were unsuccessful and the efficient preparation of 7-ACA became a major challenge.

One approach for this problem was the chemical transformation of a penicillin to a cephalosporin followed by the known removal of the penicillin side chain. Morin and coworkers believed such a mechanism must involve an oxidative step and investigated the chemistry of penicillin sulphoxides. Their successful work on the ring expansion reaction together with later developments using silyl protection for the carboxyl group (DSM, Gist- brocades) laid the foundation for the present day industrial route to 7-ADCA from penicillins.

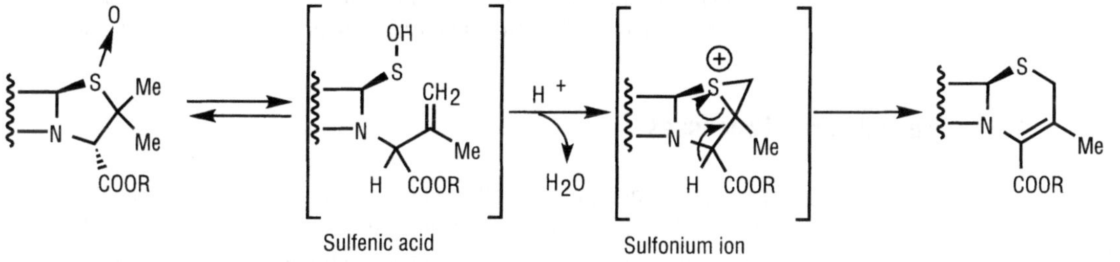

Fig. I.2 Expansion of the penicillin skeleton into the cephalosporin skeleton

In the first step of the now generally known and applied production process for 7-ADCA, penicillin G is oxidized to the corresponding S-oxide with for example peracetic acid in cold aqueous medium. The product is isolated as the free acid in excellent yield and dried for the following silylation step. This can be efficiently achieved using N,N`-bis(trimethylsilyl)urea, which in the presence of hydrogen bromide catalyst and pyridine leads to the ring expansion of penam to cephem ring (mechanism involves a sulfenic acid and a sulfonium ion shown in Fig. I.2).
Addition of water removes the silyl group and affords so-called cephalosporin G as the penultimate intermediate. Finally the latter is treated with immobilized penicillin acylase to hydrolyse the side chain and produce 7-ADCA, in an analogous manner to 6-APA. The side chain by-product can be recovered and recycled into the penicillin fermentation process. Enzymatic removal of the side chain has superseded the previous chemical route for side chain cleavage (silylation, iminochloride formation with PCl_5, conversion to iminoether with alcohol and hydrolysis, under deep-cooling conditions), where obvious disadvantages were the use of PCl_5 and deep cooling in organic solvent.
The production of 7-ADCA is presently dominated by DSM-Anti-Infectives in The Netherlands, and this technology position will soon be further strengthened by the introduction of a new fermentation process to replace the chemical ring expansion route.

7-ACA

As noted above earlier attempts to enzymatically deacylate cephalosporin C did not produce 7-ACA. However an efficient chemical cleavage of the α-aminoadipyl side chain of cephalosporin C was described by the Ciba group in 1968. Silyl ester of cephalosporin C was treated with PCl_5 to convert the peptide bond to iminochloride, which on treatment with methanol formed the iminoether. This hydrolyzed with ease to give 7-ACA.

To complete the development cycle of 7-ACA processes, we have today a two-enzyme-one-pot process for deacylation of cephalosporin C. Initially the enzyme D-amino acid oxidase catalyzes the oxidative deamination of the adipyl side chain group to the α-keto adipyl derivative. This loses carbon dioxide in the presence of oxygen and the resulting glutaryl derivative is hydrolyzed by glutaryl acylase to 7-ACA.

The early availability of these three nuclei, based on industrially viable processes, has greatly enhanced the subsequent commercial development of a wide range of semi-synthetic β-lactams. The most important examples are dealt with below.

1.3 Pre-industrial Processes

Generally speaking many of the methods used in peptide chemistry are applicable to the synthesis of semi-synthetic penicillins and cephalosporins. Selection depends on the functionality of the acylating group and on the sensitivity of the β-lactam nucleus.

Acylation of 6-APA with a variety of substituted carboxylic acids was carried out using the acid-chloride or mixed anhydride method. The new penicillins were obtained in good yield and frequently isolated as K salts using potassium 2-ethylhexanoate. The acid chlorides were reacted with 6-APA in aqueous acetone in the presence of sodium bicarbonate, or the mixed anhydrides were prepared in situ with ethyl chloroformate and triethylamine in acetone, prior to reaction with 6-APA. Such procedures were used to prepare Pen V and Methicillin.

The synthesis of Ampicillin, however, was far from being a simple acylation. In the initial Beecham approach, D(-)phenyl glycine was N-protected by reaction with benzylchloroformate to give the corresponding Z-derivative. The mixed anhydride was then prepared in acetone using ethyl chloroformate and triethylamine and, without isolation, was reacted with 6-APA in cold sodium carbonate solution. On completion of acylation, the mixture was acidified and Z-Ampicillin extracted into an organic solvent. Finally the Z-group was removed by catalytic hydrogenation with Pd catalyst to give Ampicillin. This process suffered from the disadvantage that large amounts of catalyst are required due to the presence of sulphur.

Dane and Dockner described a significantly improved method for the synthesis of α-amino substituted penicillins based on the β-dicarbonyl protecting group for the amino function. In this 'Dane salt route' named after Professor Elisabeth Dane (Munich University), D (-) phenylglycine and ethylacetoacetate are warmed in methanol/potassium hydroxide to form the corresponding enamine adduct, stabilized by hydrogen bonding. Conversion to the mixed anhydride using ethylchloroformate or pivaloyl chloride and triethylamine followed by coupling to the triethylammonium salt of 6-APA in dichloromethane afforded Ampicillin in 74 % yield.

Even at the early stage of Danes work (1962), Beecham was quick to see a potential synthesis of Ampicillin via the Dane salt route. The Beecham route utilized methylacetoacetate (the original Dane route disclosed only β-diketones) and ethylchloroformate as the mixed anhydride-forming agent. In the early version of the synthesis, 70% Ampicillin yields were obtained in a convenient process amenable to scale-up.
After these initial developments the industrial preparation of β-lactam antibiotics was governed by three themes:

a) Minimal handling and manipulation of the nucleus because of the sensitive character of the β-lactam moiety and the high prices for these molecules.
b) Side chain activation as acid chlorides or mixed anhydrides.
c) Side chain protection as the Dane salts.

Coincidentally, by the 1960's, the active field of Peptide Chemistry had likewise reached a high point, offering a formidable array of coupling methods from which to select and develop for the efficient industrial synthesis of β-lactam antibiotics.
Strangely enough, the dominant method selected for peptide bond formation in SSA's, the so-called 'Dane Salt' method, is hardly ever applied in classical peptide synthesis.

§2 First Generation of Industrial Processes

2.1 The largest products

With the market introduction of Ampicillin in 1961 (Beecham, UK), an unprecedented boom in business and process developments started. Ever since, a steady stream of new β-lactam derived molecules has reached the market with a peak period between 1970 and 1990. With the availability of 6-APA as the starting molecule, numerous derivatives with a range of substituents at the 6-amino position (and to a lesser extent at the 3-carboxyl function) were developed. Some of these, like the 6-(3-aryl-5-methylisoxazolyl)- penicillins had a modest marketing success though as a matter of fact are still on the market. It took, however, 11 years after Ampicillin before a second "blockbuster" hit the market i.e. the introduction of Amoxicillin in 1972 (again Beecham, UK). Although several additional 6-APA derivatives, with various substituents at the 6-amino position, were subsequently developed, (e.g. carbenicillin and ticarcillin) it is fair to say that with the introduction of Amoxicillin the top of the penicillin period was reached.
The ensuing development and marketing success of Augmentin (Amoxicillin in combination with the β-lactamase inhibitor clavulanic acid) has obviously greatly enhanced the enduring therapeutic benefits of the penicillins.

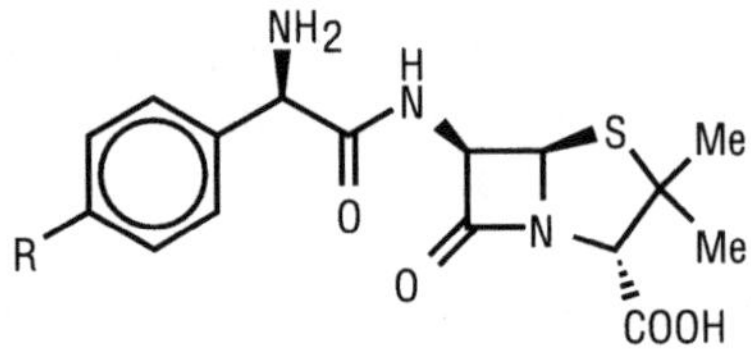

Ampicillin (R=H)
Amoxicillin (R=OH)

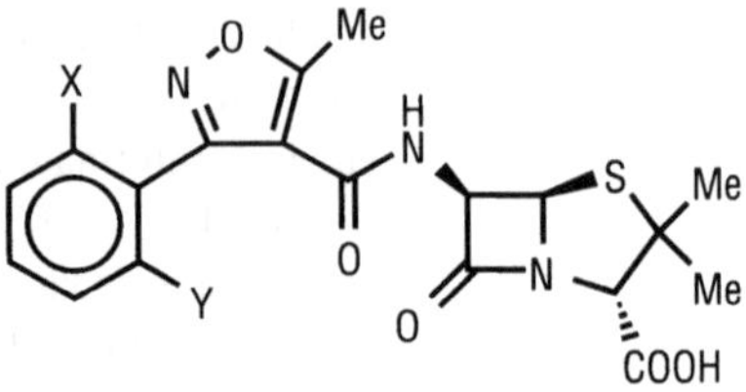

Oxacillin (X=Y=H)
Cloxacillin (X=H ; Y=Cl)
Dicloxacillin (X=Y=Cl)
Floxacillin (X=Cl ; Y=F)

Fig. I.3 The major penicillins

Even before Amoxicillin was introduced, the first cephalosporin had already reached the market, i.e. the Ampicillin congener Cephalexin (1970) developed by Lilly. After the discovery of the cephalosporins and commercial development by major pharmaceutical companies such as Glaxo and Lilly, it was the industrial process for 7-ADCA developed by Gist-brocades (now DSM Life Science Products; vide infra) that opened the door for large scale manufacturing of this class of cephalosporins. Once again a full range of substituents at the 7-amino position of 7-ADCA could be tested for therapeutically improved β-lactam antibiotic activity. Successful products such as Cefadroxil (BMS, 1977) and Cephradine (Squibb, 1972) are still with us today. It is remarkable to note, that the side chains both in the penicillins and the 7-ADCA derived cephalosporins are very much the same; arylglycines and derivatives are the common moiety.

With the detection of Cephalosporin C and the subsequent industrial development of the corresponding nucleus 7-ACA a new stream of cephalosporins arrived on the market. Cephalothin (1964) and Cefazolin (1971) developed by Lilly and Fujisawa were the earliest examples, of which Cefazolin is still a marketing success. The 3-acetoxy and 7-amino substituents in 7-ACA allowed for a much wider variation in the search for therapeutically useful molecules, than in the 6-APA/7-ADCA derived series. Heterocyclic substituents became the dominating side chains in these cephalosporins with 2-amino-4-thiazolyl acetic acid derived side chains at the 7-amino position taking the lead. Commercially still successful products like Cefotaxime (Hoechst, 1980) and Ceftriaxone (Roche 1983) are prime examples. It may be noted that optical activity is no longer an issue in the side chains employed. In fact, apart from a few isolated examples based on mandelic acid (Cefonicid: SKF, 1984; Cefamandole: Lilly, 1977) and α-sulphonylphenyl acetic acid (Cefsulodin; Takeda, 1980), all optically active side chains are derived from arylglycines.

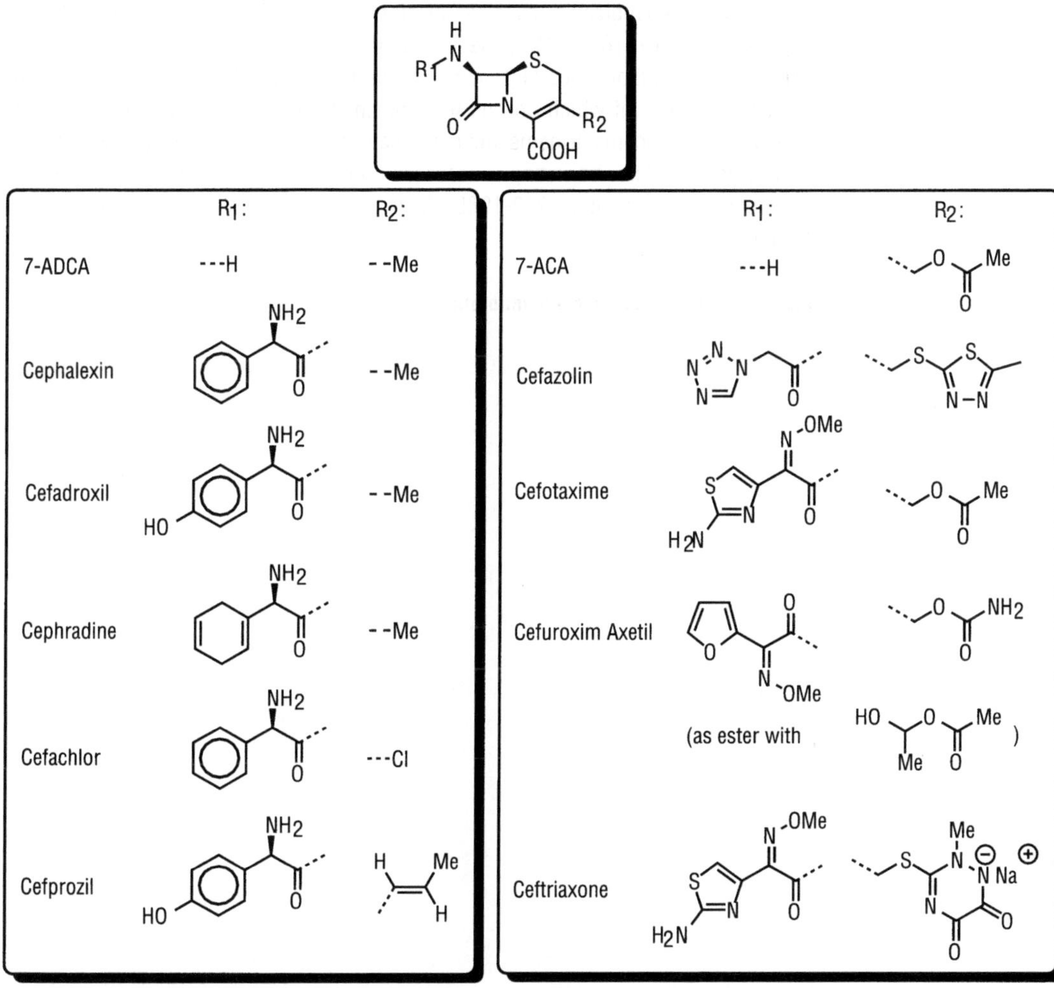

Fig. I.4 The major cephalosporins

Besides the development of cephalosporins, the available synthetic methodology of the 1970-1990 period spurred a complete overhaul of the β-lactam molecule. Several therapeutically beneficial molecules were developed, whereby, in simple terms, every molecular change in the original penicillin structure was allowed, provided the 3-amino-β-lactam moiety was left intact. Relatively simple molecules like the monobactams (i.e. Azthreonam, Squibb, 1983) were a result, as well as highly elaborated molecules such as oxa-cephalosporins (i.e. Moxalactam, Shionogi, 1981), carbapenems (i.e. Imipenem/Cilastatin, Merck, 1985) and carbacephems (i.e. Loracarbef, Lilly, 1993).

Less manipulated molecules such as Cefaclor (Lilly, 1979), Cefuroxim Axetil (Glaxo, 1987) and Cefprozil (BMS, 1992) have evidently had a greater commercial success. Although several thousands of β-lactam derived antibiotics have been developed in the last 40 years (of which dozens have reached the market), and are still being developed (i.e. penems, trinems and combinational structures of lactams with other antibacterials), the marketing successes remain mainly with the products introduced in the period 1970-1985. Table I.1 gives an overview of the largest products.

Table I.1 The largest β-lactam antibiotics

Product	Year of Introduction	Estimated market volume 1970	2000 (tons)
PenG	1941	3.000	25.000*
PenV	1953		
Ampicillin	1961	900	20.000
Amoxicillin	1972	-	
Cefachlor	1979	-	2.000
Other Penicillins		100	
Augmentin	1981	-	See Amox.
Cephalexin	1970		
Cephradine	1972	-	4.000
Cefadroxil	1977		
Cefazolin	1971		
Cefotaxime	1980		
Ceftriaxone	1983	-	3.000
Cefuroxime Ax.	1987		
Cefprozil	1992		
Other Cefalosporins	-		
Totals		4.000	45.000**

*Including usage as starting material for other β-lactams.

**Excluding PenG/V as starting material for other β-lactams.

Amoxicillin (including Augmentin) is clearly by far the largest single β-lactam antibiotic. The indicated 11-fold increase to 45.000 tons/year in the total volume of β-lactam antibiotics during the period 1970-2000 not only shows the great impact of these products on world health care, but has also been the main driving force behind continuous process improvements and process innovations.

2.2 Life cycle of Ampicillin processes

Following the initial processes using conventional peptide chemistry as described in chapter I.1, Ampicillin manufacturing was governed by two processes. The first was developed by Beecham and prompted by the discovery of the Dane salts in 1962. These salts are very readily prepared through condensation of amino acids with β-dicarbonyl compounds in alkaline methanol. Industrially, cheaply available ethyl or methyl acetoacetates are used. Yields of 95 - 98 % can easily be achieved, and moreover, the crystalline salts are stable at room temperature, easily isolated, handled, stored and shipped. The amide bond with 6-APA is formed through in-situ activation of the carboxyl group of the Dane salt as mixed anhydride followed by the addition of 6-APA and an organic base. Suitable solvents are acetone or dichloromethane. Reaction temperatures are from -30°C to - 45°C requiring sufficient cooling capacity. In the early processes ethyl chloroformate was employed to form the mixed anhydride. Many organic basis (mainly amines), often for patent reasons or yield optimization, have been used to catalyze the mixed anhydride step and to obtain a complete coupling reaction. The commonly used amine is triethylamine.

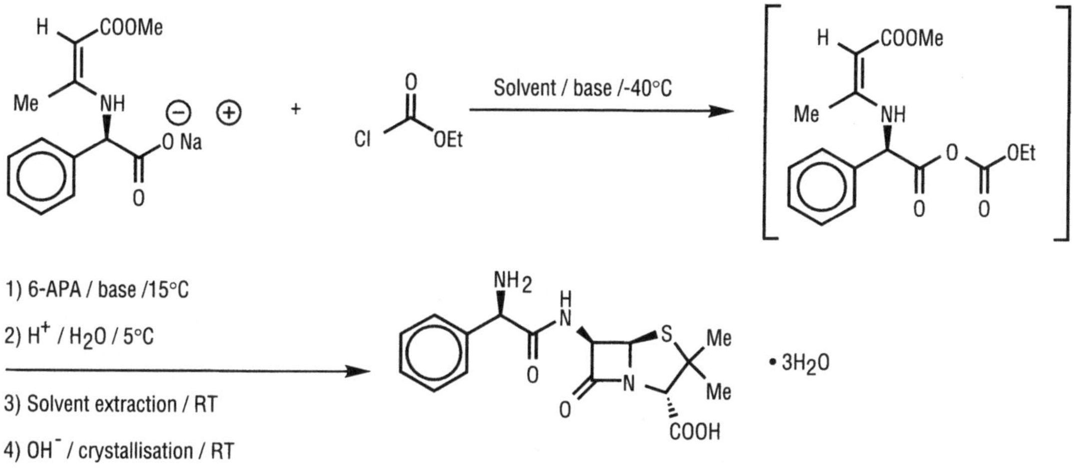

Fig. I-5 Conventional Dane process for Ampicillin.

Work-up and isolation of Ampicillin is rather straightforward through aqueous acidic removal of the Dane protection, whereby Ampicillin dissolves as the stable hydrochloride in the aqueous phase. Organics are removed by solvent extraction or phase separation, after which Ampicillin is precipitated as the crystalline trihydrate at its isoelectric point of pH 5. Best crystals are obtained by simultaneous addition of Ampi.HCl solution and aqueous base to a small amount of buffer at constant pH 5 (so called concordance crystallization). Initial yields of 80 - 85 % have over the years been improved to the range of 85 - 88 %.

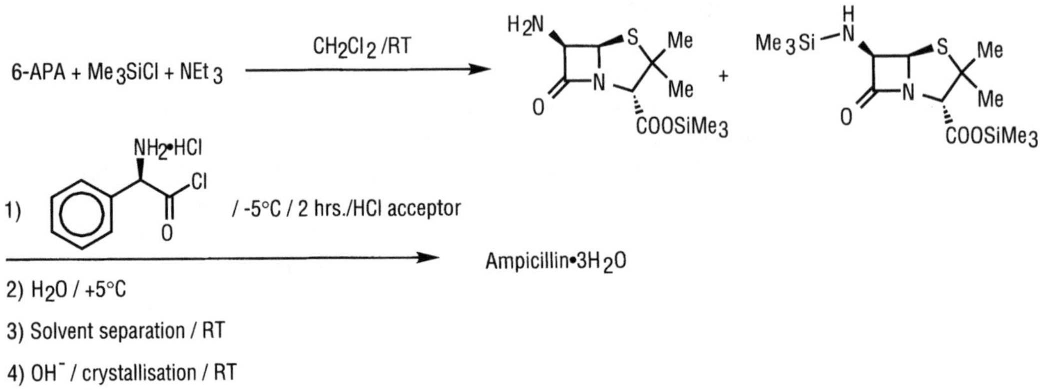

Fig. I-6 The Acid Chloride process for Ampicillin.

The second process was developed by Bristol Meyers and employed the acid chloride hydrochloride of D(-)phenylglycine as the protected/activated side chain. Both Gist-Brocades and DSM Andeno have improved the process. Manufacturing, storage and shipment of the acid chloride requires special precautions which will be described in Chapter I. 2.3. With the acid chloride in hand preparation of Ampicillin is quite straightforward, although a disadvantage compared to the Dane salt route is the requirement of silyl protection of 6-APA. Thus, a suspension of 6-APA in dry dichloromethane is treated with an organic amine (triethylamine, dimethylaniline and several others have been used) and a silylating agent such as trimethylsilyl chloride or dimethyl dichlorosilane. Also HMDS and BSU, in which amine and silyl functions are combined, have been used. Reaction temperatures are from -10°C to 0°C avoiding the need for investments in deep cooling equipment as in the Dane process. After addition of the acid chloride and further stirring, the mixture is diluted with water, the pH adjusted to 1.5 and organic solvents are removed. Isolation of Ampicillin is carried out as described for the Dane process. Yields and quality are normally slightly better than in the traditional Dane process. This simple one pot process was proven to be very robust and particularly suitable in countries where experienced plant operators are scarce and money for investments (i.e. cooling equipment) is limited. The process has greatly contributed to the start of efficient Ampicillin manufacturing in Latin America, Eastern Europe and many Asian countries. The reaction scheme, a typical procedure and a simplified plant scheme, are shown below.

Typical procedure using HMDS (Hexamethyldisilazane)

A. Silylation

A 6-APA suspension in dry dichloromethane (10 % w/v) is treated with acetamide (2.2 equiv.) and hexamethyldisilazane (1.05 equiv.). The mixture is refluxed for 1 hour (solution occurs) and brought under vacuum to remove excess ammonia gas.

B. Acylation

The silylated 6-APA is cooled to -10°C and acid chloride(1.05 equiv.) is added portion wise.The reaction mixture is stirred for $1^1/_2$ - 2 hours at 0°C - 5°C.

C. Hydrolysis / crystallization

Cold water is added with vigorous stirring and the pH adjusted to 1.5. The lower dichloromethane phase is removed, and the aqueous phase is adjusted to pH 2.5 with conc. ammonium hydroxide, keeping the temperature around 5°C.
The aqueous phase is filtered and crystallized according to the concordance crystallization method (at constant pH 5.0). The Ampicillin trihydrate crystalline slurry is stirred at 0°C for 2 hours.

D. Ampicillin Trihydrate

The product is collected and washed with cold water, acetone and dried. The yield average is ca. 88 %.

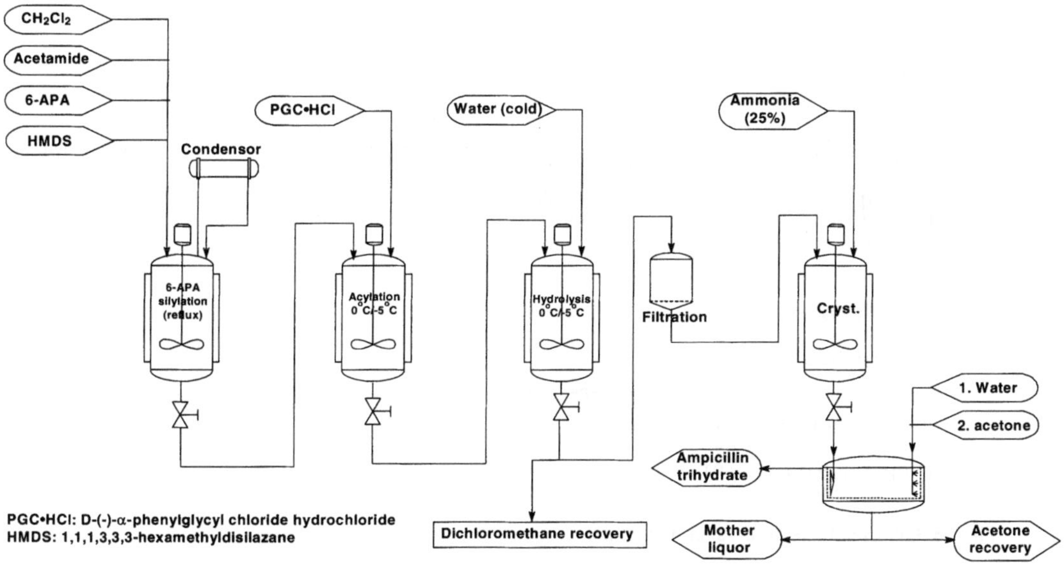

Fig. I-7 Ampicillin chloride hydrochloride process flow diagram

By 1975, when the first patents for Ampicillin expired, over 60 producers were already active, Europe and the America's being the most productive regions. The great demand, a growth of 8-10 %/year was quite common in those years, could only be met through licensing out by both Beecham and Bristol Meyers. At the same time lack of patent protection in several countries (i.e. Italy, Spain, Latin America) or protections limited to process patents only (i.e. The Netherlands) paved the way to Ampicillin becoming a generic drug 'avant la lettre'. Next to this, the growing demand from Asia, in particular India and the ASEAN countries, added another life cycle to the product. A top number of producers of around 100 was probably reached in the 1980-1990 period, whereas in the next decade the top in volume, i.e. ca. 6000 tonnes/year, was reached.

This huge worldwide demand spurred the desire in many countries to produce not only Ampicillin but also the precursors PenG, 6-APA and D (-)-phenylglycine. The activated side chain, i.e. the acid chloride hydrochloride was an obvious first candidate, quickly followed by the more expensive and more sensitive 6-APA molecule (mostly from imported PenG; local fermentation of PenG has not been very successful in many countries with China as a recent exception).

With the availability of a robust enzymatic process for the splitting of PenG to 6-APA and phenylacetic acid, local production of 6-APA became successful earlier than side chain manufacturing (see chapter I.2.5). Several attempts to produce indigenous phenylglycine acid chloride failed because of the great moisture sensitivity of the process and the product, resulting in low yields and off-color product (on hindsight, only producers which had access to the DSM Andeno technology became successful).

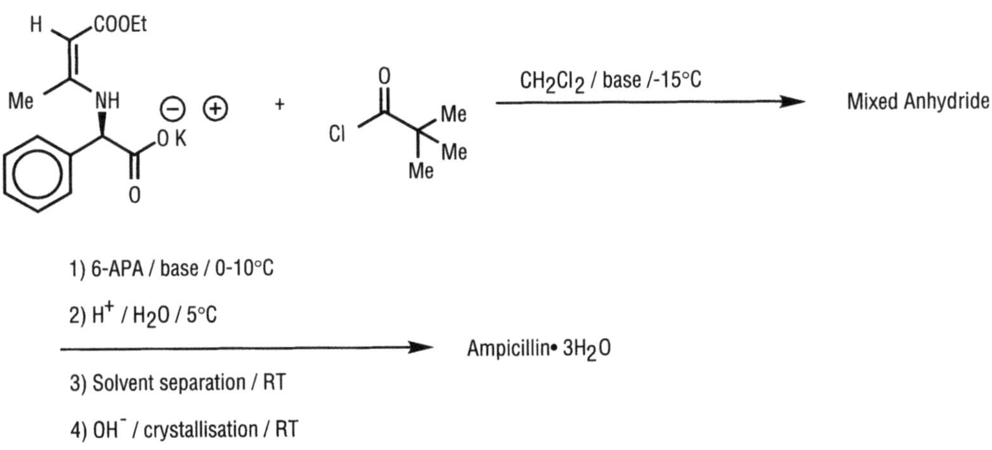

Fig. I-8 Improved Dane process for Ampicillin.

Application of the conventional Dane process was not a viable alternative because of lower yields on 6-APA and not as good a quality compared to the acid chloride route. Local competition on cost price, in particular in India, was already an issue at that time (ca. 1985). Developments in Korea by Kim led to a rediscovery of an improved version of the Dane route already indicated by professor Dane herself. Employing pivaloyl chloride instead of ethylchloroformate in the preparation of the mixed anhydride resulted in a robust and high yielding process for Ampicillin of excellent quality. Yields of 90 % or even slightly better can be achieved. A typical procedure and reaction scheme are shown in scheme I.8.

Ampicillin via the Dane salt route

A. Preparation of Mixed Anhydride
A solution of D- (-)-phenylglycine potassium ethyl Dane salt in dichloromethane is treated with an equivalent of pivaloyl chloride in the presence of a basic catalyst. Addition is at such a rate so as to maintain a suitably low temperature (e.g. below -15 deg. C.). The mixture is stirred for 1 hour to give the mixed anhydride of the Dane salt.

B. Preparation of 6-APA solution
A suspension of 6-APA in dichloromethane is solubilised with the help of an excess of triethylamine at 0/10 deg. C.

C. Preparation of Ampicillin Trihydrate
The 6-APA solution is slowly added to the mixed anhydride under deep-cooling conditions (i.e. -30 deg. C) with good stirring. After an additional few hours of stirring, the mixture is hydrolyzed with water and HCl at around 0 deg. C. The layers are separated and the aqueous phase is treated with base to give a final pH of 5. The crystallized Ampicillin trihydrate is collected, washed with water and acetone, and dried.

The drawback of low temperature was overcome by better and cheaper global availability of (equipment for) liquid nitrogen. Also the developing boom of Amoxicillin, where the Dane process is commonly used (see I. 2.4), eased the way to investments in cooling facilities. Today the Dane process is also the most commonly used procedure for Ampicillin.

Altogether the 40 year period of Ampicillin manufacturing has seen three major process variants in the coupling of side chain and nucleus. All of them employed stoïchiometric methods combined with up-to-date organic chemistry, i.e. quick (to) adaptation to silylating agents. The upcoming concern about environmental issues has mainly been met by increasing process efficiencies and several recycles of solvents and auxiliary agents. Nowadays, dichloromethane, acetone, triethylamine and pivalic acid are all recycled with high efficiencies. In fact, process improvements, optimizations and recyclings were needed because of increasing cost-price based competition and went hand in hand with the environmental issues. An interesting question is whether the life cycle of Ampicillin will be long and strong enough to see a serious industrial application of the biocatalytic processes as outlined in chapter I.3.

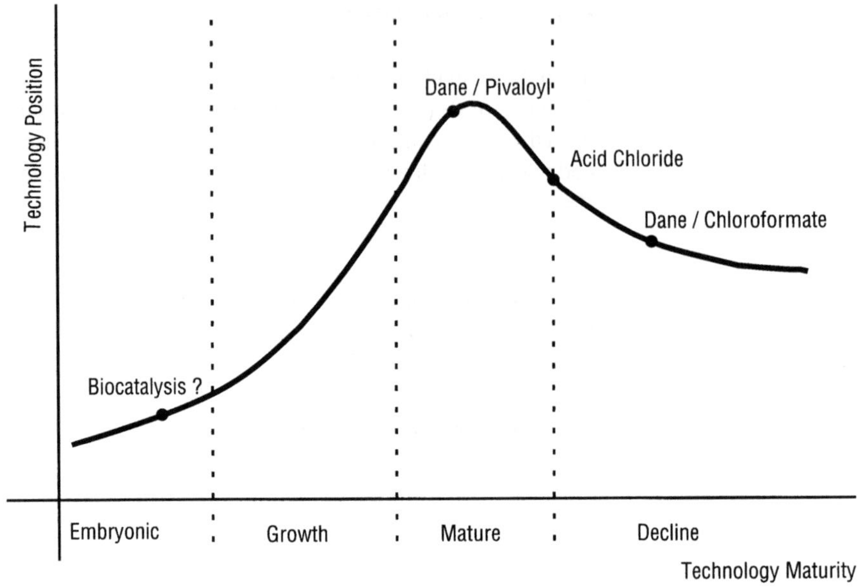

Fig. I.9 Life cycle of Ampicillin Processes

2.3 Development of a Phenylglycine product line

The well-known resolution process of DL-phenylglycine with camphor sulphonic acid quickly became the industrial process of choice. The required racemic amino acid was readily obtained through a conventional Strecker synthesis with benzaldehyde. The resolving agent was either made through sulfonation of natural camphor or resolution of the racemic acid (from synthetic camphor) with optically active phenylglycine. Both are low-yielding processes necessitating an efficient recycling of the resolving agent sometimes aggravated by short supply (and high prices) of natural camphor. For many years this process, developed by DSM Andeno, has been and is the most outstanding example of resolution on an industrial scale.

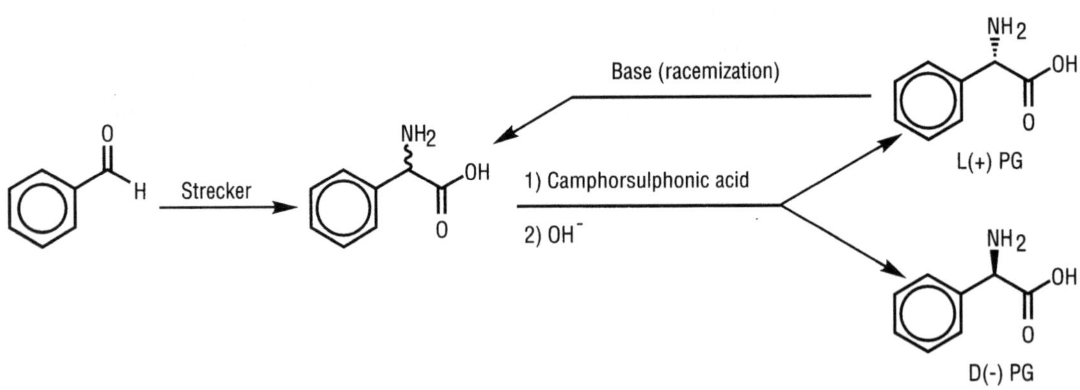

Fig. I-10 Resolution process for phenylglycine

The process runs in water, employs 0.6 eq. of resolving agent and gives only NaCl as waste. The process is very robust and can be run for several months without interruption. In-process purifications are not needed. A yield in excess of 40 % of the desired enantiomer is obtained in a single crystallization of the diastereomeric salt with an ee of ca. 99 %.

Resolving agent and racemized distomer are recycled smoothly. Racemization can be done efficiently with aqueous base at elevated temperatures. Racemization catalysts like aromatic aldehydes are not needed, thus omitting an additional recycle loop.

For many years it was impossible to prepare the undesired diastereomeric salt in a suitable crystalline form, preventing a quantification of the quality of the resolution process. Only from a carefully chosen aqueous solution of hydrochloric acid (1 ml 10 N HCl, 14 ml water, 18 gr (50 mmol) of crude (+) PG.CAS) could this salt be isolated in a low yield of 7%. Other conditions resulted in co-precipitation of L (+) PG, L (+) PG.HCL and/or mixtures thereof. With both diastereomeric salts in hand x-ray crystal structures and several physical parameters could be measured to quantify the quality of this resolution.

	p-salt	n-salt
Crystal system	Monoclinic	Orthorhombic
Space group; no	C2, Z=4; 5	$P2_12_12_1$,Z=4;19
a (Å)	24,867	6,903
b (Å)	7,037	15,262
c (Å)	11,359	17,513
V (Å3)	1933,3	1845,1
D (g. cm-3) calc.	1,317	1,380
D (g. cm-3) measured	1,313	1,378
Solubility (g/100 g; 20 (C) (in 14 ml H2O + 1 ml 10NHCl)	49,0	4,9
Solubility (g/10 g H2O; 20 (C)	>150	5,0

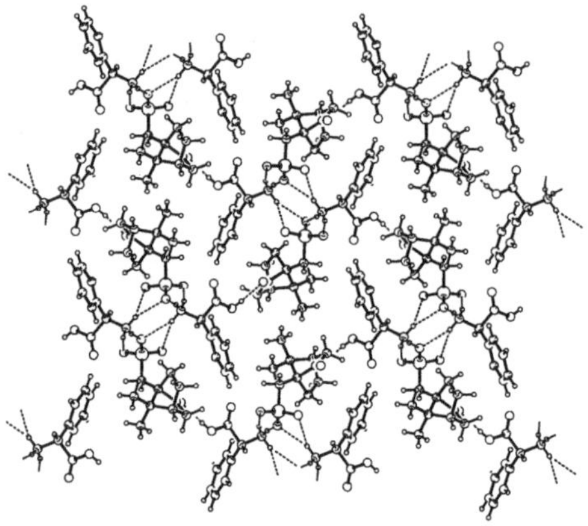

D-(-)-phenylglycine-(-)-camphorsulfonate

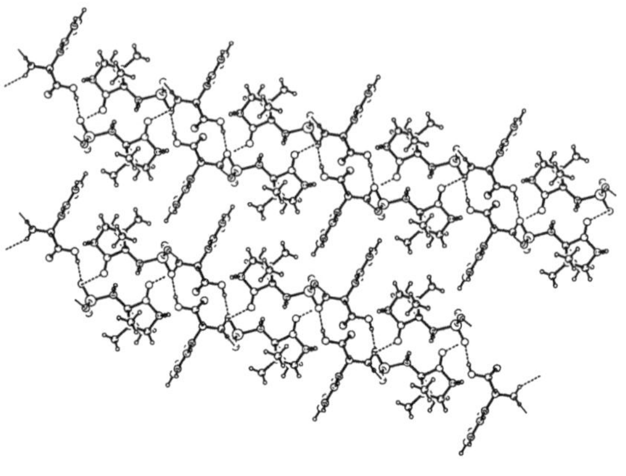

L-(+)-phenylglycine-(-)-camphorsulfonate

Fig. I.11 Physical parameters and crystal structures of D(-)PG.CAS and L(+)PG.CAS

Construction of binary phase diagrams using melting points and DSC measurements was hampered by decompositions. A ternary phase diagram in aqueous hydrochloric acid could be constructed. The eutectic point was found to be above 99 % of n-salt.

Using the formula*

$$S = k.t = \frac{1 - 2\,x_{eu}}{1 - x_{eu}} = \frac{s_p - s_n}{s_p}$$

the quality of this resolution is found to be between 0.85 and 0.98.Resolutions of similar high quality are found in the pairs dl-Ibuprofen/S-phenylglycinol (S = 0,80) and dl-mandelic acid/R-phenyl ethylamine (S = 0.90).

The high efficiency of the camphor sulphonic acid process has greatly hampered the introduction of other processes at a relevant industrial scale. An enzymatic process based on dl-phenyl hydantoin has therefore not come to industrial maturity; unlike such a process for p-hydroxyphenylglycine (see I. 2.5). For some producers supply and/or recycling of camphor sulphonic acid was a bottleneck resulting in the development in India and China of resolution processes based on dl-phenylglycine methyl ester and tartaric acid or (more efficiently) dl-phenylglycine aminonitrile and tartaric acid. The latter process is in fact an asymmetric transformation as the ammonium nitrile tartrate can be racemized in situ. Also the methyl ester based process can be turned into a asymmetric transformation provided the process is run in a ketone as solvent (acetone, MIBK) or when an aldehyde is added as racemization catalyst.

* S = quality of resolution ; $0<S<1$
k = max. yield of n-salt in an optimalized resolution; k = 1 at 50%
t = optical purity of isolated n-salt; t = 1 at 100 % de
x_{eu} = molar composition at eutectic point
s_p, s_n = solubility of p respectively n-salt

Derivatives of phenylglycine as required in the preparation of Ampicillin and Cephalexin were very readily prepared on industrial scale in the case of Dane salts. Condensation of the amino-acid in alkaline methanol or ethanol with acetoacetates affords the Dane salts in almost quantitative yields. Whereas in the early years mainly the Na/methyl Dane salt was applied, nowadays the K/ethyl salt is preferred. Better stability and good solubility for mixed anhydride preparation are probably main reasons.

Preparation of the phenylglycine chloride hydrochloride is a quite different endeavor. Whereas the mixed anhydride, prepared from the Dane salt, can only be handled in situ at low temperature, the acid chloride aims at serving the customer with an alternative that can be stored safely and handled at room temperature. Additional advantages for the customer are a protected and preactivated intermediate with a minimum molecular weight and a minimum of handlings and investments to produce Ampicillin or Cephalexin.

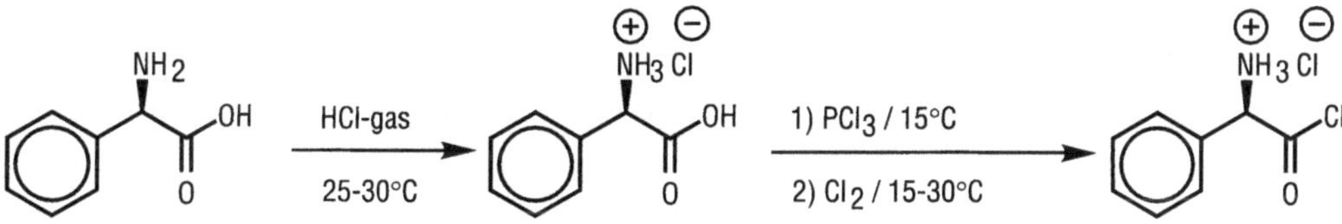

Fig. I-12 Preparation of D(-)phenylglycyl chloride hydrochloride

The acid chloride is prepared in a two-step process consisting of hydrochlorination with dry HCl-gas and chlorination with chlorine and phosphorous trichloride. Reactions are performed in dry, low boiling halogenated solvents like dichloromethane, dichloroethane, chloroform or carbon tetrachloride. All reactions are run in suspension and the end-product can readily be isolated on a centrifuge provided all steps are carried out in a dry, closed system. Pre-dried air or nitrogen is required. Solvents, hydrogen chloride and phosphorus oxytrichloride can all be recovered and re-used.

Yields well over 90 % can be obtained. The quality of the product is strongly influenced by any moisture and the quality of the recovered solvents and reagents. Off-colour product is a major risk as well as the formation of small amounts of the dipeptide PG-PG and the corresponding diketopiperazine. Loss of optical purity is not an issue. Good quality product can be safely drummed, under dry conditions, and shipped worldwide. Application of the acid chloride can best be done by employing complete drums; left-overs should not be used for a next batch and opening of drums for sampling and analyses should be avoided. Since coupling is under heterogeneous conditions, the crystal size and particle size distribution are critical specifications.

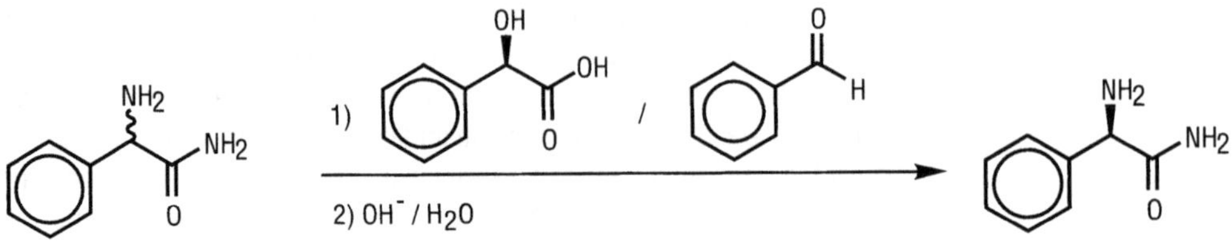

Fig. I-13 Asymmetric transformation of dl-phenylglycine amide

With the development of enzymatic couplings for Ampicillin and Cephalexin (see I.3) new derivatives of phenylglycine were needed. Although the initial process developments were done with the D(-)phenylglycine methyl ester, which is readily obtainable as the HCl-salt, the preferred product for enzymatic coupling is D (-) phenylglycine amide. The amide has the advantage of being a precursor in the phenylglycine process whereas the methyl ester is a derivative. The amide can be obtained through enzymatic kinetic resolution of the corresponding racemate, though, the asymmetric transformation developed by DSM is preferred. Treatment of dl-amide with mandelic acid and a catalytic amount of benzaldehyde in an organic solvent results in a complete transformation to the D(-)amide with 99 % ee. Solvent and resolving agent can easily be recovered. The starting amide is obtained through a Strecker synthesis with benzaldehyde. When desired the D (-) amide can be converted to the free amino acid or derivatives.

2.4 Industrial Processes for Amoxicillin

Initially Amoxicillin was considered only marginally better compared to Ampicillin and market introduction could therefore best be done when the first Ampicillin patents expired around 1975. With the discovery that blood levels of Amoxicillin were twice as high as with Ampicillin at the same dosage, the market launch was brought forward to 1972. The fast uptake of Amoxicillin into the bloodstream causes this unexpected effect. The result has been a sound marketing success with annual growths often above 10%. The subsequent development and market launch of the combination drug with clavulanic acid (Augmentin) has greatly enhanced this success. In the early 90's Ampicillin was surpassed in volume and also replaced by Amoxicillin on the WHO list for essential drugs.

The industrial production of Amoxicillin was, virtually right from the start, a process

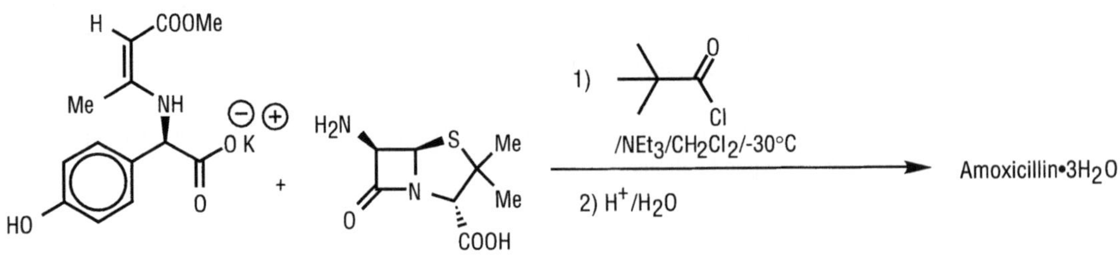

Fig. I.14 Industrial process for Amoxicillin

based on the Dane salt route. Various Dane salts have been used but the potassium/ethyl salt is presently the preferred derivative. Just as with Ampicillin ethyl chloroformate is replaced by pivaloylchloride in the synthesis of the mixed anhydride and dichloromethane is the solvent of choice. Yields can go above 90 %.

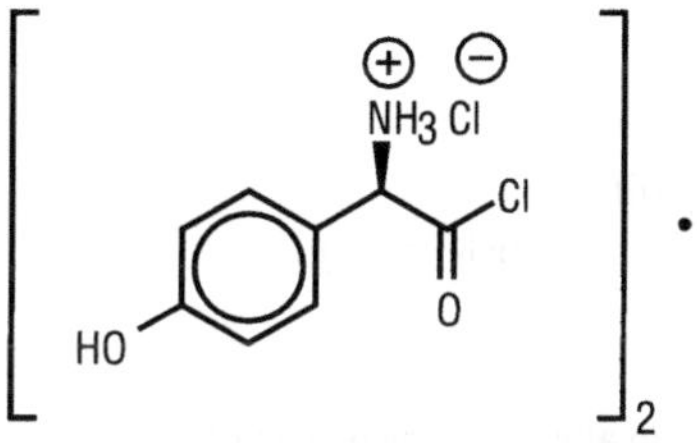

Fig. I.15 D(-)-*p*-hydroxy-phenylglycyl chloride hydrochloride

Attempts to develop an acid chloride route analogous to Ampicillin met with initial success through a process originating from Bristol Meyers. The acid chloride hydrochloride of D(-)-*p*-hydroxy-phenylglycine was found to be surprisingly stable when isolated as a hemi-solvate of 1,4-dioxane. Unfortunately the adduct can only be prepared efficiently using phosgene. Also more attractive stabilizers cannot replace the expensive and toxic dioxane, which is also difficult to recover readily. Apart from application at Bristol Meyers the acid chloride route has not come to industrial maturity.

2.5 Development of *p*-Hydroxyphenylglycine

Whereas with phenylglycine the struggle centered around the derivatives applied in the coupling with 6-APA and 7-ADCA, in the case of hydroxyphenyl glycine the focus was on the resolution of the racemate and the synthesis thereof. Virtually right from the start the process of choice for dl-*p*-hydroxyphenylglycine was the Mannich condensation of phenol, ammonia and glyoxylic acid.

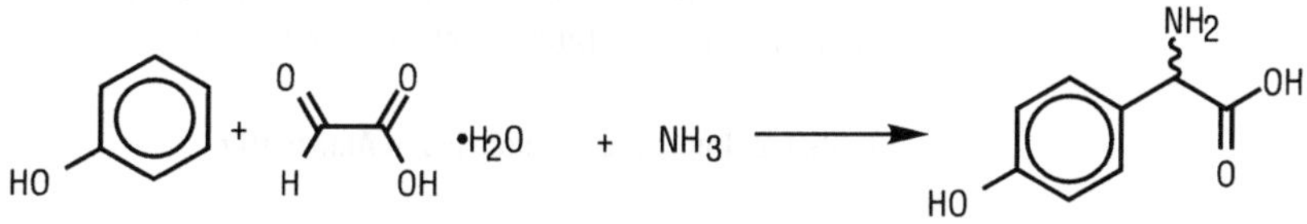

Fig. I.16 Mannich condensation to dl-*p*-hydroxyphenyl glycine

Despite the apparent simplicity of this process a still uncompleted battle was needed to improve the economics. Due to lack of regioselectivity initial yields were no better than 40 %. Through the years p/o ratios have been improved from 3/1 to 8-10/1 thereby allowing isolated yields of around 60 % (on glyoxylic acid). Although most variations were run in water, vast ranges of temperatures (20-120 °C), pH (2-12) and promoters (solvents, Lewis acids) can be found in the patent literature. Two alternatives, which have seen industrial application, need to be mentioned. The first one is a two-step process developed by Hoechst-France in which dl-*p*-hydroxymandelic acid is isolated as an intermediate. Regioselectivity is slightly better and the reaction with ammonia (in fact NH_4Cl) can be run at its own optimal conditions at 120°C. In the second process, developed by DSM, sulfamic acid is employed as amino source, also resulting in improved regioselectivity and allowing acidic reaction conditions.

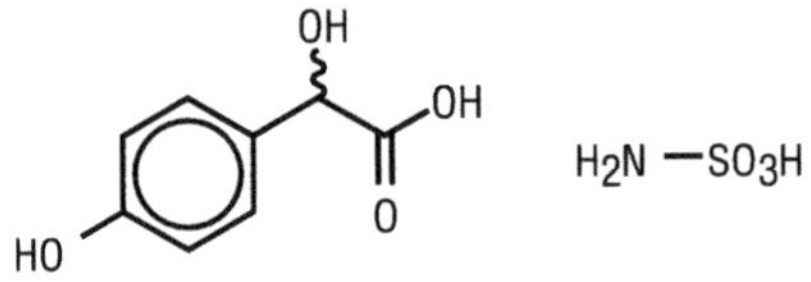

Fig. I.17 Intermediates in alternative synthesis of dl-*p*-hydroxyphenyl glycine

With the emergence of biocatalytic processes for D(-)-*p*-hydroxyphenylglycine (vide infra), employing dl-*p*-hydroxyphenylhydantoin as starting material, new parameters became available for improving the Mannich condensation; urea now being the source of nitrogen. On average slightly better selectivities and yields are reached compared to the processes producing the free amino acid.

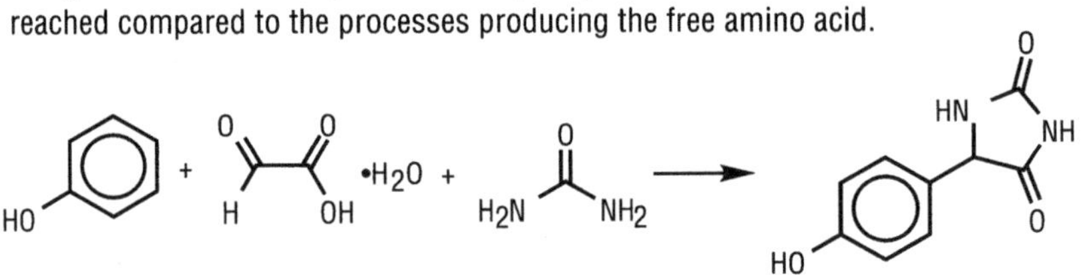

Fig. I.18 Preparation of dl-*p*-hydroxyphenyl hydantoin

Given the huge global market for Amoxicillin it is worthwhile speculating on the maximum yield of Amoxicillin in the future taking into account the improvements already achieved. Table I.2 shows the results.

Table I.2. Improved efficiencies in Amoxicillin

	1975	1990	2005
Pen.G -> Amox (% yield)	55	75	90
Glyoxylic acid -> Amox (% yield)	20	50	75
Kg waste/kg Amox	50	15	5

The industrial history of the resolution process to D(-)-*p*-hydroxyphenylglycine is a nice example of developments in chirality. After an initial process, developed by Beecham and employing entrainment resolution of dl-amino acid sulfonates, most producers embarked on resolution through diastereomeric salts, bearing in mind the successful phenylglycine/camphorsulphonate process. For a number of years the DSM Andeno process based on resolution with 3-bromocamphor-8-sulphonic acid, was the dominating process. However, unlike the phenylglycine process, recyclings were less efficient and build-up of impurities required regular in-process purifications. *p*-Hydroxybenzaldehyde and other oxidation products caused many headaches. Although milder process conditions could be achieved by employing aromatic aldehydes as catalysts in the racemization of the unwanted L(+) enantiomer, the process was superseded by the kinetic resolution of dl-*p*-hydroxyphenylhydantoin employing enzymes.

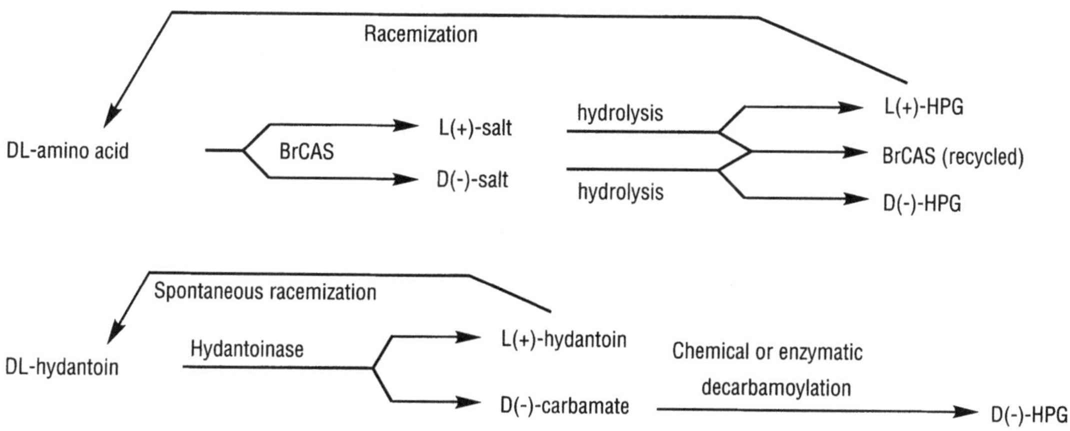

Fig I.19 Resolution processes for D(-)-*p*-hydroxyphenyl glycine

Snam Progetti in Italy developed the first biocatalytic resolutions of 5-arylhydantoins. Examples with 5-phenyl and 5-(*p*-methoxyphenyl) substituents did not achieve industrial success. With the discovery of the proper biocatalyst, *p*-hydroxyphenyl hydantoin could successfully be tackled, greatly facilitated by the spontaneous, in-situ, racemization of the unwanted enantiomer. The process was quickly brought to industrial scale in Japan/Singapore (Kanegafuchi) and Italy (Recordati) and gave virtually quantitative yields of D (-)-*p*-hydroxy-phenylglycyl carbamate.

Efficient conversion to the required free amino acid took a number of years to find. Initial chemical routes, employing sodium nitrite and acetic acid, are now completely replaced by enzymatic decarbamoylation. The two enzymes can be operated simultaneously or in separate steps.

2.6 Processes for Cephalosporins

This section summarizes the chemistry of industrial processes used for cephalosporins that have gained significance in present day clinical use, amongst them:

a) 7-ADCA derived orally active cephalosporins: Cephalexin, Cephradine and Cefadroxil.
b) 7-ACA derived cephalosporins, with or without substitution of the 3-acetoxy group.

Cephalexin, Cephradine and Cefadroxil

As with Ampicillin synthesis, there are two routes applied for the industrial synthesis of these 7-ADCA-based cephalosporins, namely the Dane salt and acid chloride hydrochloride route. The Dane salt route has two process variants depending on whether the 7-ADCA is solubilized as a salt or as a silyl ester.
The current trend, notably in India, is a clear shift from the acid chloride to the Dane salt process. This reflects the relatively high cost and difficulty in producing the acid chloride side chains under indigenous conditions as compared to the corresponding Dane salts. Indeed, Cephradine is nowadays only produced by the Dane route, although the chloride hydrochloride of D (-) dihydrophenyl glycine can be prepared in a process very similar to that for the phenylglycine acid chloride. It is estimated that over 80 % of both Cephalexin and Cefadroxil are made by the Dane route.

The Dane salt processes for Cephalexin, Cephradine and Cefadroxil are typified by the following consecutive steps:

A. 7-ADCA Dissolution
The universal solvent for all three processes is dichloromethane. Solid 7-ADCA is unreactive as such (zwitterionic), and dissolution using a strong organic base is needed to free the amino-group for reaction with mixed-anhydride activated Dane salts.
Either 1,1,3,3-tetramethylguanidine (TMG) or 1,8-diazabicyclo (5.4.0) undec-7-ene (DBU) can be used to efficiently dissolve 7-ADCA in dichloromethane at -10 to 10 deg. C. This step is considerably more expensive than for 6-APA dissolution where cheaper triethylamine can be used (this is not basic enough to dissolve 7-ADCA).
As indicated above an alternative method for dissolution of 7-ADCA is by in situ silylation using most commonly hexamethyldisilazane (HMDS) in refluxing dichloromethane. Addition of catalysts such as acids or ammonium salts increase the rate and degree of silylation. The latter is complete when evolution of ammonia by-product has ceased (ammonia is removed under vacuum prior to the coupling step). The HMDS method is predominantly applied in India, though silylation is also used by certain major producers in Europe. It is more economic than using TMG/

DBU for the 7-ADCA dissolution step, though coupling yields are some 2% lower probably due to steric hindrance caused by simultaneous silylation of the 7-amino group.

B. Dane salt activation

The synthesis of Dane salts has previously been discussed. These must be chemically activated before coupling to solubilized 7-ADCA. Pivaloyl chloride is currently the most commonly used reagent for the activation step which is catalyzed by an aromatic amine such as pyridine or a methyl pyridine. The temperature range is -30 to -70 deg. C (liquid nitrogen cooling) and activation is completed within 30-60 minutes normally.

C. Coupling

In this step 7-ADCA solution is added to the above activated Dane salt at -30 to-55 deg. C (liquid nitrogen cooling) to produce the corresponding Dane-protected cephalosporin derivative. During coupling the K or Na salt of pivalic acid is formed as a byproduct.

D. Hydrolysis

The Dane protecting group is removed by hydrolysis of the above coupling mixture with water and hydrochloric acid at pH 1.0. The product is transferred from the organic to the aqueous layer. Pivalic acid remains largely in the organic layer whilst organic bases are also transferred into the water layer.

When using silylated 7-ADCA ester instead of TMG/DBU, the by-product of hydrolysis, hexamethyldisiloxane, remains mainly in the organic layer (recovered and recycled). This oily by-product can also contaminate the aqueous product layer and addition of co-solvent (eg methanol) is needed to avoid formation of a 'sticky' crystalline end-product which could be difficult to centrifuge.

After hydrolysis the layers are separated and the aqueous layer filtered prior to crystallization. The dichloromethane can be efficiently recovered (85 %) whilst pivalic acid is normally not recovered (incinerated).

E. Crystallization

In recent years the main yield improvement in the production of 7-ADCA-based cephalosporins has come from isolation of a second crop of product (extra 5-10 %), from the mother liquors of the first main crop. This is required because of the higher intrinsic water solubility of cephalosporins when compared to SSP's. The end-product can be isolated by one of two methods: in the first case by direct crystallization at the isoelectric point (pH 5.0 finally) followed by isolation of a second crop from the resulting mother liquors via insoluble complex formation (β-naphtol is used in India for Cephalexin where total yields are 85-87 %). Alternatively, a second procedure is to isolate all the product initially as an insoluble solvate (e.g. as dimethylformamide solvate) prior to crystallization of first and second crops. Additional work-up procedures are required to recover DMF (complicated by the presence of the strong organic bases used in the dissolution of 7-ADCA) and to recycle the complexants such as β-naphtol.

Compared to these Dane salt routes the chloride hydrochloride processes can be run more simply. For example, an advantage is that the side chain acid chloride can be added as a solid directly to silylated 7-ADCA in dichloromethane, thereby simplifying the acylation step. In a typical production process, initially a suspension of 7-ADCA, acid acceptor (e.g. acetamide) in dichloromethane is azeotropically distilled to remove all traces of water prior to adding hexamethyldisilazane and refluxing further. The resulting solution of 7-ADCA silyl ester is pre-cooled to -20 deg. C to absorb the exotherm on subsequent addition of the acid chloride. The acylation is completed after a few hours stirring at -10 deg. C, and the silyl group is cleaved by acid hydrolysis (pH 1.5). The water layer is separated from dichloromethane, filtered and the product crystallized as described above.

7-ACA derived cephalosporins

A large number of semi-synthetic cephalosporins (SSC`s) have been prepared as part of structure-activity relationship studies on modifications to the side chain and 3-position of cephalosporin C. In these studies the activity of substituents could be qualitatively assigned though not potency, reflecting the complex inhibitory process in vivo. Consequently, a relatively limited number of these SSC`s have found broad clinical application and some examples are listed below.

Cephalothin

This was the first 7-ACA based SSC with high activity against penicillin resistant staphylococci. Preparation is by a modified Schotten-Baumann reaction of 7-ACA with 2-thienyl acetyl chloride in aqueous acetone/sodium bicarbonate, and isolation of the product as the sodium salt in good yield. Reaction of Cephalothin sodium salt with a slight excess of aqueous pyridine at pH 6.5/60 deg. C affords Cephaloridine (3`-pyridinium analogue of Cephalothin). Such substitution with pyridine bases proceeds better in the presence of an excess of salts such as potassium iodide. In general, since the direct substitution of the 3`-acetoxy group requires higher temperatures (60-70 deg. C; some lactam degradation), the use of iodine as an excellent leaving group is a distinct advantage. The intermediate 3-iodomethyl cephalosporins required can be prepared in situ or isolated, by the reaction of 3-acetoxymethyl cephalosporins with iodotrimethylsilane in organic solvent, followed by room temperature reaction with diverse N- and S-nucleophiles.

Cefazolin

Cefazolin is active against gram negative bacteria as a result of good stability towards β-lactamases. It has found wide clinical use as an injectable antibiotic in China. Essentially there are two alternative routes to Cefazolin, or indeed to the other SCC`s in this category, depending on whether the 3- or the 7-substituent is initially prepared. When yields are similar, then the more economic approach is to first substitute the 3`-acetoxy group of 7-ACA by 2-mercapto-5-methyl-1,3,4-thiadiazole by a method described above. Coupling of the side chain tetrazole-1-acetic acid can then follow using a conventional mixed anhydride process. Alternatively the latter 7-acylation reaction can be initially performed on 7-ACA prior to displacement of the 3`-acetoxy group by the thiadiazole. On a cautionary note the handling of the side chain can be hazardous due to the potentially explosive nature of tetrazole-1-acetic acid. Cefazolin free acid is converted to the sterile sodium salt by a salt exchange reaction using for example sodium 2-ethylhexanoate.

Cefotaxime and Ceftriaxone

The aminothiazole cephalosporins Cefotaxime (Hoechst) and Ceftriaxone (Hoffman La Roche) are two of the most important members resulting from an extensive SAR study of this series. They display an outstanding activity profile conferred in part by their stability towards β-lactamases. The original aminothiazoyl side chain was discovered accidentally by research workers at Takeda in Japan whilst substituting 7-chloroacetoacetyl-cephalosporin with thiourea. The expected side chain substituted product obtained initially, cyclised to the 2-aminothiazol-4-ylacetyl-cephalosporin. Further work on modification at C-2 of the side chain ($-CH_2-$ replaced by syn -C(=N-OMe)-) and coupling to 7-ACA gave Cefotaxime .

The side chain 2-aminothiazolyl-2-methoxyimino acetic acid can be activated for coupling either as the acid chloride hydrochloride (from PCl_5 / < -10 °C), or in situ by reaction with a Vilsmeyer reagent (DMF / $POCl_3$), or more conveniently as the active thioester of 2- mercaptobenzothiazole. The latter is probably the most suitable method for Cefotaxime and analogues since the active ester is a stable solid which reacts under mild conditions with 7-ACA ester derivates(e.g. silyl ester) resulting in high coupling yields. One drawback in the latter method is the need to efficiently remove/recover 2-mercaptobenzothiazole released during coupling, to avoid end-product contamination and reduce costs of the activated ester.

Both Cefotaxime and Ceftriaxone are injectable cephalosporins, and are converted to the salt form by reaction with sodium 2-ethylhexanoate or similar reagents. This final crystallization step usually also results in further purification in going from the free acid to the sterile salt preparation.

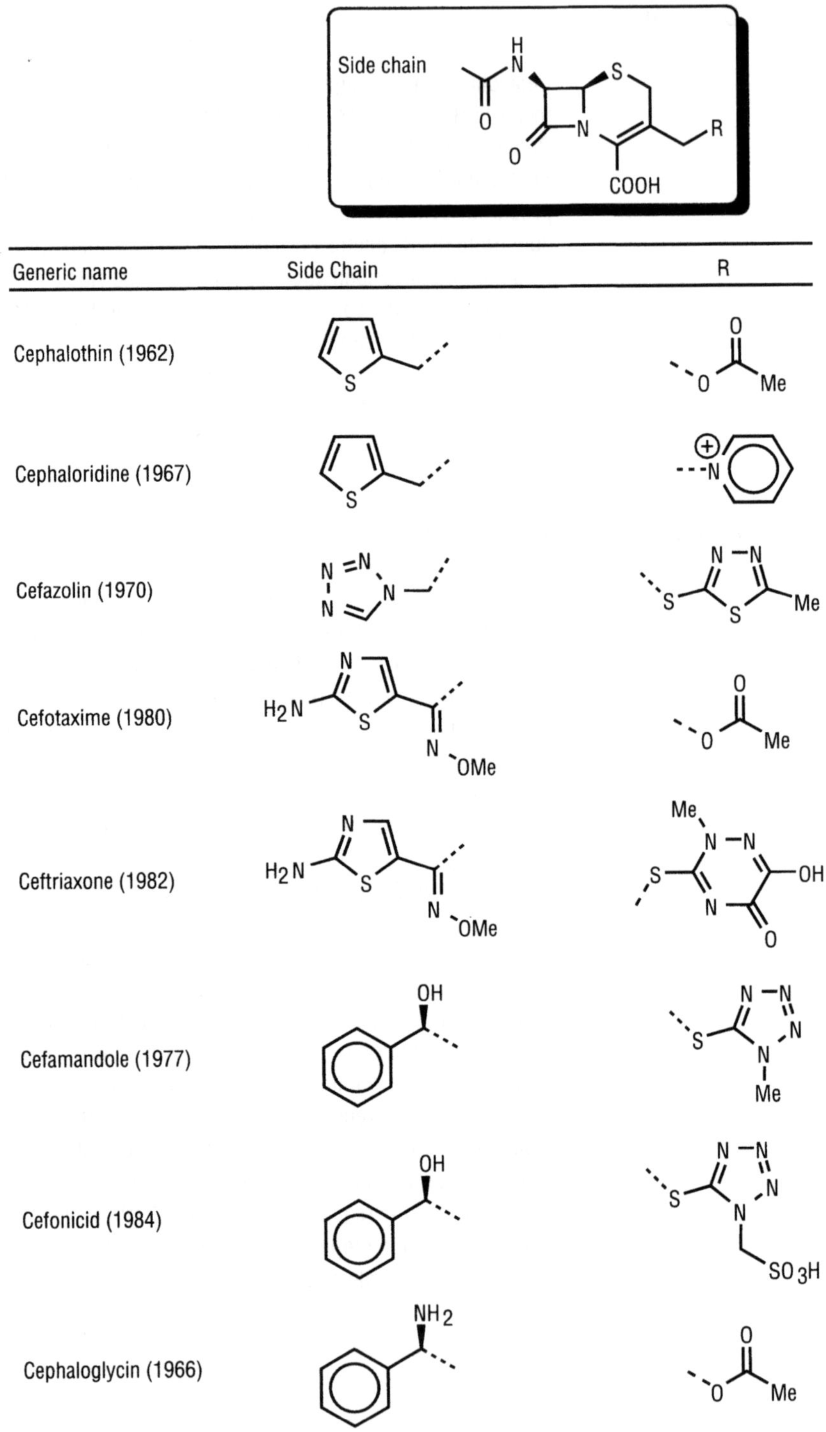

Fig. I.20 Examples of 7-ACA derived Cephalosporins in clinical use

Cefamandole and Cefonicid

Introduction of D-mandelic acid as side chain affords cephalosporins such as Cefamandole and Cefonicid with excellent activity profiles. Coupling of the O-formyl protected acid chloride of mandelic acid (DSM product) to the respective 7-ACA substituted derivatives, can be carried out efficiently under Schotten- Baumann conditions. The O-formyl group can be removed finally by treatment with base.

Cefaclor and Cephaloglycin

Acylation of 7-ACA with D(-)phenylglycine activated derivatives via the acid chloride or mixed anhydride methods analogous to Cephalexin, produces orally active Cephaloglycin. Whilst the 3-chloro-substituted analogue Cefaclor can be prepared by similar coupling procedures, the real challenge in this synthesis is not so much the final condensation step, but rather the multi-step preparation of the 3-chloro-nucleus. The latter reaction sequence admirably developed by the Lilly group involves ozonolysis and chlorination of 3-methylene cephams, and clearly determines the economics of the total process in terms of overall yield from penicillin starting material.

This brief survey has aimed to outline the industrial routes available for the synthesis of SSA`s, mainly with reference to peptide bond formation between the various side chains and β-lactam nuclei. From the numerous classical activation methods developed in the field of peptide chemistry, basically only three have been successfully/economically developed for SSA`s, namely the acid chloride, Dane salt/mixed anhydride and active ester coupling procedures. This situation has long since reached maturity, and the time is now ripe for a boost in new water based enzymatic coupling procedures to improve process economics and reduce environmental impact.

2.7 Market versus Process Development

The vast commercial success of the SSA's has quickly resulted in a change from discovery and technology driven process developments to business led strategies. With the early availability of good working processes for the required intermediates (nuclei, side chains) and the fast onset of worldwide consumption the scene was set for cost based competition. Whereas in the 1960-1980 period industrial manufacturing and competition was mainly restricted to Europe and the USA, thereafter a real global cost-competitive business emerged. Presently Western Europe, India and China are the most prominent regions for industrial production.
These factors, together with the ready availability of process know-how, the relatively low investments needed to start a commercial plant (with the fermentation of PenG or PenV as a notable exception) and the confidence in prolonging life cycles of the end products, have been and still are the drivers behind the market and

process developments of the β-lactams. Commercial success therefore, is governed by lowest cost processes preferably combined with technology leadership. Local or national policies, i.e. import duties, export incentives, tax relief, investment grants etc., might give temporarily advantages to some producers, though eventually the result is determined by continuing cost reduction through improved or innovated processes, increase of scale and integration of plants and processes.

Lowest cost and technology leadership can only be achieved through detailed market and competition monitoring, meaningful benchmarking and continuous attention to in-house process developments. At DSM a number of techniques for assessment of technologies and competitors have been used over the years. Most of these methods might be familiar to experts in marketing or economics, but will be interesting reading for workers in research and development. Moreover, the strategic power of these analyses lies mainly in its combined application over a substantial number of years (5 years or more).

2.7.1 Process profile analysis

The exploration of the pro's and con's of several process options can be done without formulas. Any chemist should be able to rank in a matrix his process alternatives against important parameters such as consumption of raw materials, equipment needed, safety and environment hazards, product quality and efficiency. Very clear process profiles (fig. I.21) are the result as shown below for three processes for a β-lactam antibiotic (fig. I.22). An enzymatic route, still in development, is compared with two environmentally safe alternatives using traditional intermediates. In all processes halogenated solvents have been eliminated. Dichloromethane, commonly used in manufacturing processes for antibiotics, is replaced by non-halogenated equivalents such as methyl acetate.

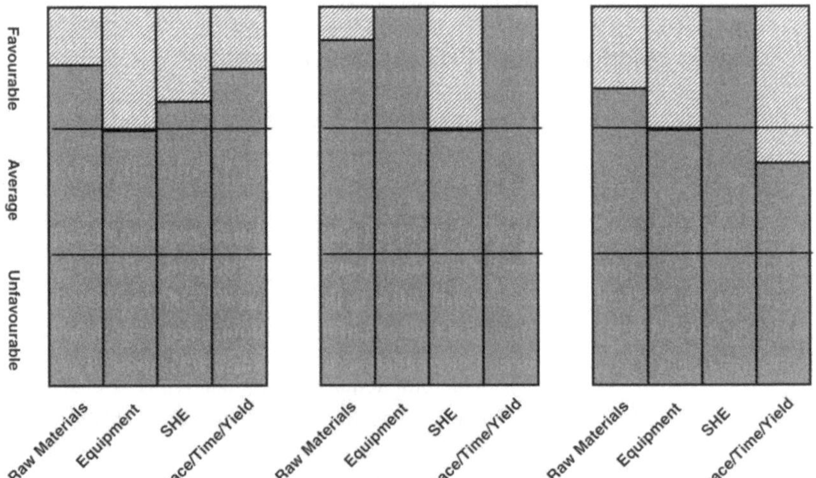

Fig I.21 Process profiles for modern routes to semi-synthetic antibiotics

In a more elaborate approach, originally developed in the pharmaceutical industry and adapted to fine chemicals by DSM Andeno, every process parameter is quantified on an individual scale. Table. I.4 shows the result for six alternatives methods to make Ampicillin measured against 14 different parameters. The best process can easily be recognized and also weak spots in alternative routes can be detected directly.

The technique of process profile analysis is very useful in:

- Choosing the best chemical routes in early stages of development.
- Comparing competitor processes.
- Assessing the right process variant for a given location.

In a broader sense the method can be used to weigh economics against ecological/ sustainability factors.

Table I.4 Detailed process profiles for 6 Ampicillin routes

	Max. Pts.	I	II	III	IV	V	VI
OPERATIONAL COSTS							
I. Raw materials	15			7			11
II. Yield: overall	5			5			3
average	5			5			4
III. Process time	10			8			6
IV. Space/Time/Yield	10			6			6
	45	28	27	31	24	28	30
CAPITAL COSTS							
V. Number of steps	10			7			5
VI. Special Equipment	10			10			7
	20	16	17	17	13	13	12
PROCESS CONTROL							
VII. Reproducibility	3			3			2
VIII. Tolerance in critical parameters	5			5			2
IX. Linear or converging proc. scheme	2			2			2
	10	6	6	10	6	6	6
INTERNAL RISKS							
X. Safety/Environment	5			5			3
XI. Health	10			10			7
	15	10	12	15	15	10	10
EXTERNAL RISKS							
XII. Availability raw materials	5			5			5
XIII. Sensitivity to government rulings	3			2			1
XIV. Patent position	2			1			2
	10	7	7	8	9	8	8
TOTAL	100	67	69	81	67	65	66

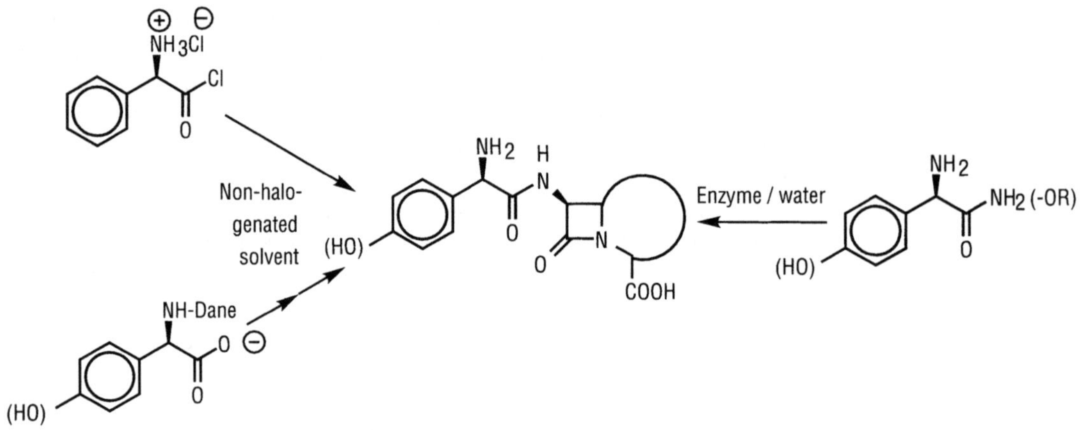

Fig. I.22 Modern routes to semisynthetic antibiotics

If desired the definition and weight of the individual parameters can be adapted. For example, in economic versus sustainability studies the parameters regarding internal and external risks can be given an impact equal to the economic factors. Also the health and environmental issues can be defined more precisely and more quantitatively (i.e. COD and BOD of waste streams, the presence of toxic metals). Profile analysis has also been used successfully in rating various synthesis methods i.e. oxidation or reduction reactions in order to define a research strategy. As in most comparative analyses 'apples should be compared with apples', i.e. all processes should be compared at the same level of development (which might be difficult for new ideas) and personal bias should be avoided by making it a group analysis or through further quantification of the points given to a specific parameter.

2.7.2 Cost curves and cost formulas

In comparing different, industrially applied routes to a single end product we have to start with the chemical and technical details. Great care should be taken to insure competing processes are compared on an equal footing with the in-house process. Also the issue of missing data on the competitor's process should be addressed, without using it as an excuse for not doing the analysis. The exercise can always be repeated when new information becomes available. Then, together with experienced process engineers, a hypothetical plant should be built for each process.

All process details and corresponding investments and costs should be compared under identical socio-economic conditions at a single location. The scale for each process should be the same and relevant to the market volume. The result, in a hypothetical case as shown in fig. I.23, could be that no discrimination between the process alternatives is found. This might be a very important conclusion; it shows that the chemical route is not the discriminating factor but other socioeconomic parameters are decisive.

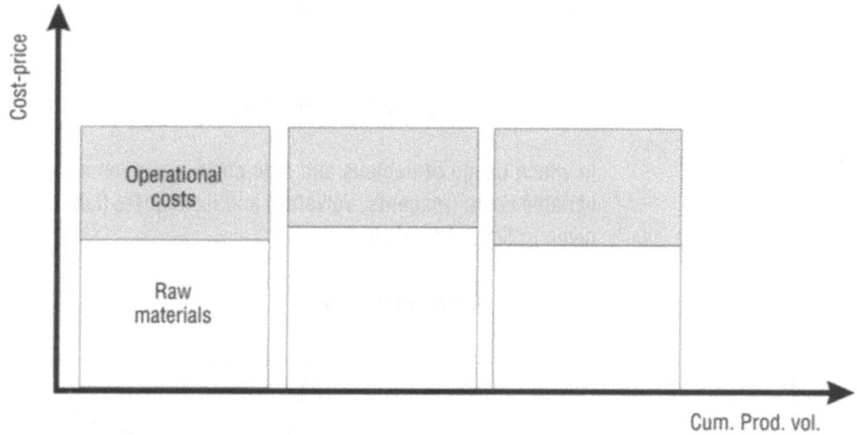

Fig. I.23 Technology assessment:
three industrial processes under identical socio-economic conditions

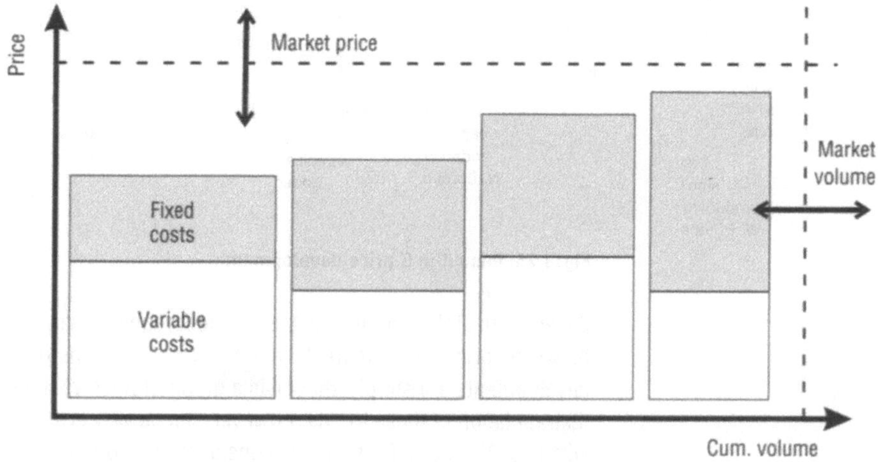

Fig. I.24 Costprice curve
(5 years after market introduction)

With these tools at hand the final step to a quantitative approach of competition analysis is easily done. Adjusting the aforementioned technology assessment to the local conditions and scale of the different competitors results in a competitor/cost price curve (as shown for 4 producers in fig.I.24). An initial investment might be needed to gather the required economic data and information on the locations involved. After that a yearly update can be quickly done. Interestingly, the power of this analysis lies in the consistency of its application over a number of years. Very often companies change their methods from year to year.

In benchmarking studies competitive positions are often expressed as cost-price formula's i.e.:

Cephalexin = 0.65 ADCA + 1.00 Dane + \$ 15,- + \$ 10,-

in which usage of nucleus and side chain are given as kg/kg Cephalexin and other variable costs (reagents, solvents) and fixed costs (labor, equipment, financing) are given in \$/kg of Cephalexin.

7.2.3 Experience curves

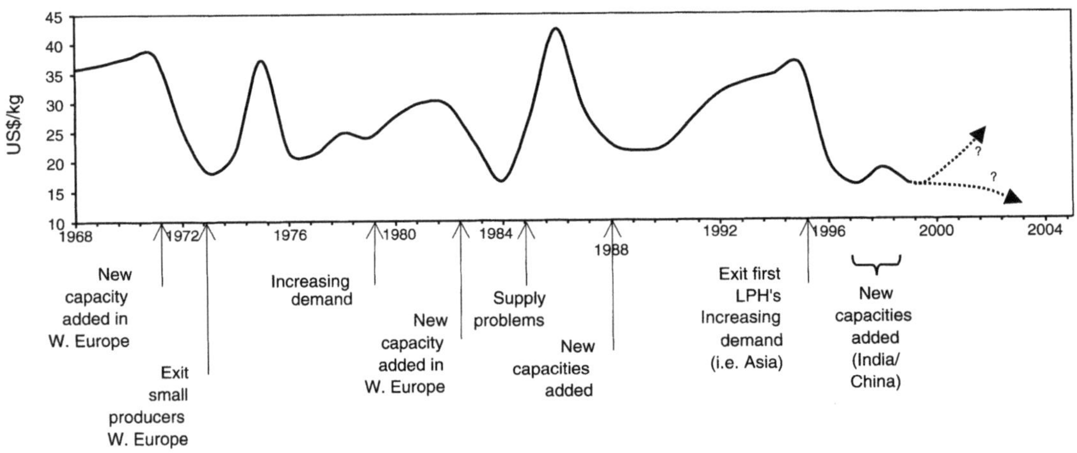

Fig. I.25 Penicillin G price development

Based on the information collected in the foregoing analyses, an interesting relation between price and market volume can be constructed. Very often price developments are simply shown as a graph of price vs. time (Fig. I.25.).

Extrapolation of these trends, however, can be very erroneous and even dangerous when used as basis for strategic considerations. Bringing in the market volume and correcting prices for annual inflation, a much more useful trend line is obtained (Fig. I.26).

Extrapolation to 2005 or 2010 by using the estimated increase in market volume gives a very useful indication of future market prices.

Knowing the future market price is of course everybody's wish. The value of this correlation, though, is much more of strategic importance. R&D can use it to assess the place of a process on the well-known S-curve for innovations and advise about investing in process improvement, developing a new process or abandoning the product altogether. In particular in products with growing markets and a global, open competition this relationship has proven to be of strategic value.

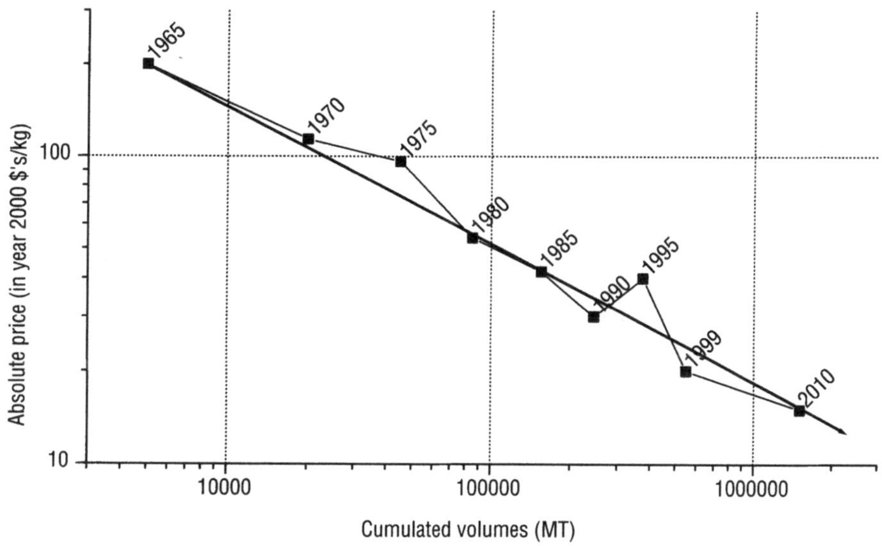

Fig. I.26 Experience curve for Pen G price

§3 Second Generation of Industrial Processes

3.1 Introduction of Biocatalysis

Over the last 10-15 years the syntheses for SSA's have been a prime playing field for the industrial development of biocatalysis. In fact, before the first large scale introductions of biocatalysis in the fine chemical industry around 1985, the utilization of enzymes in molecular transformation was already practiced in the manufacturing of some β-lactam products. During the late 1950's a suitable enzyme became available to split PenG to 6-APA and phenyl acetic acid. As early as 1960 four companies (Bayer, Beecham, Bristol and Pfizer) independently announced the discovery of these enzymes. Not many years thereafter most 6-APA producers employed the commercially readily available penicillin G acylase derived from *E.coli* as the biocatalyst of choice. For a time, however, a chemical route was highly competitive. Only when robust and immobilized biocatalyst systems came to the market did the enzyme route reach its dominant position as it has today. For 7-ADCA a similar process has been developed, whereas for 7-ACA, starting from Cephalosporin C, a two enzyme process is needed as already presented in I.1. The background and experience in fermentation of most companies involved in these products has of course been of great importance. Also, it is of no surprise that the relatively simple hydrolytic processes were the first to be carried out by biocatalysis.

Also because of historic reasons many β-lactam interested companies embarked on enzymatic synthesis of SSA's. With very little commercial success however. A selection of early attempts is shown in table I.5.

Table I.5 Enzymatic synthesis of β-lactam antibiotics

Product	Enzyme	Source	First year of publication
PenG	PA	*E.coli*	1960
Ampicillin	PA	*E.coli*	1969
Cephalexin	AH	*A.pasteurianum*	1972
Cephaloglycin	AH	Several	1972
Amoxicillin	PA	*E.coli*	1975
Cefadroxil	AH	*Ps-melanogenum*	1975
Cefamandole	PA	*E.coli*	1992
Cefazolin	PA	*E.coli*	1992
Cephalotin	PA	*E.coli*	1996

Well known, original drawbacks of enzymes in organic synthesis were at the base of this lack of success (i.e. dilute aqueous systems, expensive and unstable enzymes). In the period 1980-1990 the momentum geared towards biocatalysis for several reasons:

- Original drawbacks disappeared because cheap industrial enzymes (mainly hydrolases from the detergent industry) in robust, immobilized form became available.
- Chemical processes for SSA's had reached a high degree of maturity whereas there was an increasing need to replace traditional, stoichiometric processes in order to improve product/waste ratio's and to reach further cuts in costs;
- The unsuccessful translation of chemo-catalytic process know-how from petro-chemicals to multifunctional fine chemicals such as β-lactams.
- The ready acceptance of enzymes by organic chemists as part of their toolbox.
- The ease with which enzymatic processes could be run in existing equipment.

With these stimuli at hand a number of companies i.e. NOVO, Dobfar, BMS, DSM/ Gist Brocades, embarked on the development of biocatalytic routes to SSA's.

3.2 Enzymatic couplings with 6-APA and 7-ADCA

Cephalexin was the first successful application of biocatalysis in the synthesis of SSA's on industrial scale.

Simple thermodynamic coupling of 7-ADCA and D (-) PG was impossible because of incompatibility of the side chain Zwitter ion with the enzyme, although an early French patent (1968) claimed such a process for Ampicillin.

Kinetically controlled coupling using 7-ADCA and D(-)PG as amide or ester can be done efficiently but requires delicate choices of reaction conditions to reach optimal conversions. Complicating factors are undesired enzymatic hydrolysis of both the side chain precursor and the end product to give D(-)PG and 7-ADCA. In particular PG formation is problematic because of its very low solubility.

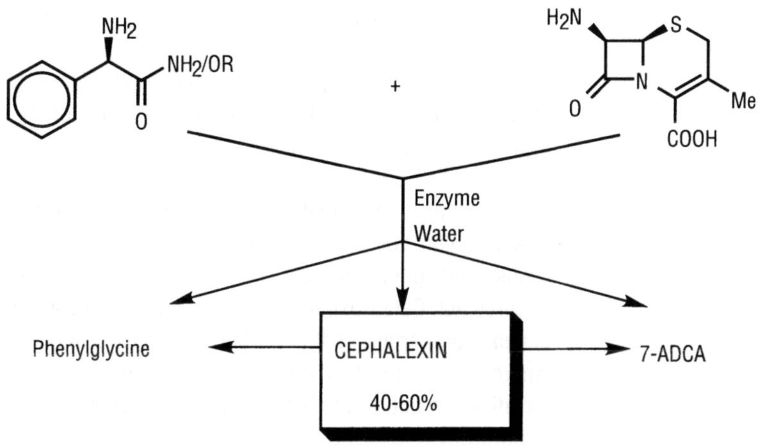

Fig. I.27 Enzymatic Cephalexin synthesis

Reactions are run in water at 0-20°C using penicillin acylase from E.coli or related sources as catalyst and varying stoichiometric ratio's of nucleus and side chain precursor depending on the down stream processing. In typical examples 40 - 60 % of Cephalexin is obtained per cycle followed by several isolation and recycle steps. Down stream processing is facilitated by running reactions in sieve-bottom reactors allowing simple separation of immobilized biocatalyst particles from product suspensions or solutions. A number of pH shifts and crystallizations allow separation and isolation of 7-ADCA (to be recycled), D (-) PG (to be reused) and the desired Cephalexin. Product quality is equal or better than traditionally obtained Cephalexin. DSM put a production plant on stream in Barcelona in 1997.

Obviously, much more work can be done to bring this first example to maturity. The following chapters will show a larger part of this research. So far, substantial gains have been made on the environmental issues as shown below (Table I.6).

Table I.6 Environmental characteristics of chemical vs. enzymatic Cephalexin

Item (per kg of Cephalexin)	**Enzymatic Process**	**Chemical Process**
Utilities		
- Electricity (Kwh.)	+50 %	
- Steam (as oil. eq.)	-60 %	
- Water (m3)	+200 %	
- Liquid nitrogen (kg.)	-100 %	
Waste Streams (kg)		
- Inorganics	0.5	0.5
- Organics (non halogenated)	0.2	1.0
- Solvents (non halogenated)	0.3	1.7
- Solvents (halogenated)	none	0.9

DSM has developed similar enzymatic couplings for Ampicillin, Amoxicillin, Cefaclor and Cefadroxil. Each process has its own, different characteristics and elaborate development programs are needed to reach optimal results. For Ampicillin the high product solubility and the lower stability of the β-lactam (compared to the cephalosporins) means efficient, downstream processing and product isolation is hindered. For Cefaclor recourse had to be taken to product recovery through complexation with aromatics (i.e. β-naphthol), leading to additional recycling loops. In Amoxicillin, advantage can be taken of the low product solubility resulting in good yields and good product quality because β-lactam decomposition is also suppressed. Cefadroxil has proven to be a particularly difficult case. Solubility and stability properties of the starting materials complicate the coupling conditions, whereas high product solubility results in complex down stream processing. Isolation through complexes with aromatics such as β-naphthol offers an opportunity.

3.3 First steps towards (Process) Integrations

Next to continuing process improvements and process innovations, the integration of manufacturing units has been a vehicle for cutting costs and increasing economics of scale. Taking Ampicillin synthesis as an example, in the period 1965-1975 almost all individual process steps were carried ou by different, individual companies. Thus in 1970, dl-phenylglycine was produced by DSM, D(-)phenyl-glycine by Ward Blenkinsop, D(-)phenylglycylchloride hydrochloride by Océ Andeno, PenG and 6-APA by Gist Brocades and Ampicillin by pharmaceutical companies such as Beecham and Bristol Meyers. Some 25 years later many names have disappeared and have been replaced by integrated conglomerates of companies. Nowadays DSM is the most completely integrated company in the market segment of semi-synthetic antibiotics.

Thus, for Ampicillin and Cephalexin, DSM is fully backward integrated starting with sugar (for PenG fermentation) and toluene (for side chain preparation). Through continuing process improvements and innovations, in particular the introduction of biocatalysis, much more efficient and integrated manufacturing positions have been reached. Fig. I.28. shows the results for Cephalexin, whereby an original 10-step process employing stoïchiometric processes only, has over the years been replaced by a 6-step process dominated by biocatalysis.

Also for Amoxicillin and Cefadroxil DSM has reached full backward integration. Starting with *n*-butane for maleic anhydride and glyoxylic acid (at DSM/Chemie Linz in Austria) and toluene for benzoic acid and phenol (at DSM Rotterdam), the synthesis of the D(-)p-hydroxy phenyl glycine side chain is completed at DSM Deretil in Spain. The urea required in the latter process is obtained from DSM, Geleen, The Netherlands.

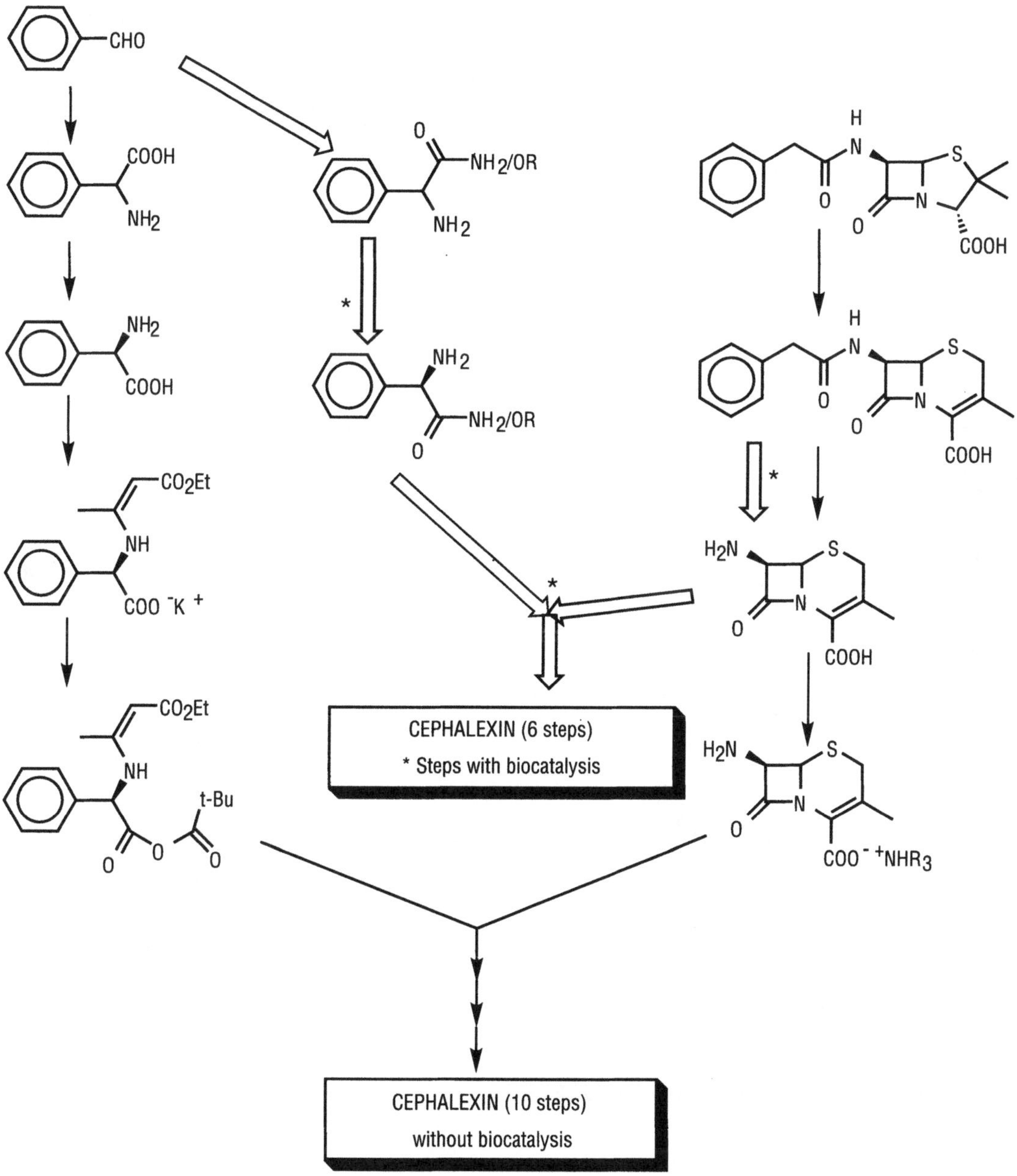

Fig. I-28 Process integration in Cephalexin synthesis

As for Ampicillin, the required PenG, 6-APA and 7-ADCA are produced at DSM-Gist in Delft, whereas couplings to the final bulk Amoxicillin or Cefadroxil are carried out at several sites of the global DSM network (i.e. Latin America, Europe, Egypt, India and China). Fig. I.29. shows the integration scheme, whereby further branching is omitted e.g. phenol to caprolactam. Market and social developments permitting, further integration through reductions in the number of manufacturing sites is an obvious next step.

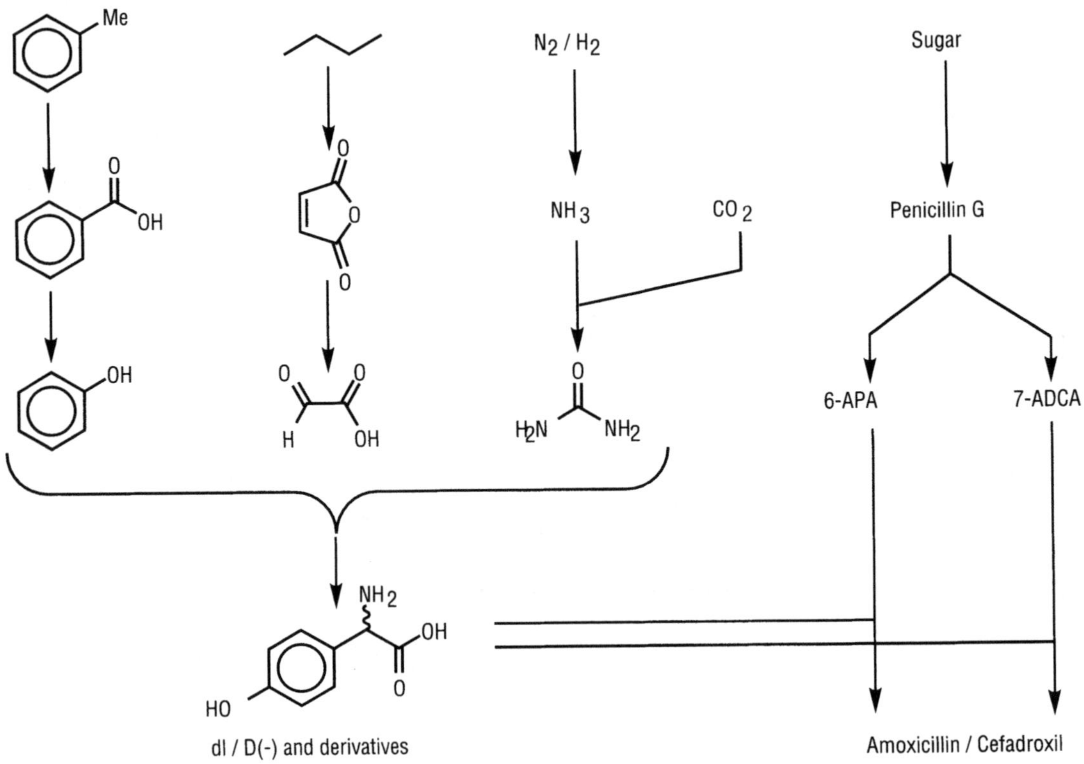

Fig. I-29 Integrated synthesis at DSM

§4 Third Generation Processes?

4.1 Marketing prospects

Any further process innovation in the synthesis of β-lactam antibiotics will be dictated by the market developments in particular the future need for new or existing anti-infectives. Although penicillins and cephalosporins are products with so far an unsurpassed therapeutic value, the call for new products is becoming increasingly louder. After a period of composure, around 1980, new infectious diseases and increasing resistance against existing products created a new impetus for research and development of new antibacterials. This lull in product development, however, implies that it is very unlikely that new antibiotics with great market impact (i.e. a new 'blockbuster') will emerge before 2010. A pipeline analysis shows some 150 molecules are in development; one third in clinical trials and two thirds in preclinical phases. Three broad lines of attack against the present infections can be recognized. First of all several new molecules are screened for which the leads are obtained from newly detected natural antibiotics (from micro-organisms, plants, animal or human sources). Also increased insight in infection mechanisms and their relations to the genetic structures (at molecular and macro-molecular levels) allows for new leads and strategies. The pipeline analysis shows that ca. 70 new molecules are under study at this moment as a result of these approaches. Several studies mention the oxazolidinones developed by Upjohn as the most promising result. Inhibition of micro-organisms with the most notorious resistance against the present arsenal of ca. 300 antibacterials is at the basis of this positive outlook (i.e. methicilline resistant staphylococcus aureaus, MRSA, and vancomycine resistant enterococcus, VRE).

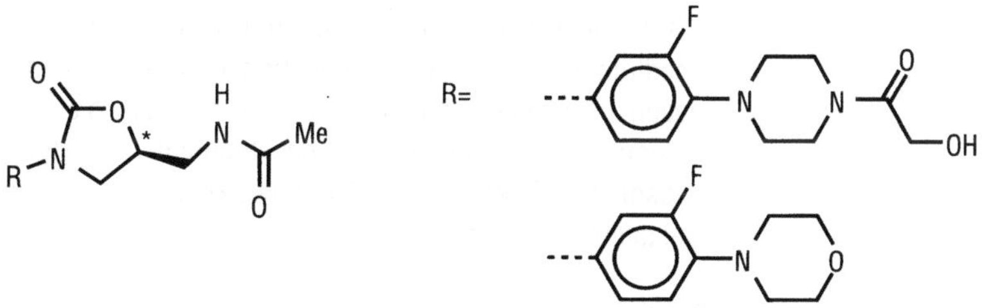

Fig. I-30 Oxazolidinones, examples of new antibacterials

A second line of approach stems from the increasing insight in the working mechanisms of the present antibiotics including the mechanisms for the development of resistance.
Three major mechanisms applicable to most of the present products are recognized: efflux pumps (pumping incoming antibacterials out of the cell), intracellular protection (in vivo modification of the target of the antibiotic) and chemical modification (inactivation of the antibiotic through specific enzymes, i.e. β-lactamases). All three mechanisms are used as leads for new products. Although completely new molecular structures might be the result of these studies, modifications and combinations of existing molecules is a more likely outcome.
Several new β-lactams, macrolides, substituted tetracyclines, aminoglycosides and glycopeptides can be found in the pharmaceutical development programs. A total of ca. 75 products are in development presently of which ca. 30 are in clinical trails. Still, there is a long way to go before products with substantially better therapeutic value will reach the market place.

The third line of approach is from a bulk drug manufacturing point of view the most interesting one: fighting development of resistance against the present drugs through a more rational approach in usage and prescription, and thereby lengthening the life cycle of existing drugs. This is gaining momentum in several countries. Increasing restrictions of antibacterial use in a prophylactic mode, i.e. in animal feed, household products or cosmetics, is a first step in which several countries have already taken legal measures. This is just one line of attack. A second one is emerging in various European countries, i.e. Finland and UK, in which doctors and hospitals are urged to prescribe antibiotics in a more rational way. Elimination of antibiotic use for viral infections such as in the common cold is an easy first target. Several studies show that up to 50 % of antibiotics consumption is unnecessary. Other studies mention the possibility of fighting most present bacterial infections with only 50 % of the present product portfolio, when drug prescription is done in a strict rational and rotational scheme. From a social and ethical point of view this might be a bridge too far.

The third line of attack, however, might need some social and political pressure. This refers to the problem of incomplete courses of treatment, which are common in several developing countries. Next to political measures it can be seen as a challenge to the pharmaceutical industry to develop a delivery and pricing system in which the incentive to stop a cure after 1 or 2 days is taken away.

Although these lines in overcoming resistance to present drugs might in the short or medium term lead to some reduction in market demand, it will be a key condition to reach a sustainable business position for today's antibacterials.

Taking into account the long development times for completely new drugs it is reasonable to expect an enduring market presence of existing products such as the β-lactam antibiotics.

4.2 Process Prospects

From a scientific point of view it is relatively easy to foresee the next steps in process improvements and innovations. Whereas the β-lactams were the first products to feel the beneficial effects of biocatalysis, they will very probably also be the first to see full biosynthetic manufacturing processes. A first example is already being put into practice.

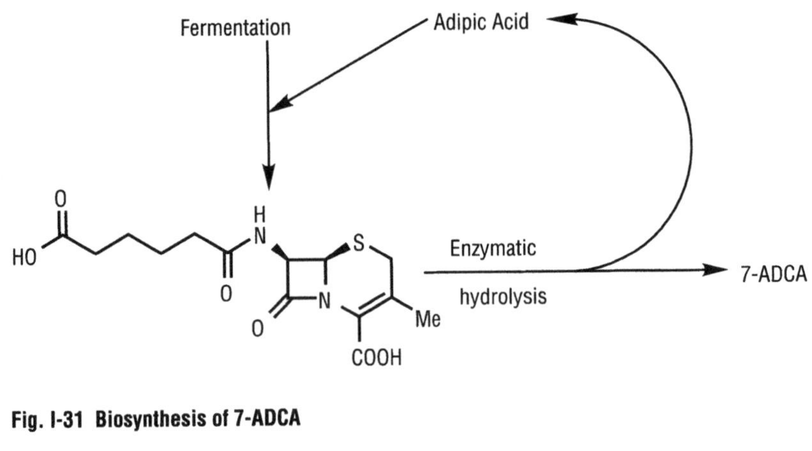

Fig. I-31 Biosynthesis of 7-ADCA

With the aid of rational metabolic pathway engineering, DSM Gist has developed a fermentative process for 7-ADCA. Thus the elegant, stoïchiometric chemical conversion of the five-membered ring in the penicillin skeleton into the six-membered ring in the cephalosporins can now be performed within a micro-organism. Genetically engineered *P.chrysogenum* does the trick. The process performs best when the traditional phenylacetic acid side chain precursor is replaced by adipic acid. The latter can readily be removed by enzymatic hydrolysis using a glutaryl acylase related enzyme. Advantages of this process are similar to the enzymatic processes given before i.e. higher purity of end product, increased efficiencies and elimination of halogenated solvents and silylating agents.

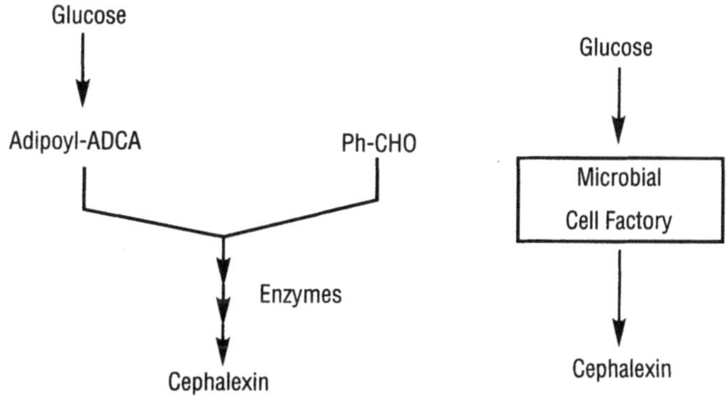

Fig I-32 Cascade catalysis and direct fermentation of Cephalexin

Although full biosynthesis of all major β-lactam end products might seem an attractive future, fermentation of non-natural side chain molecules might prove too difficult. On the other hand, several enzymes capable of recognizing these types of structures are already available from the biocatalytic processes described in the previous sections. A likely development will be a further streamlining and short-cuts in the present routes, whereby integrated processes at a single location could be a result. Market prospects will strongly affect intensity and direction of these endeavors, particularly so for the fully fermentative approaches. It is tempting however to envisage synthetic schemes as shown in Fig. I.32. for Cephalexin.

§5 Further Reading

5.1 Books

1. J.H. Sheenan, 'The Enchanted Ring, The untold story of Penicillin'. MIT Press, Cambridge, Massachusetts; London, England (1982).
2. E.H. Flynn (ed.), 'Cephalosporins and Penicillins: chemistry and biology'. Academic Press, New York and London (1972).
3. J. Baddiley and E.P. Abraham, 'Penicillin Fifty years after Fleming'. The Royal Society, London (1980).

5.2 Reviews

1. Reports from Michael Barber and Associates (UK): Chemica 1, Penicillin (1986), Chemica 2, 6-APA, 7-ADCA (1986), Chemica 3, Semisynthetic Penicillins (1987), Useful Book (1991), The Manufacture of Penicillins and its derivatives (1999), The Penicillins Business (1996) and several other MBA publications.
2. J. Wilson, 'Antibacterial products and markets', Scrip Reports, UK (1997).
3. S.C. Stinson, 'Drug Firms restock Antibacterial Arsenal', Chem. & Eng. News, Sept. 1996, 75-100.
4. C.H. Henry, 'Antibiotic Resistance', Chem. & Eng. News, March 2000, 41-58.
5. M.I. Page and A.P. Laws, 'The Mechanism of catalysis and the inhibition of β-lactamases', Chem. Commun. 1998, 1609-1617.
6. A. Bruggink, E.C. Roos and E. de Vroom, 'Penicillin Acylase in the Industrial Production of β-lactam Antibiotics', Organic Process Research & Development, 2, 128-133 (1998).
7. E.J.A.X. van de Sandt and E. de Vroom, 'Innovations in cephalosporin and penicillin production: Painting the antibiotics industry green'. Chemica Oggi, May 2000, 72-75.
8. A. Bruggink,'Green solutions for chemical challenges, biocatalysis in the synthesis of semi-synthetic antibiotics', in B. Zwanenburg et. al. (eds.), Enzymes in Action, 449-458 (2000), Kluwer Academic Publishers (The Netherlands).

Chapter II Molecular Precision in the Chemistry of Cephalosporin type Antibiotics *57*

§1 Introduction ***57***

§2 Clathrate type complexation of cephalosporin antibiotics *60*
2.1 Complexation with β-naphthol *60*
2.2 Induced fit phenomena in clathrate formation *66*
2.3 Prediction of clathrate formation by molecular modeling *78*
2.4 Efficiency of naphthalene derived complexing agents *78*
2.5 Green complexants for the clathration mediated enzymatic synthesis of Cephradine *83*
2.6 Synthesis of cephalosporins using a clathration induced asymmetric transformation *86*

§3 Synthesis of amino acid side chains ***89***
3.1 Relevant literature *89*
3.2 Regiochemistry of the Mannich type condensation to hydroxy-phenylglycine (HPG) *91*
3.3 Synthesis of *p*-hydroxy-phenylglycine from p-benzoquinone *92*

§4 Modification of the β-lactam nucleus of cephalosporins ***95***
4.1 Synthesis of 3-carboxycephems from 7-aminocephalosporanic acid (7-ACA) *95*
4.2 Synthesis of 3-halocephems from 3-carboxycephems *96*
4.3 Wittig reactions with 3-formylcephalosporins *98*
4.4 Concluding remarks *99*

§5 References ***100***

Chapter II Molecular Precision in the Chemistry of Cephalosporin type Antibiotics

B. ZWANENBURG, G.J. KEMPERMAN, J. ZHU, R. DE GELDER, F.J. DOMMERHOLT, G.T.M. TITULAER, R. KELTJENS AND A.J.H. KLUNDER (DEPARTMENT OF ORGANIC CHEMISTRY, NSR INSTITUTE FOR MOLECULAR STRUCTURE, DESIGN AND SYNTHESIS, UNIVERSITY OF NIJMEGEN)

§1 Introduction

A characteristic feature of the second generation of industrial processes for the production of semi-synthetic penicillin and cephalosporin type antibiotics is the extensive application of biocatalysts for various molecular transformations. As outlined in Chapter 1 (section 3.1) in the passed 30 years several β-lactam antibiotics have been synthesized using enzymes. At present, the original drawbacks of enzymes being expensive, difficult to handle and only applicable in dilute aqueous solution, have been overcome thanks to the introduction of more robust and immobilized biocatalysts. Moreover, the biased opinion of many organic chemists regarding the use of bioreagents and biocatalysts for synthetic purposes no longer exists. On the contrary, the attractiveness of bioreagents and biocatalysts in organic synthesis is now widely recognized.

In this chapter we will focus mainly on (the synthesis of) cephalosporin type antibiotics. For the coupling of the β-lactam nucleus to the amino acid side chain effective use of enzymes can be envisaged. This enzymatic coupling, which is illustrated for Cephalexin in scheme II.1, does not require protective groups, activating agents, or low temperatures and can be conducted in water. Hence, this process may have an enormous benefit for the environment. Although immediate economical benefit is difficult to obtain, this process leaves more opportunities for future cost reduction than the conventional chemical coupling.

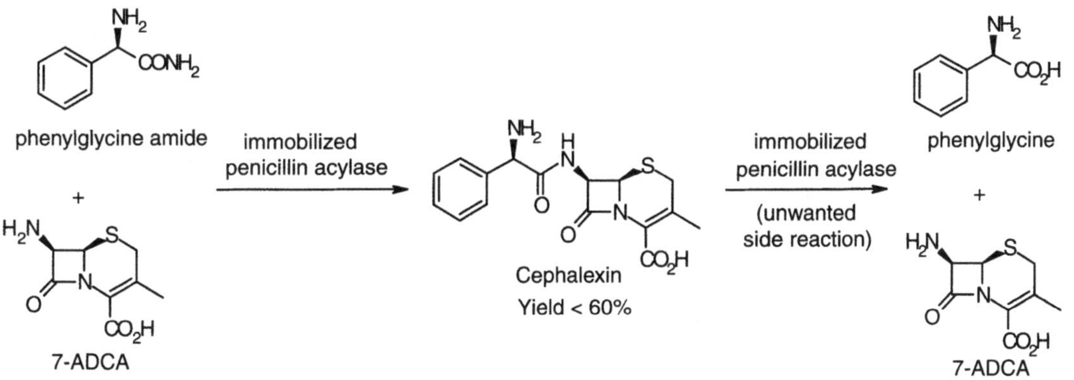

Scheme II.1 The enzymatic synthesis of Cephalexin.

This enzymatic process has still several drawbacks of which two, the major ones, are explained below.

i. Low coupling efficiency due to secondary hydrolysis of the product.
Due to the unfavorable equilibrium constant of the thermodynamic coupling[1], the enzymatic reaction has to be carried out in a kinetic fashion. This means that the amino acid side chain must be offered to the enzyme as an ester or amide derivative. In that form the enzyme can couple the side chain to the nucleus via an intermediate enzyme-acyl complex[2]. The thus formed product however, also serves as a substrate for the enzyme and as a consequence the product is hydrolyzed to nucleus and amino acid side chain. The side chain resulting from this secondary hydrolysis is no longer present in the form of an ester or amide but obtained as the amino acid, which cannot be coupled again. As a result the coupling efficiency and thus the yield of the enzymatic reaction are low. A second source of unreactive side chain stems from the undesired enzymatic hydrolysis of the activated side chain.

ii. Difficult down-stream processing.
The reaction mixture resulting from the enzymatic coupling is an aqueous solution containing starting materials, product and side products, which are all amino acid derivatives. From this complicated reaction mixture the desired antibiotic must be isolated in a pure form. Conventional methods for the isolation of the cephalosporins from aqueous solutions cannot be applied as they require higher concentrations.

By eliminating these drawbacks important cost reductions can be achieved. A solution for the problems during the down-stream processing was given in a patent of NOVO Nordisk[3]. In this patent it is claimed that after the addition of β-naphthol to an aqueous solution of Cephalexin, a precipitate is formed which contains Cephalexin and β-naphthol in a 2:1 ratio. The applicability of this concept of complexation was investigated both for the isolation of Cephalexin from aqueous solutions and by direct incorporation during the enzymatic synthesis[4]. Complexation with β-naphthol during the enzymatic synthesis, which is depicted in scheme II.2, appeared to work surprisingly well. After Cephalexin is assembled enzymatically, it immediately precipitates as a complex with β-naphthol. As in this way the product is withdrawn from the solution, it is not susceptible to secondary hydrolysis anymore, which results in a substantially higher yield of Cephalexin. In addition, after completion of the reaction the product can readily be separated from the reaction mixture by filtration of the complex. Subsequent hydrolysis of the complex yields pure Cephalexin, while β-naphthol can be reused.

Besides the application in the enzymatic synthesis, complexation of Cephalexin with β-naphthol can also be used for the isolation of a second crop of product from aqueous waste streams resulting from the chemical coupling. In this way a considerable increase of the yield can be realized.

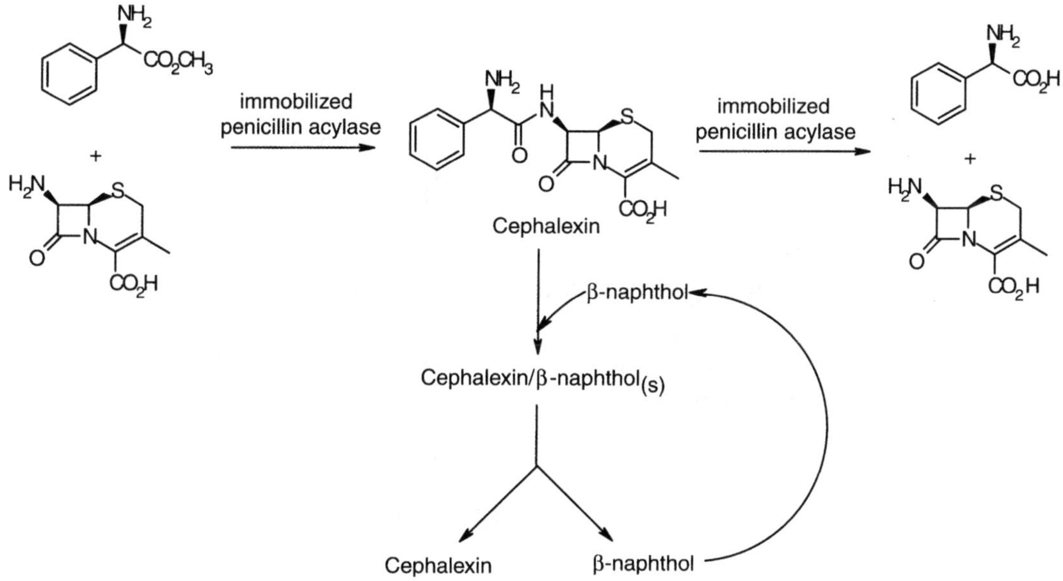

Scheme II.2 Complexation with β-naphthol during the enzymatic synthesis of Cephalexin.

The enzymatic coupling procedure in the production of cephalosporin type antibiotics is a typical example of greening of chemical processes, *i.e.* the use of environmentally benign operations employing biocatalysis. Currently, green chemistry receives worldwide much attention as its principles will considerably reduce the amounts of waste, which are inherent to the conventional chemical processes[5]. Moreover, the public image of chemistry will be improved enormously by including the concepts of green chemistry in the chemical process industry, especially in the production of fine-chemicals.

Despite the industrial relevance of the aforementioned complexation for an efficient down-stream processing of cephalosporin antibiotics, the molecular structure of the β-naphthol complexes had not been elucidated when we started our research in this area. In this chapter the structure of these complexes will be discussed in detail. In addition, the search for other complexing agents will be described. Moreover, a study of the efficiency of some complexing agents for the removal of cephalosporins from aqueous solutions and the compatibility of such agents with the enzymatic coupling conditions will be presented. It will also be demonstrated that selective complexation of cephalosporins can be utilized to achieve an asymmetric transformation of these β-lactam antibiotics.

The second topic treated in this chapter deals with the synthesis of the amino acid side chain *p*-hydroxy-phenylglycine, which is present in the important antibiotics Amoxicillin and Cefadroxil. For this amino acid four different approaches can be envisaged, as outlined in Scheme II.3

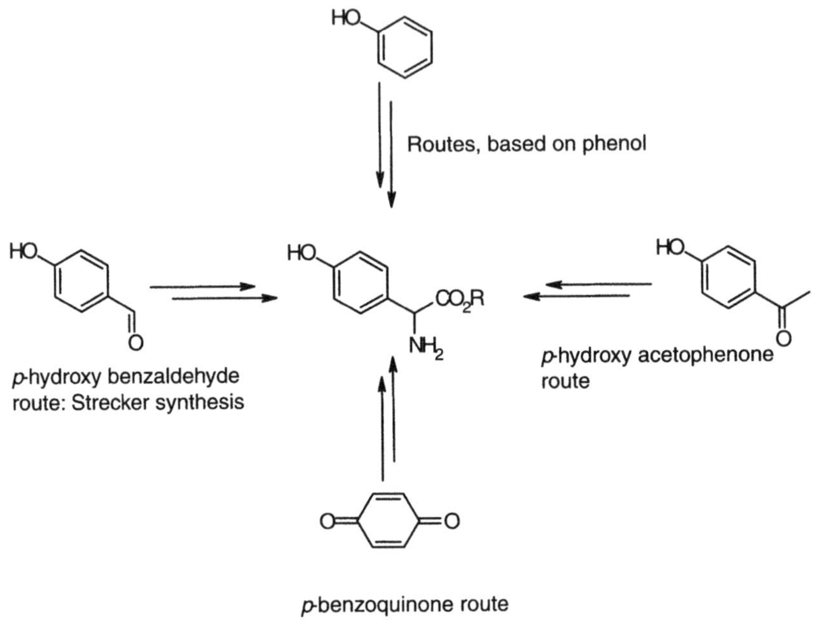

Scheme II.3 Different approaches for the synthesis of *p*-hydroxy-phenylglycine

New insights in the route starting from phenol will be described and furthermore, a novel synthesis using benzoquinone as the starting material will be presented.

A third issue to be treated in this chapter, concerns structural modifications of the β-lactam moiety, especially that present in cephalosporin derived antibiotics. Modification of the substitution pattern of the β-lactam nucleus opens avenues for the synthesis of new β-lactam type antibiotics. This approach is nicely exemplified by the series of cephalosporin derived antibiotics with structural variations at the C-3 carbon of the cephem skeleton (see Chapter 1, section 2.6). It is important to note that even relatively minor structural variations of these antibiotics may have a considerable effect on the therapeutic performance. Fine-tuning of the substituents in cephalosporins is still an important issue in the current development of this class of antibiotics. In this chapter some synthetic approaches to structural variation in cephems will be described.

§2 Clathrate type complexation of cephalosporin antibiotics

2.1 Complexation with β-naphthol[6]

The cephalosporins **1**, **2**, **3** and **4** selectively form complexes with β-naphthol[3], which precipitate from an aqueous solution. In this crystalline state, the cephalosporins are less susceptible to degradation. Moreover, complexation provides a method for the isolation of the product, since the precipitated complex can be separated from the reaction mixture by simple filtration. This facilitates down-stream processing in the large-scale production of these cephalosporins, which indicates the industrial relevance of this method. After decomplexation, which can be achieved by acidification of an aqueous suspension followed by extraction with an organic solvent, the cephalosporins can be obtained in pure state from the aqueous phase, by neutralization and crystallization.

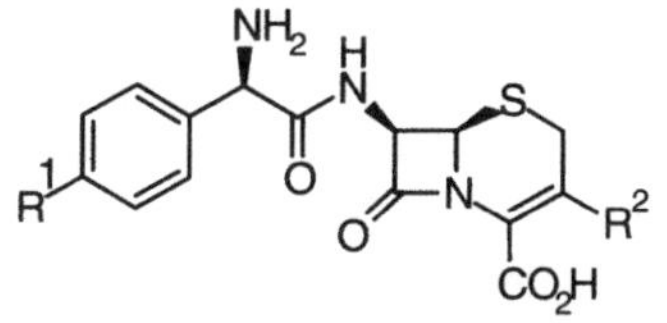

1 Cephalexin $R^1 = H$, $R^2 = CH_3$
3 Cefaclor $R^1 = H$, $R^2 = Cl$
4 Cefadroxil $R^1 = OH$, $R^2 = CH_3$

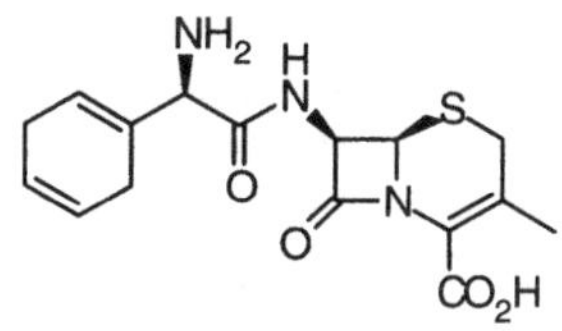

2 Cephradine

Despite the industrial relevance, the molecular structures of these β-naphthol complexes have not been elucidated prior to our study. Only the ratio of the cephalosporin and β-naphthol has been reported, as well as the water content.[3] Crystal structure analysis of complexes of β-naphthol with β-cyclodextrin[7] and with androsta-1, 4-diene-3, 17-dione[8] revealed that in the former case β-naphthol is the guest molecule, whereas in the latter case no clear distinction can be made between host and guest molecules.

An intriguing question is whether β-naphthol serves as the host or acts as a guest molecule in the complexes with cephalosporins. Complexes of cephalosporins with β-naphthol are usually obtained as precipitates upon treatment of an aqueous solution of these antibiotics with β-naphthol. For the preparation of single crystals for the X-ray analyses however, methanol was used as a co-solvent.

Remarkably, the crystal structures of the complexes of Cephalexin, Cephradine and Cefaclor with β-naphthol are isomorphous, as can be deduced from the powder diffraction patterns shown in Figure II.1.

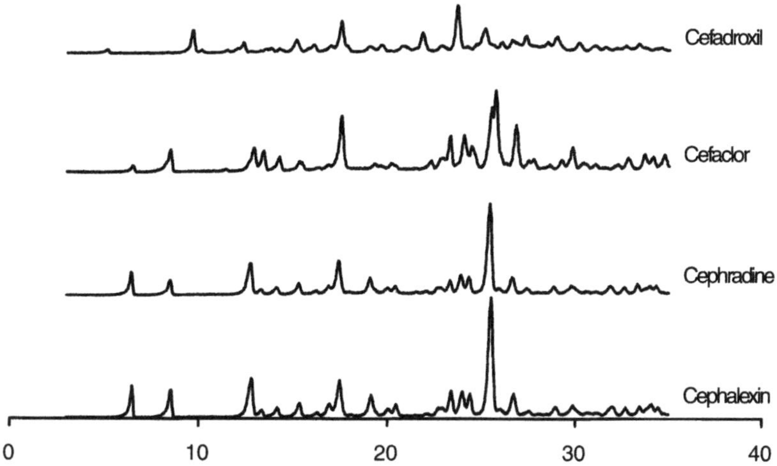

Fig. II.1 The powder diffraction patterns of the β-naphthol complexes of Cephalexin, Cephradine, Cefaclor and Cefadroxi

Although the van der Waals radius of chlorine is comparable with that of a methyl group and partial saturation of a phenyl ring hardly results in structural changes, in general, such subtle molecular alterations may cause drastic deviations in crystal structures. The β-naphthol complexes of cephalosporins **1**, **2** and **3** are of the clathrate type in which the cephalosporin serves as the host and β-naphthol as the guest molecule. In these clathrates, the hosting molecules form 2-dimensional layers, in which they are held together by hydrogen bonding and electrostatic interactions. The 2-dimensional layers are packed in such a manner that a 3-dimensional structure is formed. The remaining cavities are filled with β-naphthol and water molecules. In contrast, Cefadroxil complexes with β-naphthol as a clathrate with an entirely different 3-dimensional structure, lacking the two-fold symmetry that is present in the complexes of **1**, **2** and **3**. The large difference between the structure of the complex of Cefadroxil and the complexes of Cephalosporins **1**, **2** and **3** is demonstrated in the powder diffraction patterns shown in Figure II.1.

Figure II.2 shows the crystal structure of the Cephalexin/β-naphthol complex viewed in the direction of the b-axis. The cephalosporin molecules are in the zwitter-ionic form. Within the 2-dimensional layers, the host molecules are assembled in a head to tail fashion. One carboxylate group has both hydrogen bonding and electrostatic interactions with two ammonium groups and vice-versa. This results in 4-point junctions at which the host molecules have strong non-covalent interactions with each other. In the crystal structures of the Cephalexin, Cephradine and Cefaclor complexes, the antibiotic molecules adopt an arrangement in which from a 4-point junction, two molecules go up and two go down resulting in 2-fold symmetry along the b-axis, as is visualized in Figure II.3a. Building up these 4-point junctions leads to the formation of 2-dimensional layers of cephalosporin molecules. In addition, this arrangement has the consequence that the 2-dimensional layers contain holes as is shown in Figure II.3b. Basically, this is the cause of the formation of channels when the 3-dimensional cephalosporin framework is built-up from these 2-dimensional layers.

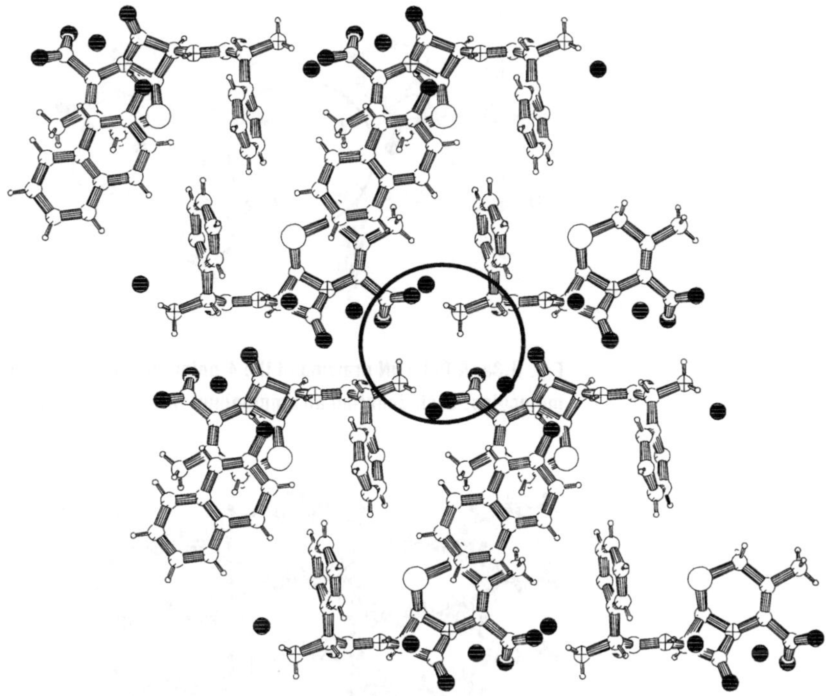

Fig. II.2 PLUTON drawing of the structure of the Cephalexin/β-naphthol complex viewed along the b-axis.

The channels shown in Figure II.3b, are filled with water and β-naphthol. The water molecules are positioned at the polar regions inside the channels, whereas β-naphthol is sandwiched between apolar parts of the antibiotic molecules. The water molecules are involved in multiple hydrogen bonding with both cephalosporin and other water molecules. The amino- and amide- hydrogen atoms are pointing towards the oxygen atoms of the water molecules. In addition, the hydrogen-oxygen distance is about 2 Å, which is the appropriate distance for effective hydrogen bonding (Figure II.3c). Although the positions of the hydrogen atoms of the water molecules could not be determined by X-ray analysis, it is assumed that they also participate in the formation of the hydrogen bonding network. This assumption is justified as follows. The distance of the water molecules towards the carbonyl- and carboxylate groups is about 3 Å. Taking into account the length of the oxygen-hydrogen bond in water of 0.95 Å , it is reasonable to assume that indeed a hydrogen bond is present. Except for the hydrogen bonding and electrostatic interactions at the 4-point junctions, all other hydrogen bonding interactions between host molecules involve water molecules. These interactions give the 2-dimensional layers a substantial additional strength. This suggests that the water molecules play the role of cement in the crystal. Recently it was shown, that water incorporated in crystal structures often serves as a gluing agent[9].

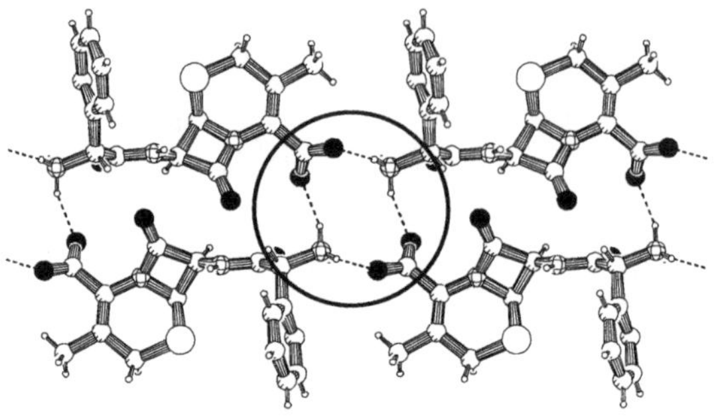

Fig. II.3a A PLUTON drawing of the 4-point junction formed by four Cephradine molecules, each donating an ammonium- or a carboxylate group.

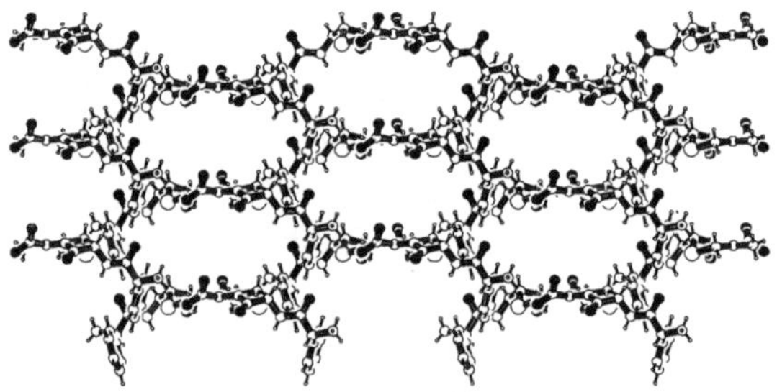

Fig. II.3b The 2-dimensional layer viewed in the direction of the c-axis. The holes which are present, are a consequence of the arrangement of the cephalosporin molecules at a 4-point junction.

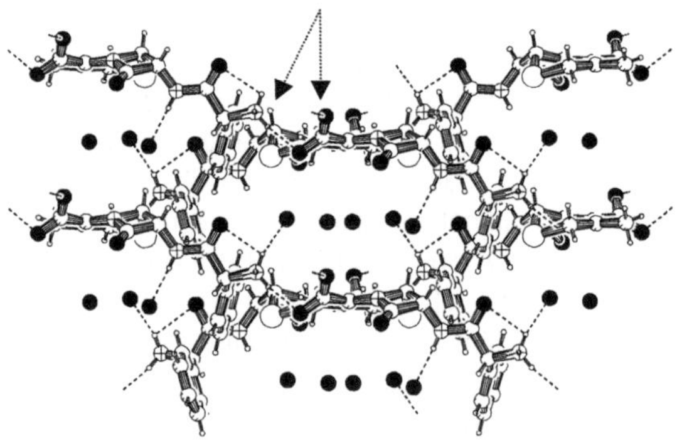

Fig. II.3c The water molecules (black balls) are positioned within the 2-dimensional layers and form hydrogen bonds with the cephalosporin molecules. The hydrogen bonds between the hydrogens of water and acceptor atoms of the cephalosporins are not shown because the positions of the hydrogens could not be exactly determined. The dotted arrows are pointing at an amino and a carboxylate function which are interacting at a 4-point junction.

The water molecules surround β-naphthol at two sides, allowing a hydrogen bond to be formed between the hydroxyl group of β-naphthol and a water molecule. This implies that the β-naphthol molecules are indirectly bound to the cephalosporin molecules via water.

The isomorphism found for the β-naphthol complexes of Cephalexin, Cephradine and Cefaclor, which was initially deduced from powder diffraction experiments, was confirmed by single-crystal X-ray diffraction. The structures of the complexes of Cephradine and Cefaclor are very similar with practically identical interaction patterns.
The crystal structure of the Cefadroxil/β-naphthol complex, differs considerably from that of the other cephalosporin complexes. This difference must be attributed to the replacement of a hydrogen on the phenyl ring by a hydroxyl group. This replacement influences the molecular structure both sterically and electronically and has dramatic consequences for the recognition properties of the molecule. This orthorhombic crystal structure contains Cefadroxil, β-naphthol and water in a ratio of 2:1:8. Thus, it contains three molecules of water per unit cell more than the cephalosporin complexes described above. Also the Cefadroxil/β-naphthol complex forms a crystal structure of the clathrate type. Cefadroxil serves as the host and β-naphthol is the guest (fig. II.4).
In the Cefadroxil clathrate structure, 4-point junctions similar to those observed for cephalosporins **1**, **2**, and **3**, are present. However, the 2-fold symmetry is lacking in the case of Cefadroxil. This deviating arrangement results in an essential difference between the Cefadroxil complex and the complexes derived from the cephalosporins **1**, **2** and **3**, namely the dimensionality of the hydrogen bonding network formed by the host molecules. While the cephalosporins **1**, **2** and **3** form 2-dimensional nets of hydrogen bonds, Cefadroxil forms a 3-dimensional network. Although the phenolic hydroxyl-group is involved in hydrogen bonding interactions, its function is not essential for the formation of the 3-dimensional network. The 3-dimensional network is constructed by interactions of only the ammonium and carboxylate groups of the Cefadroxil molecules.

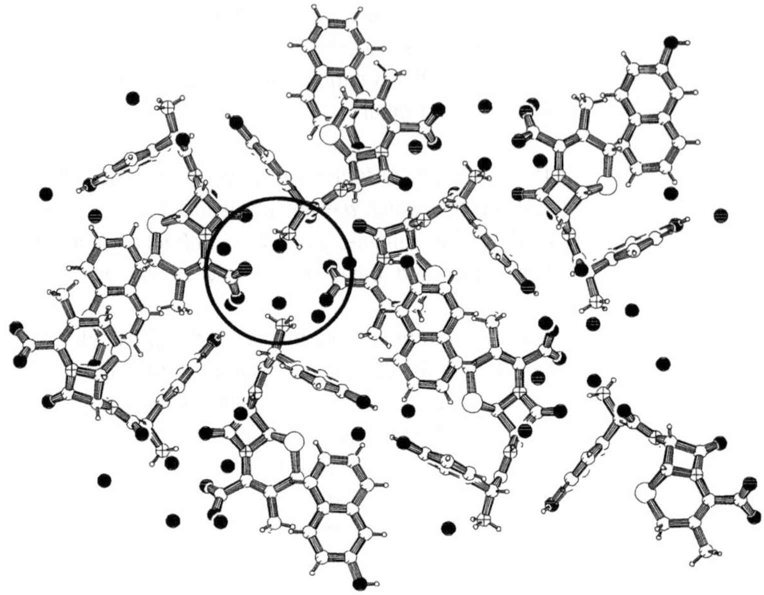

Fig. II.4 The crystal structure of the Cefadroxil complex with β-naphthol.

The essential feature of the clathrate type complexes described above is that the cephalosporins serve as the hosts and β-naphthol as the guest. The third constituent is water, which fulfils the role of cement in the crystal. The structures of Cephalexin, Cephradine and Cefaclor are isomorphous and their complexation behavior is essentially the same. The introduction of a hydroxyl, as in Cefadroxil, has a profound effect on the clathrate structure, although the same guest molecule can be accommodated. Remarkably, the cavity structure of the four Cephalosporin complexes is very similar, in which β-naphthol fits while hydrogen bonded to a water molecule. Clearly, this clathration is an example of molecular interactions requiring great precision.

2.2 Induced fit phenomena in clathrate formation[10].

Having elucidated the crystal structures of the four clathrates of cephalosporins with β-naphthol, the interesting question arises whether other guest molecules than β-naphthol can be accommodated in these hosting antibiotic frameworks. An attractive prospect of identifying other guest molecules is that such guests may open avenues for a more effective removal of cephalosporin from aqueous solutions. For this study, a series of compounds was selected, which on a molecular level show structural similarity with β-naphthol. This approach, based on the concept of molecular similarity, resembles that often used for the design of substrates for biological targets, such as receptors and enzymes.

The selected set of molecules having similar structural features as β-naphthol was subjected to a molecular modeling study using docking, implying that the β-naphthol molecule was taken out of the lattice and the new molecules were fit into the remaining cavity. This procedure of selecting potential guests leads to a series of substituted naphthalenes and two other ring aromatics, listed in Table II.1 (entries 1-16). This list was extended by a series of potential guests on more intuitive grounds, viz. entries 17-22 in Table II.1. The compounds listed in Table II.1 were all tested in clathrate formation experiments. Cephradine, Cephalexin and Cefadroxil were examined with all potential guests, while for Cefaclor a limited number of experiments was performed. The Cephalosporins **1-3** gave complexes with all compounds tested, whereas with Cefadroxil **4** only clathrate formation was observed in a limited number of cases. These results are compiled in Table II.1. The crystal structure of the clathrates formed by **1-3** with β-naphthol is referred to as type A, whereas that of **4** with β-naphthol as type B.

Table II.1 Isolated complexes of Cephradine, Cephalexin, Cefaclor and Cefadroxil with varying complexing agents.

Complexing agent	**Cephradine**	**Cephalexin**	**Cefaclor**	**Cefadroxil**
1. β-naphthol	A	A	A	B
2. β-naphthol	A	A	A	B
3. quinoline	A	A	A	*
4. naphthalene	A	A	-	*
5. 1,2-dihydroxynaphthalene	A	A	-	*
6. 1,3-dihydroxynaphthalene	A	A	-	*
7. 1,4-dihydroxynaphthalene	A	A	-	*
8. 1,5-dihydroxynaphthalene	A	A	-	*
9. 1,6-dihydroxynaphthalene	A	A	A	B
10. 2,3-dihydroxynaphthalene	A	A	A	*
11. 2,6-dihydroxynaphthalene	A	A	-	B
12. 2,7-dihydroxynaphthalene	A	A	-	B
13. coumarin	A	A	-	*
14. 8-hydroxyquinolin	A	A	A	*
15. indole	A	A	-	*
16. indene	A	A	-	*
17. 1-acetonaphtone	A	-	-	*
18. 2-acetonaphtone	A	A	-	*
19. 1-chloronaphthalene	A	-	-	*
20. 1,2,3,4-tetrahydro-1-naphthol	A	-	-	*
21. 1,5-dihydroxy-1,2,3,4-tetrahydro naphthalene	A	-	-	*
22. 2,2'-bipyridyl	A	A	A	*

* no complex formation, - not tried, "A" refers to the clathrate of cephalosporins **1-3** with β-naphthol and "B" to the complex of Cefadroxil with β-naphthol.

The X-ray powder patterns of the clathrates derived from Cephalexin **1**, Cephradine **2** and Cefaclor **3** reveal that they are all isomorphous with type A, while the few complexes obtained from Cefadroxil **4** are all isomorphous with type B. These observations indicate that the cephalosporin host molecules strongly dictate the basic lattice in which the guest molecules are being accommodated. It is highly relevant to notice that for several potential guests, which were selected on intuitive grounds and for which docking experiments suggested that fitting into the β-naphthol cavity cannot be achieved, the crystallization experiments show the opposite. *Especially the type A complexes show a remarkable tolerance for guest molecules, much more so than the type B complex.* This observation for the type A clathrates clearly suggests that there must be a considerable flexibility in the accommodation of guests, much more than can be envisaged by a straightforward replacement of β-naphthol.

Several crystal structures of both type A and type B clathrates have been solved by single-crystal X-ray diffraction. These structures revealed an interesting unpredicted role of the water molecules in these clathrates. Initially, it was thought that the water molecules in the β-naphthol clathrates are essential for the stability of the basic host framework and were therefore regarded as fixed parts in these complexes. The X-ray structures clearly showed that some guest molecules are able to remove water molecules from the cavity, whereas in other cases extra water molecules are incorporated. Such a possibility of having a variable amount of structural water was not taken into account during the docking studies, thus explaining the disagreements between prediction and experiments. The phenomena just mentioned are nicely illustrated in Figure II.5a-c for three type A clathrates derived from Cephradine. Similar observations involving a varying number of water molecules were made for type B complexes as can be deduced from Figure II.6a-c. This phenomenon of varying amounts of structural water in the clathrates has some analogy with biological systems, in which often the scope of suitable substrates for a given receptor can be enlarged by removal of water molecules from the binding site.

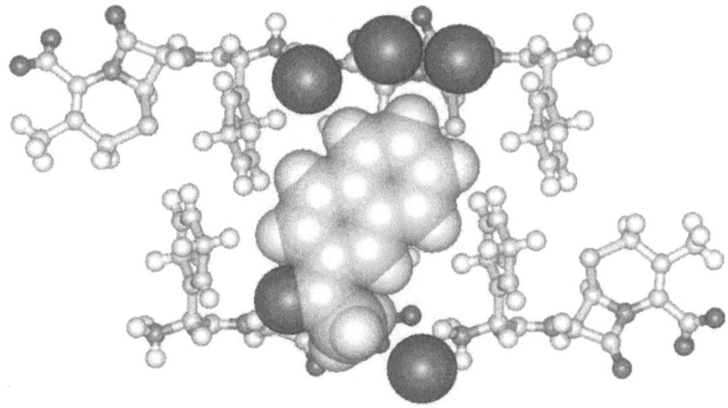

a. Cephradine/2-acetonaphthone, host/guest/water is 2:1:4.

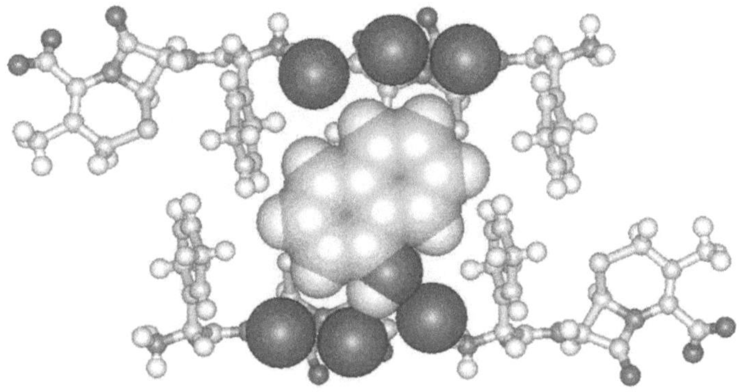

b. Cephradine/α-naphthol, host/guest/water is 2:1:6.

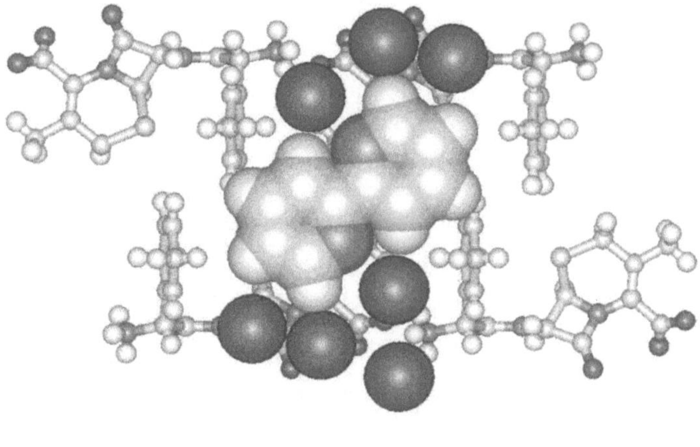

c. Cephradine/2,2'-bipyridyl, host/guest/water is 2:1:7.

Fig. II.5 Three Cephradine complexes with a varying number of water molecules.

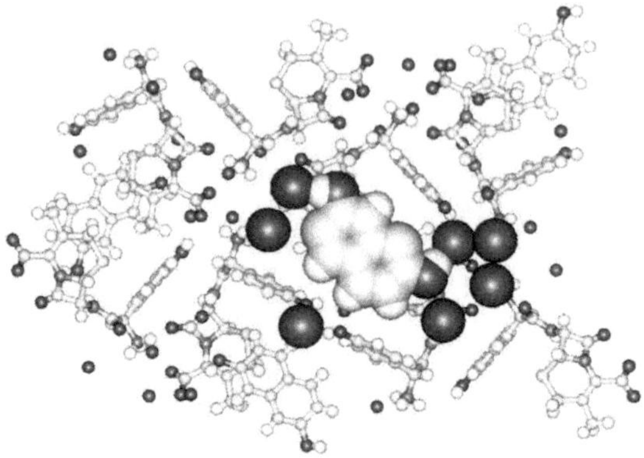

a. Cefadroxil/2,7-dihydroxynaphthalene, host/guest/water is 2:1:7.

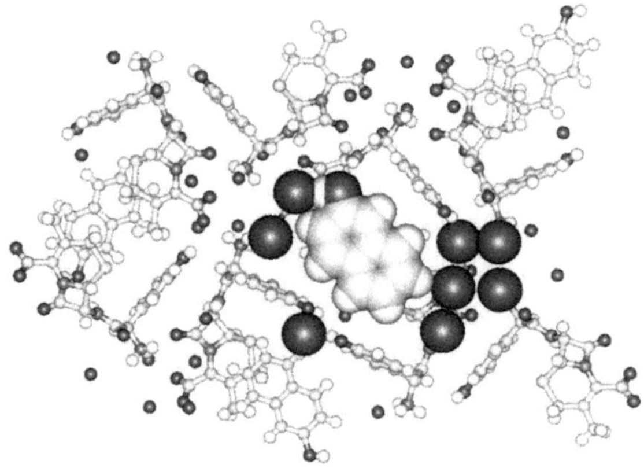

b. Cefadroxil/β-naphthol, host/guest/water is 2:1:8.

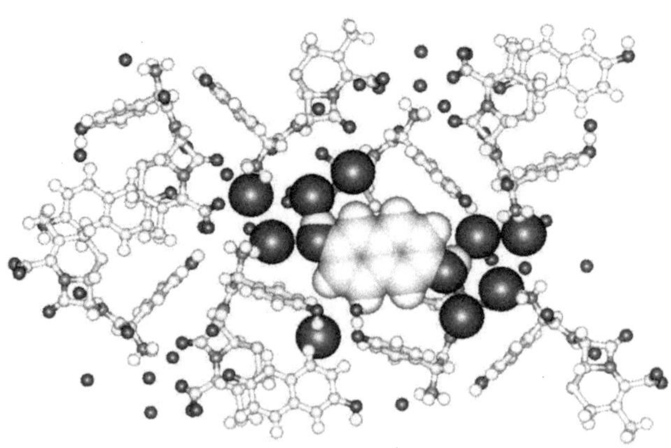

c. Cefadroxil/2,6-dihydroxynaphthalene, host/guest/water is 2:1:9.

Fig. II.6 Three complexes of Cefadroxil with a varying number of water molecules.

The role of water as a cementing agent in these clathrates derived from cephalosporins, allows a certain flexibility in accepting guest molecules in these complexes. However, the higher guest tolerance in type A clathrates cannot be reconciled by solely varying the amount of water. More information became available by a detailed analysis of the structures of a series of type A clathrates, as determined by single-crystal X-ray diffraction. The dimensions and shapes of the hosting cavities in these complexes were compared. The distances between three sulfur atoms, which form three corners of a parallelogram (see Figure II.7) were measured. The S_1-S_2 distance is a measure for the length of the type A cavity as can be deduced from Figure II.3a-c and II.7, because the line between S_1 and S_2 parallels the longest dimension of the guest molecule. For the type A structure, the distance d between two two-dimensional hydrogen bonded layers of Cephalosporin molecules can be defined. In addition, the relative slip of two layers with respect to each other can be determined, as is indicated in Figure II.7.

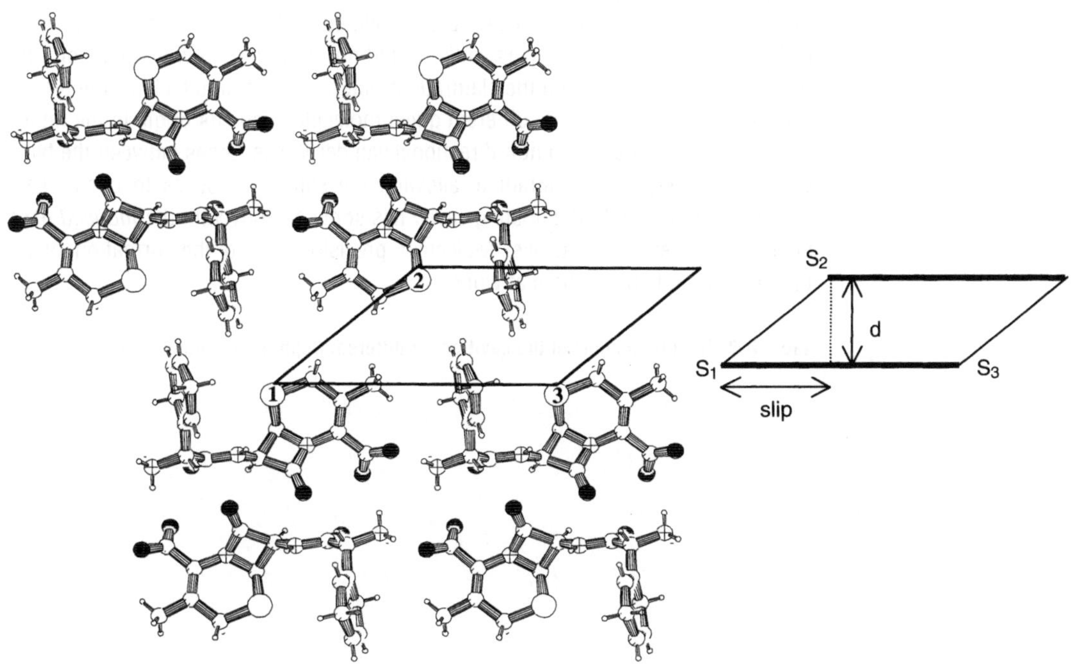

Fig. II.7 A Pluton drawing of the host-framework formed by Cephradine viewed in the direction of the b-axis. Three sulfur-atoms have been numbered to define the distances listed in Table II.2.

The sulfur-sulfur distances, the slip and the distance d are collected in Table II.2. These data reveal that the dimensions of the cavity vary with the guest accommodated in the complex. The S_1-S_2 distance, which is in first approximation proportional with the size of the cavity, decreases from the large guest 2,2'-bipyridyl to the smaller guests naphthalene, quinoline and β-naphthol. Apparently, the hosting framework is able to adjust the dimensions of the cavity to match the size of the guest, in order to achieve the most favorable crystal packing. The slip and the distance d are measures for the extent that the hosting framework is using its flexibility to adjust the size and shape of the cavity. For the Cephradine complexes the distance d between the two-dimensional layers varies only marginally, *viz.* 0.20 Å. On the other hand, the slip shows a considerable decrease (1.07 Å) going from the largest to the smallest guest. These observations lead to the conclusion that the adjustment of the size and shape of the hosting cavity mainly takes place by varying the slip rather than the interlayer distance. The consequence of this adjustment of the cavity to the nature of the guest molecule is that a wider range of guest molecules than predicted can be accommodated in the Cephradine framework, including those guest molecules which would not fit in the cavity arising from the removal of β-naphthol from the clathrate structures. The adaptability of the type A clathrate framework towards different guest molecules finds its origin in the rather weak interactions based on non-directional van der Waals forces between the two-dimensional layers of Cephradine, allowing the slipping process to occur. This fitting of the guest into the hosting cavity has some analogy with the *induced fit* of substrates in enzyme cavities. Molecular precision during the formation of an optimal crystal lattice is here at work.

Table II.2 The dimensions of the cavities for different Cephradine complexes.

Complex	S_1-S_2 (Å)	S_1-S_3 (Å)	S_2-S_3 (Å)	Slip (Å)	d (Å)
Cephradine/2,2'-bipyridyl	8.69	11.51	6.62	7.13	4.97
Cephradine/2-acetonaphthone	8.18	11.69	7.22	6.48	5.00
Cephradine/β-naphthol	8.12	11.74	7.33	6.39	5.01
Cephradine/naphthalene	8.06	11.73	7.33	6.34	4.97
Cephradine/quinoline	8.04	11.71	7.30	6.34	4.94
Cephradine/β-naphthol	7.73	11.71	7.54	5.98	4.90

The situation with clathrates derived from Cefadroxil is entirely different. This cephalosporin only forms complexes with a few guest molecules (Table II.1), which is in strong contrast with the other three antibiotics. The four Cefadroxil clathrates are isomorphous with the type B structure of its β-naphthol complex, as was established by powder diffraction analysis. This type B structure is three-dimensional in nature and is therefore lacking the adaptability by means of a slipping process as was observed for the type A clathrates. The type B framework is very rigid due to highly directional and strong hydrogen bonds, and, as a consequence, the flexibility in accommodating guest molecules is rather limited. Precise fitting is actually a prerequisite, again a matter of precision on a molecular level. In fact, only the adjustment of the number of water molecules accounts for the minimal adaptability. In other words Cefadroxil conforms more to the *lock and key* model of fitting substrates in enzyme cavities.

A drawback of the aforementioned naphthalene derivatives as complexing agents for cephalosporins is their toxicity and the inherent environmental image problem associated herewith. Hence, the search for new complexing agents was directed towards benzene derivatives, many of which have more acceptable characteristics with regard to toxicity and environmental image. Examples are biphenyl **5** (preservative E230), 2-hydroxybiphenyl **6** (preservative E231), Aspirin **7**, and methyl 4-hydroxybenzoate **8** (preservative E218). The boundaries of the adaptability of the hosting framework formed by Cephradine **2** were therefore further explored[11]

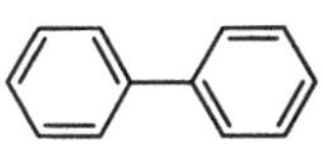

5

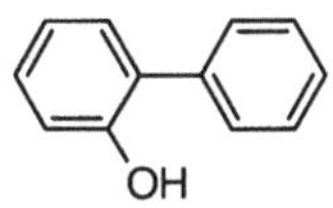

6

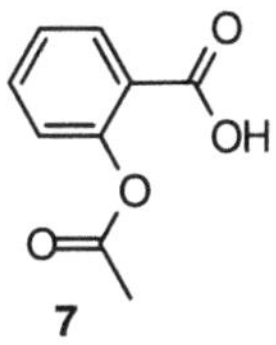

7

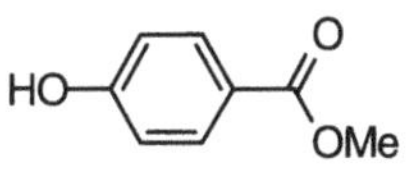

8

Thus, a series of benzene derivatives and also some other molecules, *e.g.* fluorene and benzilic acid, were subjected to complexation experiments with Cephradine. Crystalline complexes obtained were analyzed by X-ray powder diffraction in order to establish whether the structure has the *C*2 cavity structure observed for the complexes of cephalosporins **1-3** with naphthalenes. The powder diffraction technique proved to be highly valuable for this study, as non-isomorphous structure types could be readily recognized. In a number of cases the deviations caused by the induced-fit process are so large that the resulting structure no longer shows the typical X-ray powder pattern of the aforementioned *C*2 cavity structure. In these cases, the crystal structure of the complex was elucidated by single-crystal X-ray diffraction.

A large number of new guest molecules for Cephradine complexes was identified. Most of these complexes have the *C*2 cavity type structure, however, several other structure types were discovered as well. The adaptability of the hosting framework of Cephradine appeared to be high, as benzene derivatives, e.g. benzoic acid, as well as much larger molecules, *e.g.* fluorene and carbazole, could be accommodated. Some of the new structure types are clathrates, like the naphthalene complexes, containing discrete cavities. In addition, also layered and channel type structures can be created for certain guest molecules. The various types of complexes that have been prepared and characterized are summarized in scheme II.4, classified by their space group and their structure type, *i.e.* cavity, layer or channel type. The complexing behavior of the guest molecules is rather capricious, as subtle structural changes in the guest molecule can result in substantial changes of the overall structure of the Cephradine complexes, as is evident from scheme II.4.

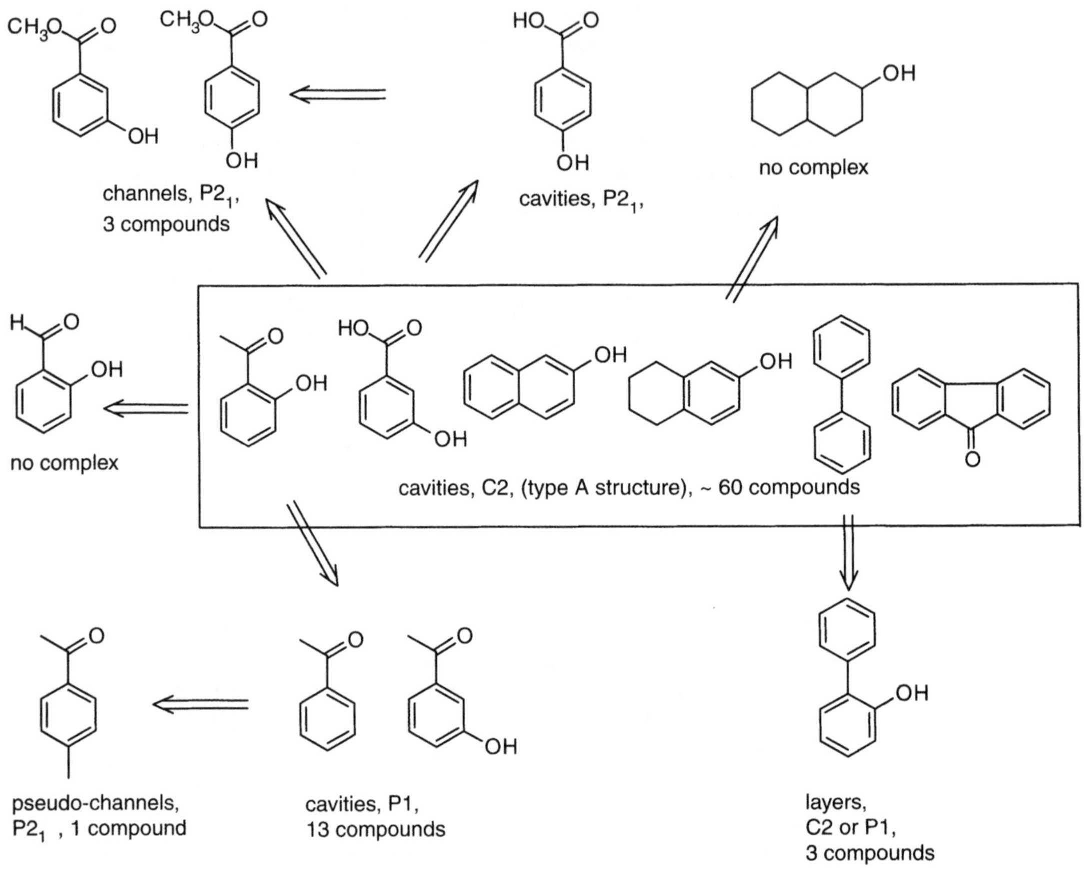

Scheme II.4 Guest molecules in the complexation with Cephradine and the type of complexes derived thereof. For each structure-type the number of compounds that form that type of complex with Cephradine is shown.

The conformational changes of the Cephradine molecule in the various complexes were analyzed by comparing the five torsion angles in the antibiotic in the respective crystal structures[11]. In spite of the fact that considerable conformational differences were observed, in all variants of the Cephradine complexes the head to tail interactions of the zwitterionic antibiotic molecules, which in most cases are present as 4-point junctions, remained unaffected. This essential assembling feature, which is encircled in figure II.8a, is in fact a supramolecular synthon consisting of two carboxylate and two ammonium groups provided by four individual Cephradine molecules. The adaptability for the large number of different guest molecules can be accounted for by induced-fit mechanisms as already encountered for the naphthalene type guests. For the *C2*-cavity structure the adaptability can mainly be attributed to the slipping mechanism along the crystallographic *a*-axis. Two other modes of induced-fit are, slipping along the *b*-axis and variation of the interlayer distance. The three ways of inducing induced-fit in the hosting framework, together with the conformational changes of the Cephradine molecules, can account for the observed high adaptability of guest molecules. A typical example is the layer type complex of Cephradine with 2-hydroxybiphenyl. The nonplanar complexing agent will not easily fit in the *C*2 cavity structure, which only tolerates practically flat molecules. The crystal structure of the complex reveals that the layered structure formed by the hosting antibiotic is identical to that in the *C*2 cavity type complexes, while the conformation of the Cephradine is also very similar. The dissimilarity between the *C*2 cavity and the structure shown in Figure II.8-a,b is the enormous difference in the interlayer distance, namely an elongation of the *c*-axis with 3.7 A.

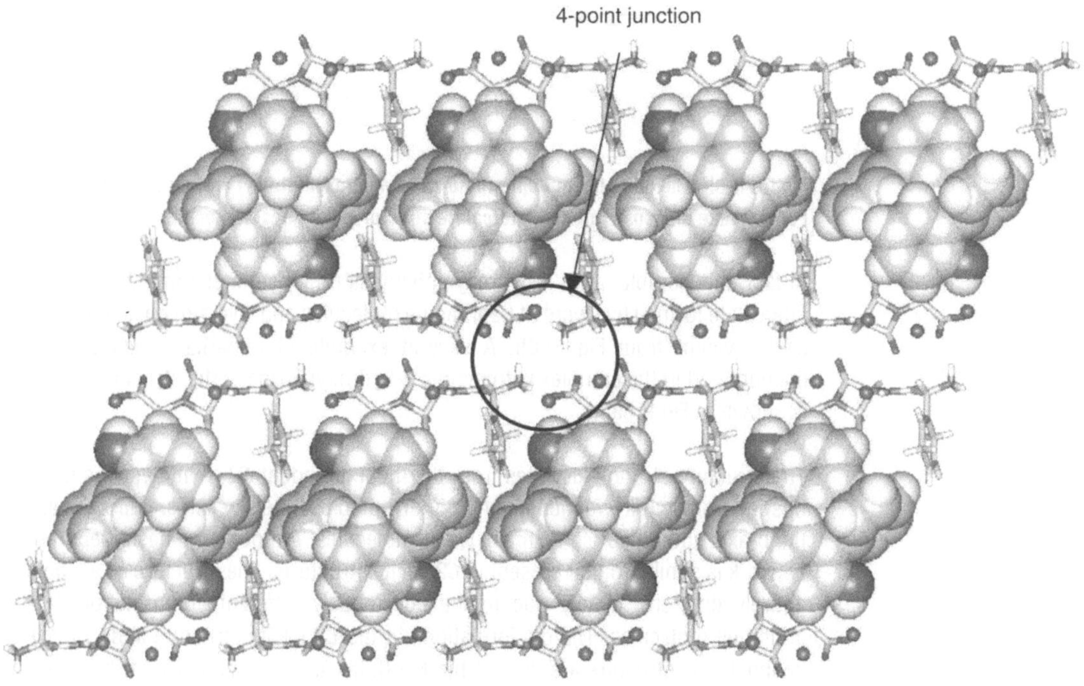

Figure II.8a The complex of Cephradine and 2-hydroxybiphenyl viewed along the *b*-axis.

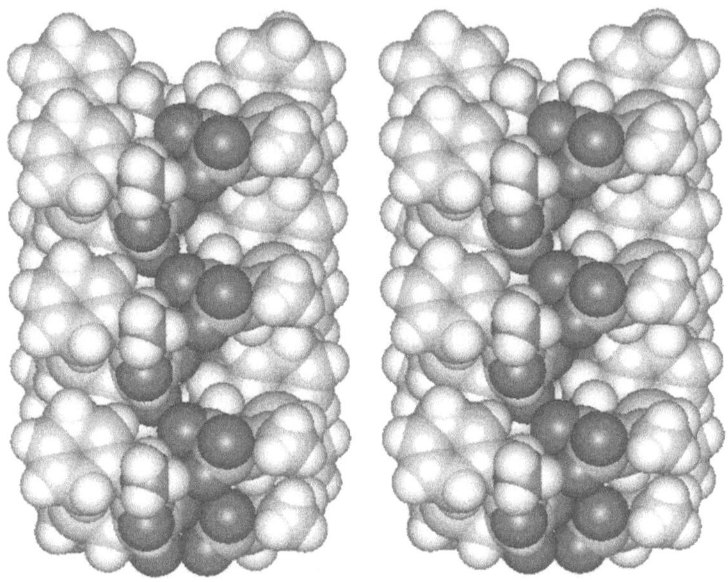

Figure II.8b The complex of Cephradine and 2-hydroxybiphenyl viewed along the *a*-axis. For the sake of visualizing the intercalate structure, the guest molecules are omitted.

In fact, this complex with 2-hydroxybiphenyl is no longer a clathrate but rather an intercalate formed by layers of Cephradine molecules and layers of guest molecules, as is evident from Fig. II.8b. A second example of remarkable adaptability is encountered in the complex with methyl 3-hydroxybenzoate, the structure of which is shown in Fig II.9a.

The structure strongly deviates from the most abundant *C*2 cavity type, as it has a hosting framework with genuine non-interrupted channels which are directed along the b-axis and do not contain any water molecules. Although the basic Cephradine network is similar to that present in *C*2 cavity structures, the structure as a whole is entirely different, mainly due to the enormously different conformation of the antibiotic molecules. The conformations in the *C*2 cavity structure and in the $P2_1$ channel type structure are depicted in Fig II.9b. As a consequence of the different conformations in $P2_1$ channel type structures, one of the four molecular bonds of the 4-point junction, which is present in all structure types, is disrupted. An amide NH of Cephradine serves as a substitute for this disrupted bond, resulting in a cyclic array of hydrogen bonding and electrostatic interactions consisting of five interaction points as shown in fig. II.9c.

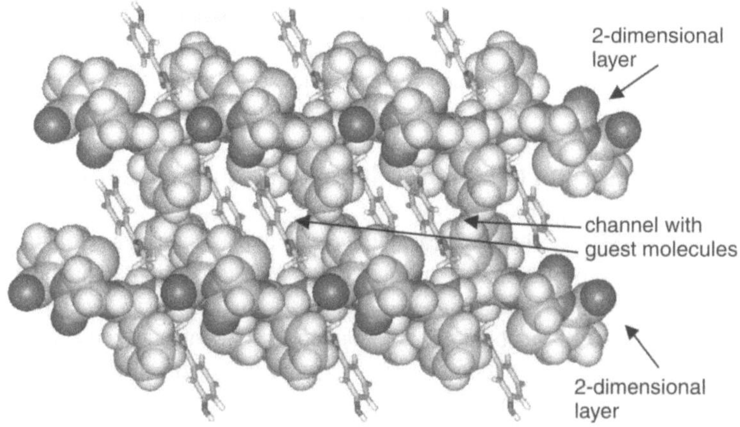

Figure II.9a The complex of Cephradine and methyl 3-hydroxybenzoate viewed along the *b*-axis.

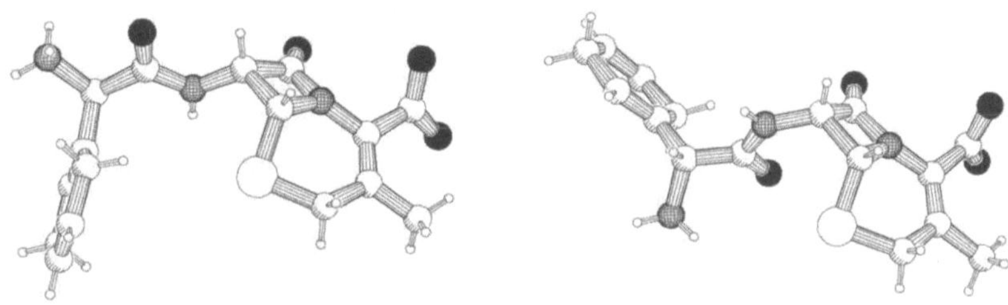

Figure II.9b A Pluton drawing of the conformation of Cephradine in the *C*2 cavity type and the $P2_1$ channel type complex, respectively.

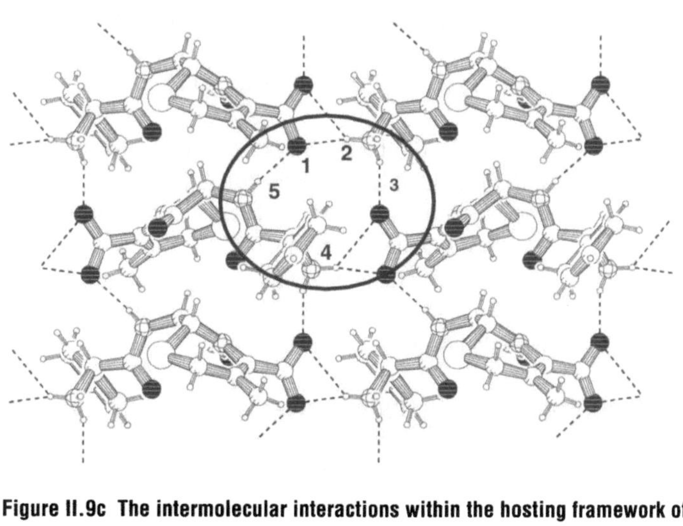

Figure II.9c The intermolecular interactions within the hosting framework of the $P2_1$ channel type structure viewed along the *c*-axis.

This study revealed that subtle variations of the structure of the guest molecules may have an enormous impact on the structure of the molecular complex with Cephradine, however the head to tail zwitterionic interactions of the antibiotic molecules remain essentially the same.

2.3 Prediction of clathrate formation by molecular modeling[12]

The empirical identification of new complexing agents is lacking any rational. It was therefore considered desirable to develop a computer-based model, which would allow to predict whether a compound is a complexant for cephalosporin type antibiotics. Attempted docking of potential complexing agents in the cavity present in the hosting lattice formed by Cefadroxil, did not lead to the identification of new complexants. This failure of the docking search is most likely attributable to the variable number of water molecules incorporated in the complexes. These water molecules affect the exact size and shape of the hosting cavity and consequently an essential prerequisite for a docking search, *i.e.* that the cavity must be clearly defined, is not fulfilled. Therefore, another approach, which makes use of the molecular similarity principle, was investigated for the prediction of complexation. The advantage of this approach is that it is based on properties of complexant molecules, thus avoiding deviation in the hosting cavities. From the calculated molecular similarities of a set of 120 molecules, comprising both complexing agents and molecules that do not form a complex with Cephradine, a predictive model was derived using Genetic Algorithms, Linear Discriminant Analysis (LDA), and various validation techniques. The way, in which the molecular similarity was calculated, appeared to be crucial. Furthermore, in order to derive a useful model, LDA has to be applied on the molecular similarities between molecules in the data set and a limited number of guiding compounds rather than on the complete similarity matrix. In this manner, a model was obtained that predicts whether a molecule forms a complex with Cephradine with a predictive score of 86%. This model is based on the similarity of a potential complexant with only three guiding compounds and is expressed by the equation:

$$D = -0.35xS_1 - 0.44xS_2 - 0.83xS_3 + 0.36$$

wherein S_1, S_2 and S_3 refer to similarity indices of a molecule under examination with the guiding compounds 2-methylbenzoic acid, methyl 3,4-dihydroxybenzoate and 3-methoxyaniline, respectively, and whereby $D<0$ means complexation and $D>0$ no complexation . Application of the model resulted in the identification of new, especially environmentally and toxicologically acceptable, complexing agents. It is worth noting that the procedure used to derive this model has general relevance for fitting guest molecules in a cavity.

2.4 Efficiency of naphthalene derived complexing agents[13]

As outlined in the introduction, the complexation with cephalosporins can be utilized in the enzymatic synthesis of these antibiotics. For an effective use, the complexation efficiency is a very important issue. This efficiency was examined for

the complexants of the naphthalene series. The influence of the hydroxyl function in β-naphthol was studied by comparing its efficiency with that of naphthalene. Conceivably, a hydroxyl function can serve as a hydrogen bond donor and acceptor to the surrounding water molecules[6] and hence it can contribute to the overall stability of the crystal, which may result in a more efficient crystallization. The behavior of β-naphthol has been compared with that of α-naphthol to investigate the influence of the position of the hydroxyl function at the naphthalene skeleton on the complexation. The influence of an additional hydroxyl function and its relative position has been studied using a series of dihydroxynaphthalenes as complexing agents. Quinoline and 8-hydroxyquinoline have been examined as guest molecules in order to shed light on the effect of a strong hydrogen bond acceptor in the aromatic system.
The complexation was studied by measuring the decrease of the concentration of cephalosporin with time after addition of the complexing agent. Most complexing agents form complexes very rapidly, in fact the remaining concentration of antibiotic did not change anymore after 90 minutes. To facilitate the dissolution of the complexant in water a small amount of methanol was used as co-solvent. In Table II.3 the relevant concentration data, the equilibrium constant of complexation and the Gibbs free energy of complexation are collected.

Table II.3 The equilibrium concentrations of Cephradine [2] and complexing agent (CA) and the equilibrium constant (K) and Gibbs free energy of complexation (ΔG)derived therefrom.

Complexing agent	**[2] mM[a]**	**[CA] mM**	**K**	**ΔG (kJ/mole)[c]**
β-naphthol	2.7	5.1	370	-14.7
naphthalene	3.7	0.2	270	-13.9
α-naphthol	1.3	7.7	769	-16.5
1,2-dihydroxynaphthalene	8.0	[b]		
1,3-dihydroxynaphthalene	2.6	11.4	385	-14.7
1,4-dihydroxynaphthalene	9.8	5.4	102	-11.5
1,5-dihydroxynaphthalene	1.7	1.0	588	-15.8
1,6-dihydroxynaphthalene	1.1	[b]		
2,3-dihydroxynaphthalene	1.8	15.9	4406	-20.8
2,6-dihydroxynaphthalene	3.0	6.9	333	-14.4
2,7-dihydroxynaphthalene	3.9	17.0	1967	-18.8
quinoline	2.9	16.5	2684	-19.6
8-hydroxyquinoline	3.6	2.7	278	-14.1

[a] The complexant was dissolved in methanol prior to addition to the Cephradine solution.
[b] The solubility was not reported in the literature.
[c] $\Delta G_{complexation} = -RT\ln K_{complexation}$.

By comparing the efficiencies of the complexing agents for Cephradine shown in Table II.3, the intriguing question arises whether the overall stability of the complexes can be correlated with the structure of the complexant. It is important to note that the complexation efficiency, which is directly related to the ΔG of the complexation reaction, depends on the stability of the clathrate formed and on the energy of solvation of both the complexing agent and the cephalosporin in water. Prior to complexation with Cephradine, the complexant has to be desolvated from water, which costs energy. This intermediary stage can be represented by the molecule in the gas phase deprived of all intermolecular interactions. The Gibbs energy of complexation ($\Delta G_{complexation}$) is the difference between the Gibbs energy of stabilization ($\Delta G_{stabilization}$) and the Gibbs energy of solvation ($\Delta G_{solvation}$). The energy diagram of the complexation reaction is depicted in Scheme II.5.

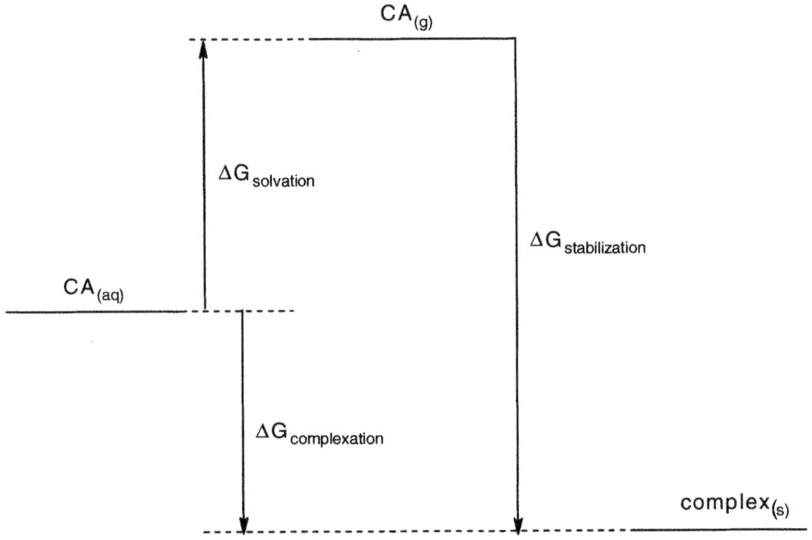

Scheme II.5 The energy diagram for the complexation reaction. CA stands for complexant.

For five complexants the thermodynamic quantities of complex formation have been determined (Table II.4). The data in Table II.3 reveal that the efficiency of complexation strongly depends on the complexing agent used. A considerable improvement with respect to the reference compound β-naphthol can be achieved. For cephalosporins **1-3**, the highest efficiency is obtained for β-naphthol as the complexant. In the case of Cephradine only 1,6-dihydroxynaphthalene is slightly better than β-naphthol, however, this complexant performs much worse in case of Cephalexin and Cefaclor.

The complexation data of Cefadroxil indicate that β-naphthol behaves the worst in this series, in strong contrast to the results obtained with the cephalosporins **1-3**. The best performing complexing agent for Cefadroxil is 2,7-dihydroxynaphthalene. The difference in behavior between the cephalosporins **1-3** on one hand and Cefadroxil **4** on the other, as far as the best complexant is concerned, can be attributed to differences in the hosting cavities of the respective antibiotics.

Table II.4 The Gibbs energy of complexation (ΔG_{compl}), solvation (ΔG_{solv}) and stabilization (ΔG_{stab}) for the complexation of Cephradine with five complexing agents (in kJ/mole).

complexing agent	$0.5 \times \Delta G_{solv}$	ΔG_{compl}	ΔG_{stab} [a]
naphthalene	-1.4	-13.9	-15.3
α-naphthol	-11.6	-16.5	-28.1
β-naphthol	-12.7	-14.7	-27.4
quinoline	3.5	-19.6	-16.1
8-hydroxyquinoline	-11.4	-14.1	-25.5

[a] ΔG_{stab} is calculated by $\Delta G_{stab} = 0.5 \times \Delta G_{solv} + \Delta G_{compl}$ (the ratio complexant:Cephradine is 2:1).

The data in Table II.3 show some remarkable results, confirming that the complexation efficiency is not entirely controlled by the effect of the complexing agent on the complex stability. As the cavity in the crystal lattice formed by Cephradine has a two-fold symmetry, it was expected that the dihydroxy-naphthalenes arising from applying a two-fold symmetry operation on α- and β-naphthol would be more efficient than their monohydroxy analogues. However, the data in Table II.3 show that 1,5-dihydroxynaphthalene performs worse than α-naphthol and 2,6-dihydroxynaphthalene performs worse than β-naphthol, despite the fact that the second hydroxyl function could be used for additional stabilization through hydrogen bonding. The dihydroxynaphthalenes in general, with 1,6-dihydroxynaphthalene as an exception, perform worse than β-naphthol indicating that the second hydroxyl function has no beneficial effect on the complexation efficiency. Similarly, 8-hydroxyquinoline performs worse than quinoline. For the subset of five molecules shown in Table II.4 the $\Delta G_{stabilization}$ can be correlated with the molecular structure of the guest molecules, thereby revealing the essential interactions responsible for complex stabilization. The observation that $\Delta G_{stabilization}$ for quinoline is only marginally higher than that for complexation with naphthalene, at first sight seems illogical. It was expected that quinoline would contribute much more to the stabilization energy than naphthalene by hydrogen bonding with water in the cavity, using nitrogen as a hydrogen bond acceptor. The crystal structure of the complex of Cephradine and quinoline was determined very accurately, even the hydrogen atoms of the water molecules could be located. The structure of this complex, pictured in Figure II.10a, reveals that the nearest water molecule is positioned with its oxygen toward the nitrogen atom of quinoline, which represents a repulsive interaction energy. In contrast, naphthalene can form a weak hydrogen bond ($C_{ar}H\text{-}O_{water}$) with a neighboring water molecule (Figure II.10b), which contributes favorably to the stabilization energy. In comparison with naphthalene, solvation phenomena in α- and β-naphthol contribute much more to the stabilization energy due to the hydrogen bonding ($OH\text{-}O_{water}$) with these guest molecules. In addition to its role as a hydrogen bond donor, the hydroxyl function in the last mentioned complexants can also serve as a hydrogen bond acceptor for a water molecule or as in the case of the β-naphthol complex, the amide proton of Cephradine. The difference in ΔG_{stab} between the complexes derived from naphthalene and α- or β-naphthol, respectively, of ca. 12.5 kJ/mole must be attributed to the presence of the hydroxyl function in the naphthols. The difference in $\Delta G_{stabilization}$ between the complexes of quinoline and 8-hydroxyquinoline is significantly smaller (9.1 kJ/mole). From the crystal structure of the Cephradine/8-hydroxyquinoline complex is evident that the hydroxyl-group of 8-hydroxyquinoline does not serve as a hydrogen bond donor toward water, but only plays a role as a hydrogen bond acceptor, implying that its contribution to the stabilization of the complex must be smaller than that of the hydroxyl function of α- and β-naphthol in complexes derived therefrom. This study of the correlation between the stabilization energy and the structural features of these five complexes, reveals that the complex stability can be enhanced by hydroxyl groups present in the guest molecule via hydrogen bonding. However, stabilization of the complex does not necessarily result in more efficient complexation, due to the profound influence of the solvation energy of the complexant in some cases. The term $\Delta G_{solvation}$ may become of comparable importance as the $\Delta G_{complexation}$ in the equation $\Delta G_{stabilization} = 0.5 \times \Delta G_{solvation} + \Delta G_{complexation}$. Thus, a structural variation of the complexant designed to stabilize the clathrate complex may be accompanied by an uncorrelated contribution to the energy of solvation of the complexant, which may be even larger and thus contraproductive to the complexation efficiency. An anticipated beneficial effect of polar groups in the complexant may be entirely counterbalanced by an increased energy of solvation. Such phenomena are of general importance in host-guest chemistry.

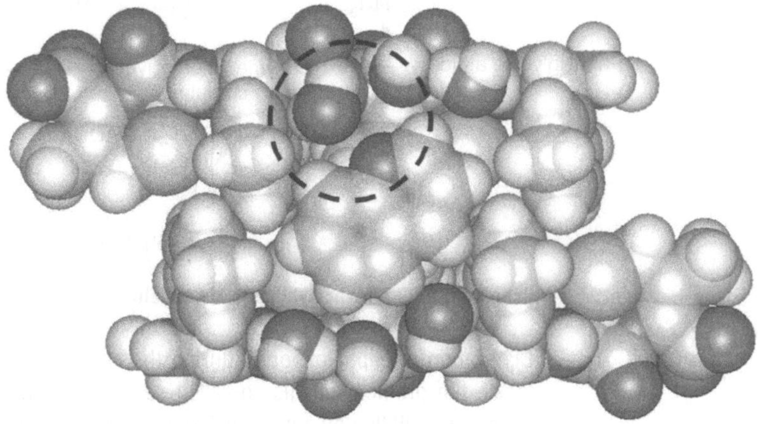

Figure II.10a The complex of Cephradine and quinoline.

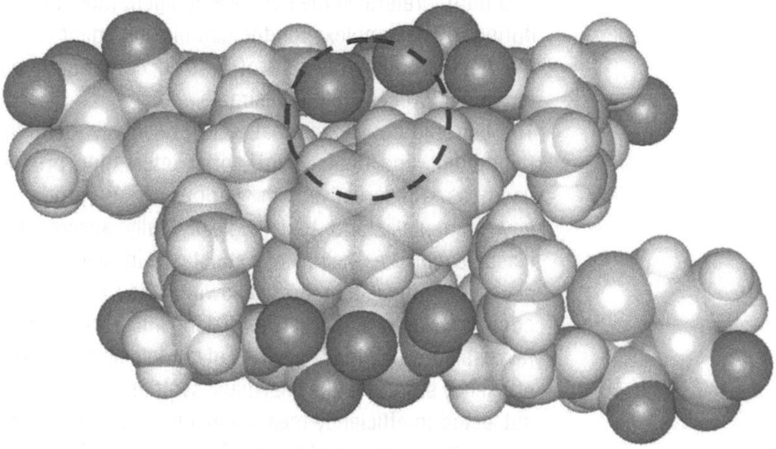

Figure II.10b The complex of Cephradine and naphthalene.

2.5 Green complexants for the clathration mediated enzymatic synthesis of Cephradine[13]

As has been outlined in the introduction, the enzymatic coupling of the β-lactam nucleus with an appropriate derivative of the side chain suffers from secondary hydrolysis. In the first step in the enzymatic synthesis of Cephalexin from phenylglycine amide an enzyme-phenylglycyl complex is formed (Scheme II.6). This complex can undergo either hydrolysis with water (k_2) resulting in the formation of phenylglycine or reaction with the β-lactam nucleus 7-ADCA (k_3) to give Cephalexin. The cleavage of Cephalexin (k_{-3}) reverts to the enzyme-phenylglycyl complex and 7-ADCA. The competing reaction of the enzyme-phenylglycyl complex with water leads to a loss of product.

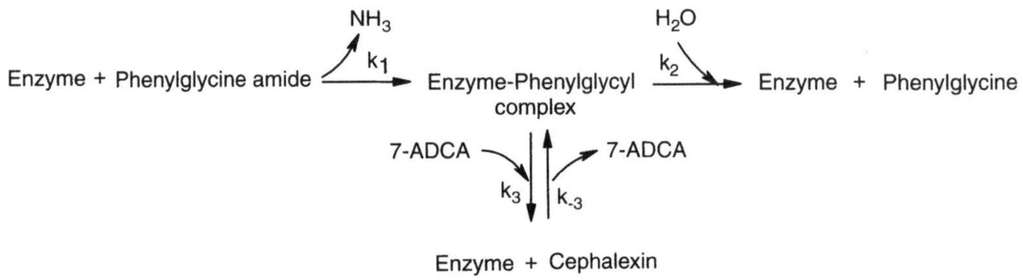

Scheme II.6 Enzymatic synthesis of Cephalexin.

Clathration of cephalosporins during the enzymatic synthesis may be suitable to tackle the problem of this so-called *secondary hydrolysis*. Several requirements have to be fulfilled by a complexing agent before application during the enzymatic synthesis of cephalosporins becomes feasible. First, the clathration process must be sufficiently effective under the conditions used for the enzymatic coupling. Second, the complexing agent used must not (irreversibly) inactivate the enzyme. And third, preferably the complexing agent must be non-toxic. Despite the fact that during the decomplexation the complexing agent can be completely removed from the final product, the use of a toxic compound, *e.g.* β-naphthol, in a 'green' enzymatic process is not desirable.
The benzene derivatives that were previously identified as complexing agents for Cephradine and Cephalexin[11], offer interesting prospects as several of them are used as preservatives in various food products. Such complexants are environmentally and toxicologically fully acceptable and according to their R classification, i.e. the risk of a substance in case of a particular exposure such as via skin, eyes, inhalation or swallowing, they have a toxicity index similar to other commonly used preservatives. Taking into account the complexing capacity and the favorable toxicity index a series of complexing benzene derivatives was selected for efficiency studies in the enzymatic synthesis of Cephradine. These compounds were subjected to efficiency measurements under conditions that resemble those of the enzymatic synthesis of Cephradine (pH= 7.2, T=5°C). The results obtained for the most effective complexants under the justmentioned conditions are collected in Table II.5. For an acceptable complexant the residual concentration of antibiotic should be below 10 mM. Only five compounds listed in this table met this criterion, *viz.* entries 1, 2, 6, 7, and 8. These five compounds were investigated in enzyme inhibition experiments using Assemblase® as the biocatalyst[14].

Table II.5 Residual concentration of Cephradine (pH = 7.2 and T = 5°C) using benzene derivatives as complexants.

Entry	Complexant	[Cephradine] (mM)
1	**2-aminobenzoic acid**	**6.6**
2	**methyl 2-aminobenzoate**	**3.7**
3	2-aminobenzamide	16
4	2-methoxyacetophenone	15
5	4-aminoacetophenone	16
6	**2-hydroxybiphenyl**	**1.3**
7	**methyl 4-hydroxybenzoate**	**5.5**
8	**methyl 3-hydroxybenzoate**	**6.9**
9	methyl 3,5-dihydroxybenzoate	14

The activity of the enzyme in the presence of a complexing agent was compared with that in the absence of complexing agent. The rate of hydrolysis of D-phenylglycine amide (the side chain precursor of Cephalexin) was taken as a measure for the enzyme activity. In all cases the complexants lowered the activity of Assemblase®. Only for compounds in entries 7, 8 and 9 reversible inhibition was observed which did not deactivate the enzyme permanently. This implies that the enzyme can be reused for next batches. This is an important observation as the enzyme accounts for a substantial part of the total cost of enzymatically prepared cephalosporins. It should be noted that the complexants of entry 2, 6 and 7 are also effective in the complexation with Cephalexin and accordingly also suitable for the clathration mediated enzymatic synthesis of this antibiotic. In Table II.6 the residual activity of Assemblase® when exposed to either of these three complexing agents as a percentage of the enzyme activity in the absence of complexing agent is listed. These residual activities are similar to those measured after exposure to β-naphthol. A decrease of the enzyme activity as such is not desirable but it may have a positive effect on the diffusion limitations normally present when Assemblase® is used as the biocatalyst. It may thus be concluded that green alternatives for β-naphthol have been identified, which may be applied in future cephalosporin production.

Table II.6 Residual activity of Assemblase® in the presence of complexants.

complexing agent	residual activity (%)
2-hydroxybiphenyl	30
methyl 4-hydroxybenzoate	30
methyl 2-aminobenzoate	40

2.6 Synthesis of cephalosporins using a clathration induced asymmetric transformation[15]

In all current syntheses of cephalosporins the β-lactam nucleus is coupled with a derivative of enantiopure D-amino side chain. The use of racemic amino acids would be an attractive and cheap alternative. However, coupling of the β-lactam nucleus with an appropriate racemic amino acid derivative, leads to an epimeric mixture of cephalosporins of which only the derivative from the D-amino acid exhibits the desired antibiotic activity. Therefore, epimerization of the stereogenic center in the amino acid side chain of the epi-cephalosporins is a must to make the coupling starting from racemic amino acids an economically feasible one.

Amino acids are usually optically stable and can only be racemized under harsh conditions.[16] Their rate of racemization can be greatly enhanced by prior conversion into their Schiff bases.[17,18] This racemization becomes even more effective when conducted in the presence of an organic acid. In general, any amino acid can be racemized by heating in acetic acid, using a catalytic amount of aldehyde.[19] A generally accepted mechanism for the racemization involves isomerization of the Schiff base.[18]

Despite these major improvements of the racemization methods of amino acids, the conditions described above cannot be used for the synthesis of cephalosporins. Due to the presence of the β-lactam nuclei they are susceptible to degradation. Exposure to a pH above 8 or to strong acidic conditions causes severe breakdown of the cephalosporins. Also the temperatures required for the methods described above are too harsh for the β-lactam antibiotics. Therefore, the epimerization of the cephalosporins was investigated in aqueous solution under mild conditions. It was found that both salicylic aldehyde and pyridoxal could be used at much milder conditions for the Schiff base promoted epimerization, whereby the latter showed a faster reaction. A catalytic amount of aldehyde (10%) appeared to be sufficient. Gratifyingly, the *epimerization*, can be performed under these very mild conditions. A pH between 7 and 7.5 appeared to be appropriate to obtain an acceptable rate of epimerization in water at room temperature. At equilibrium 36%, 32% and 37% of the epi-cephalosporins is formed from Cephalexin, Cephradine and Cefadroxil, respectively[15a].

The asymmetric transformation has been studied using mixtures of cephalosporin diastereomers obtained during the epimerization experiments described above in which pyridoxal is still present. The cephalosporins can be efficiently withdrawn from aqueous solutions by clathration with aromatic compounds. For Cephalexin and Cephradine α-naphthol is the complexing agent of choice, while for Cefadroxil 2,7-dihydroxynaphthalene is preferred.[20] It was shown that this clathration process is highly selective. Only the cephalosporins derived from D-amino acids were found in the precipitate, while their epimers remained in solution. During the asymmetric transformation the pH was maintained by adding acid to the reaction mixture. The formation of the cephalosporins from their epimeric mixtures was followed with time using HPLC. It was found that in 24 h more than 90% of the cephalosporin with the correct diastereomeric structure had precipitated as a clathrate with the appropriate naphthalene derivative.

This successful asymmetric transformation opens avenues for the clathration induced asymmetric synthesis of cephalosporins starting from racemic amino acid side chains. For this purpose, the coupling of the β-lactam nucleus and the appropriate amino acid side chains is conducted via the so-called Dane salt route, as is depicted in Scheme II.7.[21] For the preparation of the Dane salt, the racemic amino acid is heated with ethyl acetoacetate in the presence of potassium hydroxide. The thus obtained Dane salt is then activated by conversion into the mixed anhydride of pivalic acid by reaction with pivaloyl chloride. This anhydride is subsequently used in the actual coupling reaction with the β-lactam nucleus, 7-ADCA. In order to make 7-ADCA more soluble in dichloromethane its tetramethylguanidine salt is prepared prior to its addition to the mixed anhydride. After hydrolytic removal of the Dane protecting group, an aqueous solution containing an epimeric mixture of the cephalosporin was obtained in yields ranging from 80 to 88%. The lowest conversion was obtained for Cefadroxil and the highest for Cephradine. It is important to note that the Dane salt route is also used in the industrial production of cephalosporins.

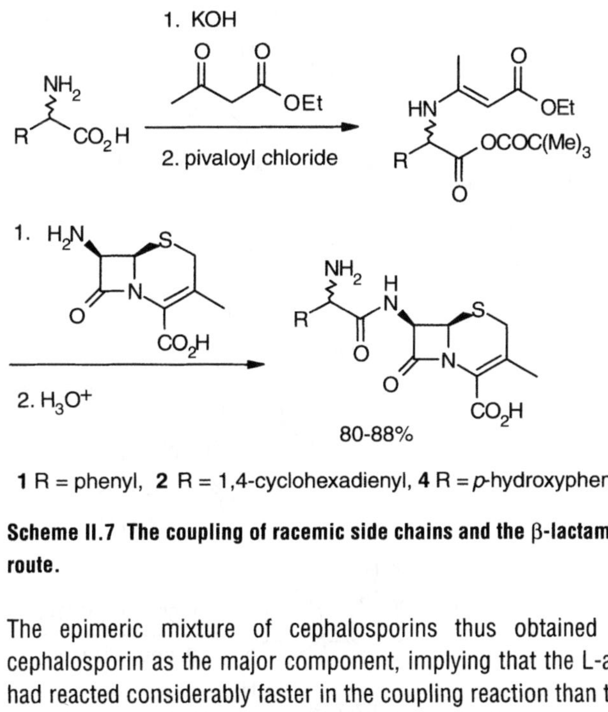

1 R = phenyl, **2** R = 1,4-cyclohexadienyl, **4** R = *p*-hydroxyphenyl

Scheme II.7 The coupling of racemic side chains and the β-lactam nucleus via the Dane route.

The epimeric mixture of cephalosporins thus obtained contained the *epi*-cephalosporin as the major component, implying that the L-amino acid derivative had reacted considerably faster in the coupling reaction than the corresponding D-form. This remarkable observation has the interesting consequence that in principle a higher conversion can be achieved in the coupling of the racemic side chains compared with that of the commonly used enantiopure side chains. The required epimerization of the *epi*-cephalosporin (Scheme II.8) could be readily accomplished under the conditions mentioned above, *viz.* by treatment of the aqueous solution of epimers with 10 mole percent of pyridoxal at a pH of 7.5. The next step is to drive the equilibrium of epimeric cephalosporins to the side of the desired epimer having the antibiotic activity. Selective clathration with a complexing agent could be accomplished for Cephalexin and Cephradine by adding α-naphthol to the mixture of equilibrating epimers. For Cefadroxil, 2,7-dihydroxynaphthalene gave the best performance in this selective clathration. After one night, the complexed cephalosporin had crystallized and could simply be collected by filtration. Decomplexation was achieved by hydrolysis of the complex with diluted aqueous acid and subsequent removal of the complexing agent by extraction with ethyl acetate. Neutralization of the highly concentrated aqueous solution of cephalosporin resulted in precipitation of the ultimate product. The overall yields, after decomplexation, based on 7-ADCA, amounted to 68%, 73% and 57%, for Cephalexin, Cephradine and Cefadroxil, respectively. The products were analyzed by X-ray powder diffraction to ascertain that the desired cephalosporin monohydrate had been obtained.

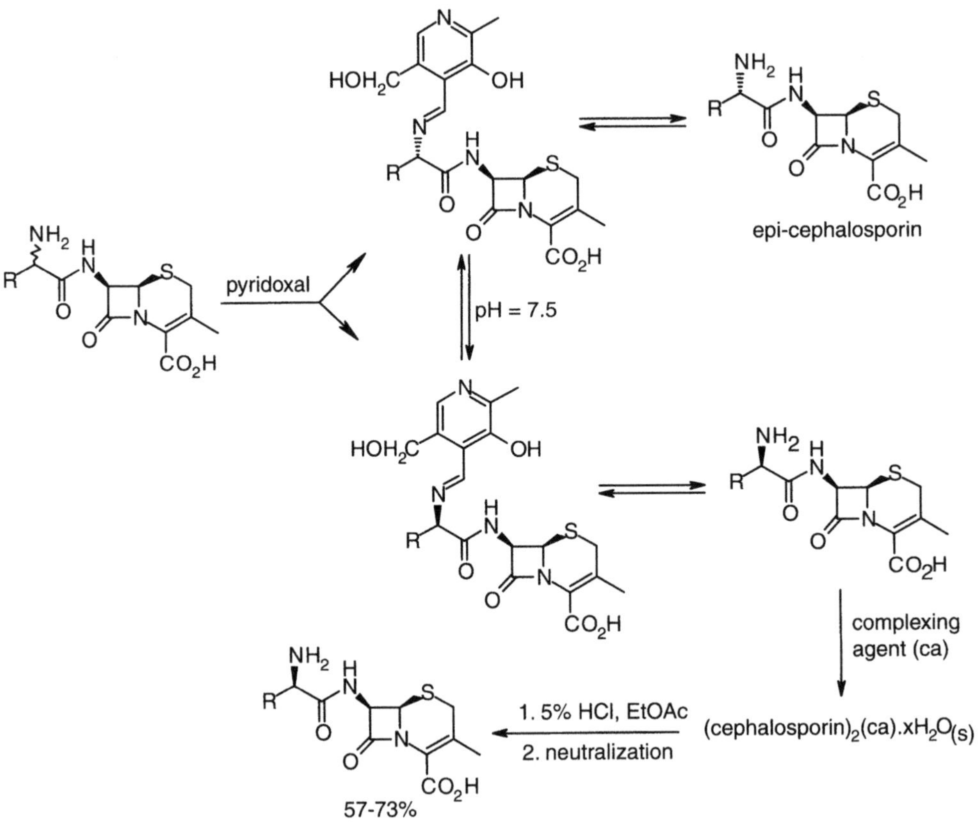

1 R = phenyl, **2** R = 1,4-cyclohexadienyl, **4** R =*p*-hydroxyphenyl

Scheme II.8 Asymmetric transformation of *epi*-cephalosporins to the antibiotics 1-3.

In summary, by using the clathration induced asymmetric transformation, the objective to prepare the cephalosporin antibiotics Cephalexin, Cephradine and Cefadroxil from racemic side chain precursors has successfully been accomplished.

This new approach to the synthesis of cephalosporins employing racemic amino acid as the starting material for the side chains may have attractive prospects for cost price reduction in the manufacture of these live saving antibiotics.[15c]

§3 Synthesis of amino acid side chains

3.1 Relevant literature

Phenylglycine is readily prepared by the Strecker synthesis with benzaldehyde, as already mentioned in Chapter I section 2.3. In this section the entire focus is on the synthesis of *p*-hydroxyphenyl glycine (HPG, **9** R=H). As indicated in the introduction (§1) a conventional Strecker synthesis[22] with *p*-hydroxy benzaldehyde **10** would be a logical option for the synthesis of HPG (scheme II.9). A modified version, the so called Bucherer-Berg method[22], has been reported in which potassium cyanide and ammonium carbonate in aqueous alcoholic solution is used and whereby 5-(*p*-hydroxyphenyl)hydantoin **11** is obtained (scheme II.9) Both methods suffer from the formation of considerable amounts of by-products and moreover the starting material is commercially not readily available.

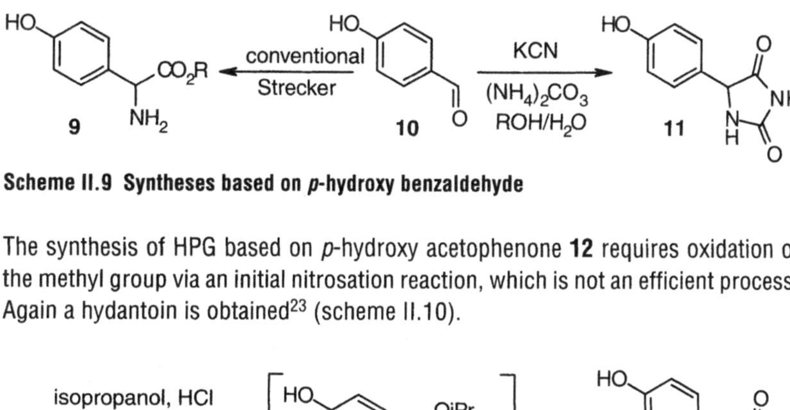

Scheme II.9 Syntheses based on *p*-hydroxy benzaldehyde

The synthesis of HPG based on *p*-hydroxy acetophenone **12** requires oxidation of the methyl group via an initial nitrosation reaction, which is not an efficient process. Again a hydantoin is obtained[23] (scheme II.10).

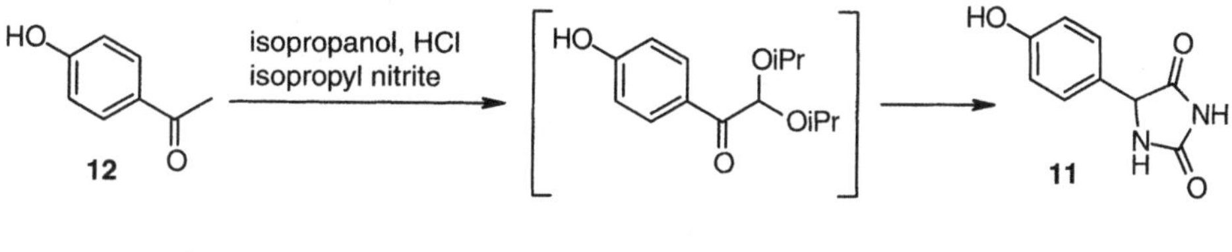

Scheme II.10 Synthesis of the hydantoin, based on *p*-hydroxy acetophenone

The industrially most used synthesis of HPG is based on phenol, in spite of the fact that it has some serious drawbacks. The synthesis involves a Mannich type condensation of phenol with urea and glyoxilic acid (**13**, R=H)[24], or its methyl ester hemiacetal (**13**, R=Me)[25], in the presence of an acid catalyst. The product is a mixture of *para* and *ortho* hydroxyphenyl hydantoin with the *para* compound **11** as the predominant isomer (scheme II.11).

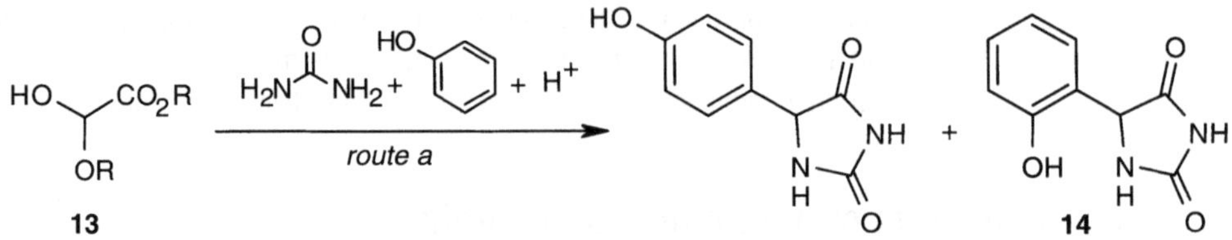

Scheme II.11 Mannich type condensation of phenol

The *ortho* : *para* ratio amounts to 1:4 in standard processes, although several improvements have been reported. Nevertheless, the co-formation of the *ortho* product is one of the major problems in this route. Modifications, such as changing the sequence of addition of the reactants as well as the reaction conditions did not solve this problem. Two alternatives, *viz* condensation of urea with glyoxilic acid (**13**, R=H) or its hemiacetal (**13**, R=Me) prior to the addition of phenol, involving an allantoin or an urethane as an intermediate, are included in Scheme II.12 as route b[26] and c[27], respectively. Also routes in which phenol first condenses with glyoxilic acid or a derivative thereof resulting in a mixture of *ortho* and *para* hydroxyphenyl mandelic acid, followed by treatment with urea or ammonia, have been considered.

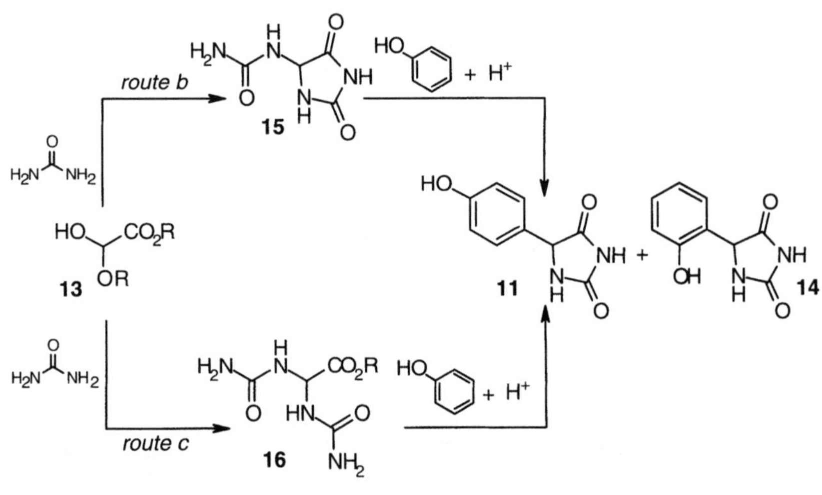

Scheme II.12 Modified Mannich type condensations of phenol

However, in this approach the *o:p* ratio is even less favorable[28]. The hydantoin products shown in the schemes II.11 and II.12 are important intermediates towards HPG as they allow the enzymatic dynamic kinetic resolution[29] of the racemate, as depicted in Scheme II.13

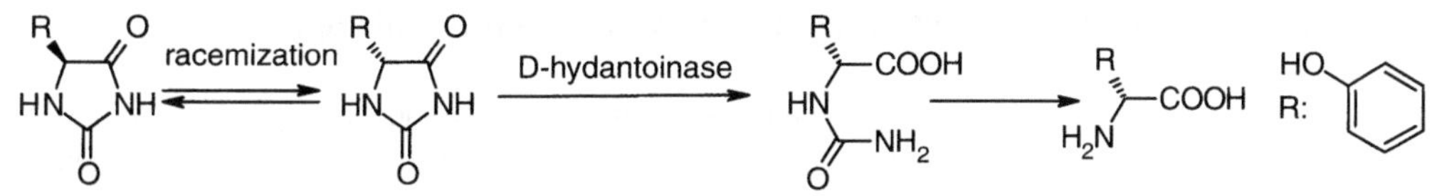

L-hydantoin D-hydantoin *N*-carbamoyl-D-HPG D-HPG, **17**

Scheme II.13 Conversion of racemic hydantoin into optically pure D-HPG

3.2 Regiochemistry of the Mannich type condensation to hydroxy-phenylglycine (HPG)

The co-formation of *ortho*-hydroxyphenyl-hydantoin **14** is one of the major problems in the Mannich type condensation of phenol, urea and a glyoxilic acid derivative. To shed more light on the regiochemistry of this condensation, urea was replaced by a series of amides with R substituents of varying steric demand, as shown in Scheme II.14. The results clearly reveal that increasing the steric size of the R substituent has in fact a contraproductive effect on the *o*:*p* ratio. Thus, other avenues must be considered to improve this ratio.

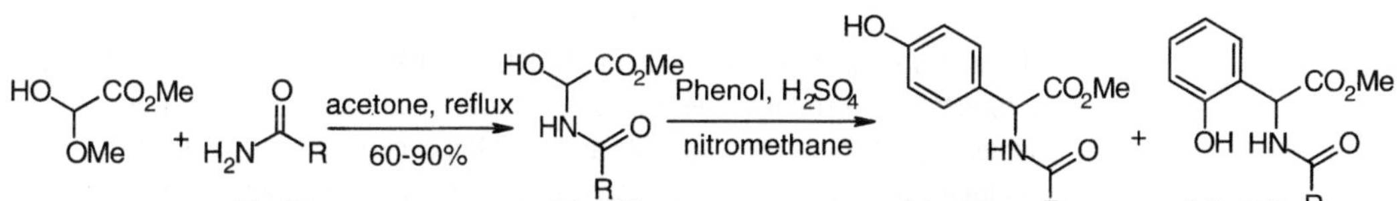

17 - 23 **24 - 30** **31a-37a** **31b-37b**

Reactant	Intermediate	R	Product	ortho/ para
17	24	Me	**31a,b**	1/3
18	25	Et	**32a,b**	1/4
19	26	iPr	**33a,b**	1/3
20	27	cyclo-Hex	**34a,b**	1/3
21	28	tBu	**35a,b**	1/1
22	29	CH_2F	**36a,b**	1/3
23	30	Ph	**37a,b**	5/6

Scheme II.14 The Mannich condensation with various amides

It was found that the various initially formed species such as **15** and **16** (scheme II-12) from the reaction of glyoxilic ester hemiacetal and urea have a profound (positive) effect on the regiochemistry of the Mannich condensation with phenol. This approach is still under active investigation.

3.3 Synthesis of *p*-hydroxy-phenylglycine from *p*-benzoquinone[30]

p-Benzoquinone would be an interesting starting material for the synthesis of HPG, assuming that the carbonyl groups can be selectively converted into the functionalities present in this amino acid. In this route the *ortho/para* problem is in fact circumvented. The first step of this novel approach to HPG is a 1,2-addition of methyl lithioacetate to benzoquinone[31] and subsequent reaction of the thus formed oxy anion with chloroformate to give product **38** in high yield. (Scheme II.15). The oxy anion can also be quenched with 5% HCl. It is of interest to note that the then obtained compound is the natural product *jacaranone*, which has been isolated from *Jaracanda caucana* Pittier and which has a significant anti-tumor activity against P388 lymphocytic leukemia in vivo.

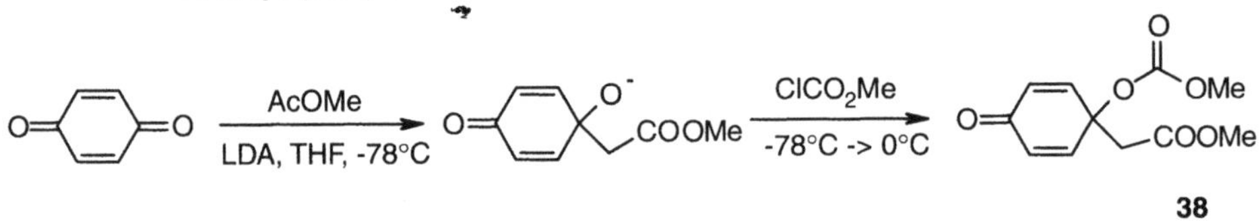

Scheme II.15 Addition of lithioacetate to benzoquinone

The key step in this approach to HPG is the base-induced β-elimination of the carbonate to produce the quinone methide intermediate **39**, followed by a nucleophilic addition of a suitable amine to the α-carbon atom (Scheme II.16). This quinone methide is a highly reactive species which escapes isolation under the conditions of this reaction sequence (however, *vide infra*). A one-pot approach in THF as the solvent in which the reactive intermediate **39** is trapped with an amine in an elimination-addition reaction proved to be very successful. Six different amines have been introduced using this procedure (see Table II.7).

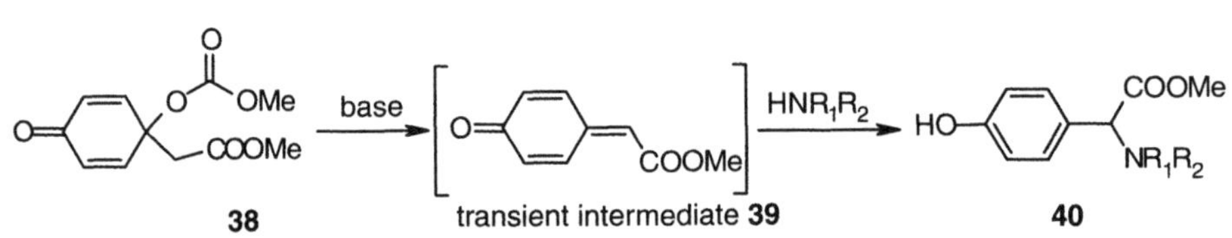

Scheme II.16 Elimination-addition reaction to HPG and congeners

Table II.7 Results of the one-pot Elimination-Addition reaction

entry	R_1	R_2	equiv	base	Yield (%)
a	H	H	excess	NH3	80-90
b	CH_2Ph	H	10	DBU	95
c	$(CH_2)_3CH_3$	H	10	DBU	100
d	C_6H_{11}	H	10	DBU	99
e	$C(CH_3)_3$	H	10	DBU	92
f	piperidine		10	DBU	87

For the synthesis of HPG compound **38** was dissolved in THF and added to liquid ammonia at -30° C. The use of THF as the co-solvent appeared to be essential, because without THF no addition product was formed. Apparently, THF has a beneficial effect on the nucleophilicity of the amine. For the synthesis of HPG no additional base was required to eliminate the carbonate group. For the other amines listed, the reactions were carried out at 0° C with 10 equivalents of the amines in the presence of a small amount of DBU (5 - 10 mol %) which is needed to initiate the elimination of the carbonate. These elimination-addition reactions all went to completion in a few hours time in high yield. Using lesser amounts of amine caused a decrease in yield.

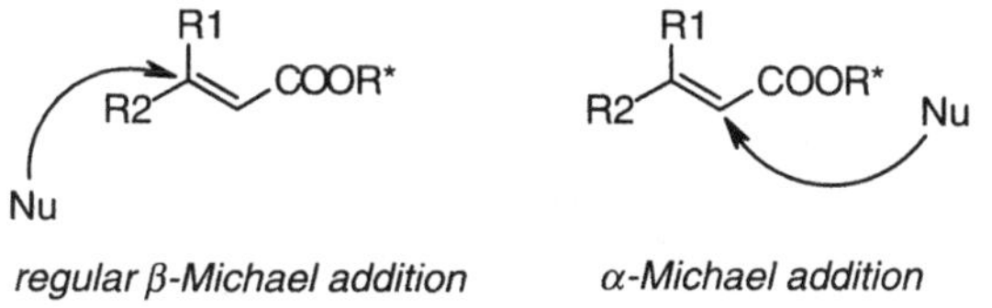

Scheme II.17 α-Michael addition vs β-Michael addition

In principle, this elimination-addition reaction can be accomplished in a diastereoselective manner when chiral acetate esters are used as the substrate. It should be noted that this is a special case of the Michael addition (Scheme II.17). Here the nucleophile comes in at the α-position instead of β-position as encountered in the regular Michael addition. This implies that the chiral auxiliary is closer to the reaction center than in a regular Michael reaction. This anomalous Michael behavior may be attributed to the driving force of the aromatization of the dienone group during this process. To study the diastereoselectivity, esters of three different chiral alcohol have been investigated (Scheme II.18) following essentially the same sequence as depicted in Scheme II.16. The results for two amines, viz ammonia and benzyl amine are included in the scheme II.18. The bornyl group did not show any diastereoselectivity, which is in line with the rather poor inducting capability of this group. The menthyl group however gave a diastereoselectivity of 15-20 %. By far the best results were obtained for the 8-phenyl-menthyl group (de 60%). Further fine-tuning of the chiral auxiliary may result in the desired improvement of the diastereoselectivity.

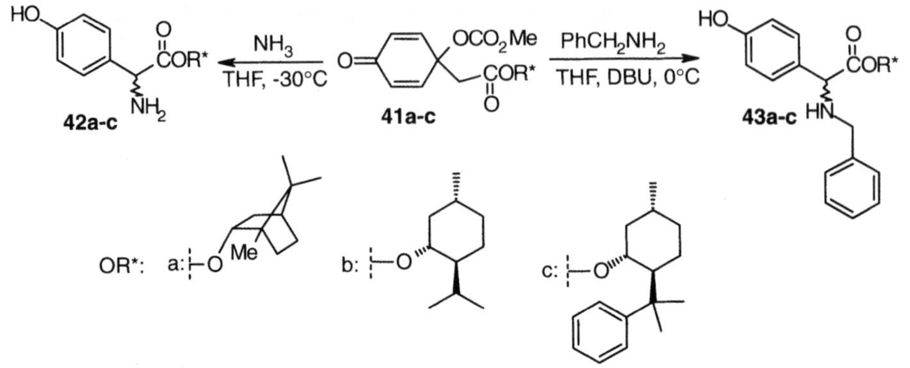

	yield (%)	de (%)
42a	55	0
42b	40	15-20
42c	57	60
43a	90	0
43b	90	15-20
43c	90	60

Scheme II.18 Elimination-Addition reaction of 41 using chiral auxiliaries

The transient intermediate **39** shown in Scheme II.16 is a highly intriguing species. Therefore steps were taken to prepare this compound by an independent route. The successful sequence leading to the actual preparation and isolation of this *p*-quinone methide is depicted in Scheme II.19.

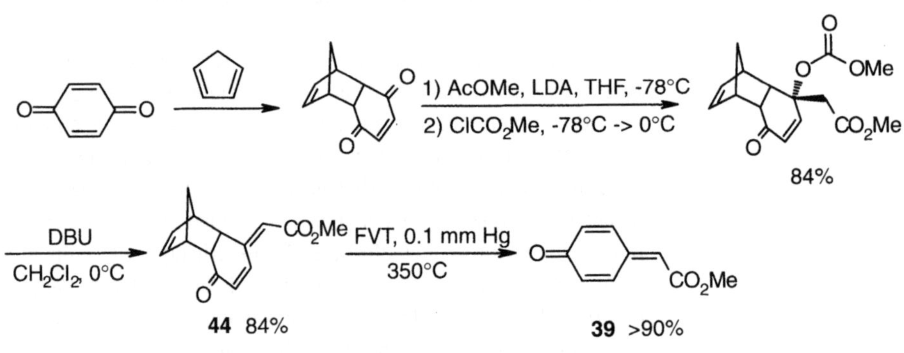

Scheme II.19 Synthesis of quinone methide 39

First one double bond of benzoquinone is temporarily protected in a Diels-Alder adduct with cyclopentadiene. Then, addition of methyl lithioacetate and subsequent reaction with chloroformate, followed by treatment with DBU, leads to product **44**, which in fact is a synthetic equivalent of the desired quinone methide **39**. Removal

of the protecting cyclopentadiene unit using Flash-Vacuum-Technology at 365° C resulted in the target compound **39** in high yield. This *p*-quinone methide **39** is a yellow solid which is reasonably stable in solution. Appealing support for the chemistry shown in Scheme II.16 was obtained when this methide was treated with amines, as shown in Scheme II.20. Indeed, as expected, the addition of amines readily took place to give HPG methyl ester and its *N*-benzyl derivative, respectively, in high yields.
The chemistry described in this section is a typical example of molecular precision in synthesis.

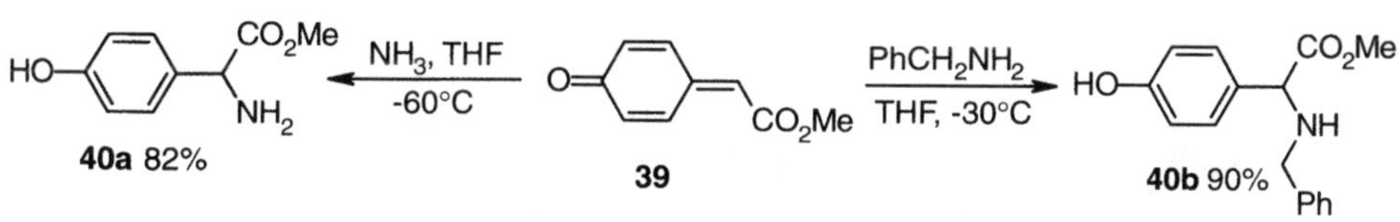

Scheme II.20 Addition of amines to quinone methide 39

§4 Modifications of the β-lactam nucleus of cephalosporins

4.1 Synthesis of 3-carboxycephems from 7-aminocephalosporanic acid (7-ACA)[32]

As mentioned in the introductory section (§1) variation in the substitution pattern of cephalosporin type antibiotics is an important approach to develop new antibiotics of this class. In this section the focus is on the synthesis of 3-carboxycephems starting from 7-ACA, **45**. This 3-carboxy compound can serve as a starting material for the preparation of other structural variants of cephems.
The general strategy to the three 3-carboxycephems **57**, **58**, and **59** shown in Scheme II.21 involves protection of the C7-amino function and the C4-carboxyl group in the starting material 7-ACA, to allow synthetic elaboration at C10. The amino function was protected as a phenylacetamide group because this moiety can be removed enzymatically. For the C4-carboxyl group the *tert*-butyl ester was used, mainly because reconversion of this ester into the free carboxylic acid can readily be accomplished. Introduction of these protecting functions gave the Δ^2 isomer **48**. This is surprising since usually mixtures of Δ^2 and Δ^3 isomers are formed. After saponification of the acetate at C10, alcohol **51** was oxidized to aldehyde **54** with 1-hydroxy-1,2-benziodoxol-3(1H)-one 1-oxide (IBX). Other well-known oxidation reagents gave lower yields or did not react at all. This is typically an example of precision in synthesis. The final step, the oxidation of aldehyde **54** to the carboxylic acid **57**, was accomplished with sodium chlorite in THF/phosphate buffer and cyclohexene as the chlorine scavenger. These subtle conditions resulted in a high yield of acid **57**.

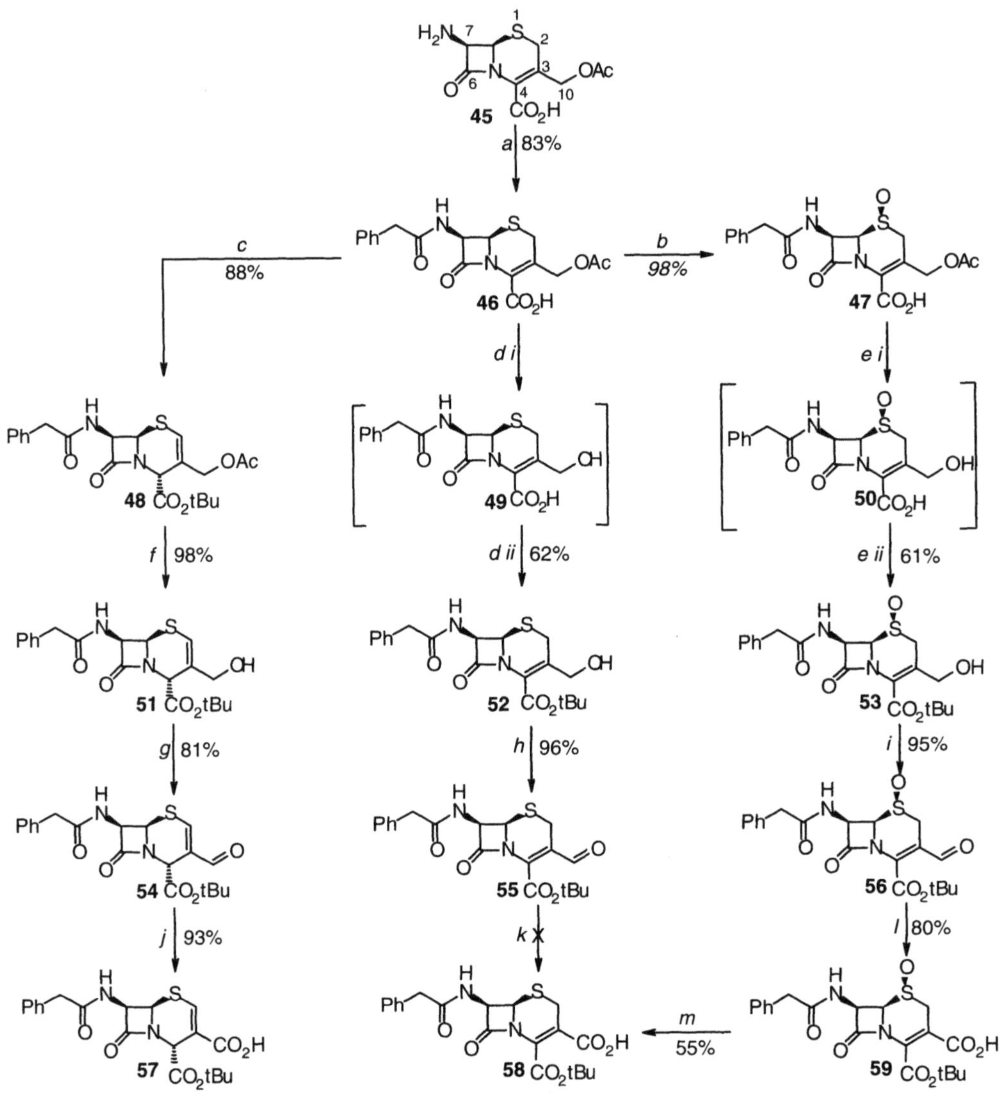

a) phenylacetyl chloride, H_2O pH 8.0, rt, overnight, 83%; b) m-CPBA, CH_2Cl_2, 0 °C, 1.5 h, 98%;
c) tert-BuOH, DCC, DMAP (cat.), CH_2Cl_2, -30 °C - rt, o/n, 88%;
d)i cephalosporin acetyl hydrolase (CAH), H_2O pH 7.5, 4-8 h, ii isourea(DCC, tert-BuOH), THF, o/n, 61%;
e)i CAH, H_2O pH 7.5, 4-8 h, ii isourea(DCC, tert-BuOH), THF, overnight, 62%; f) NaOH, MeOH, -20 °C, 4 h, 98%;
g) IBX, DMSO/THF, rt, 30 h, 81%; h) IBX, DMSO, rt, 15', 96%; i) IBX, DMSO, rt, 15', 95%;
j) $NaClO_2$, cyclohexene, THF/phosphate buffer, 0° C, overnight, 93%;
k) $NaClO_2$, cyclohexene, THF/phosphate buffer, 0 °C, 0%; l) $NaClO_2$, cyclohexene, THF/phosphate buffer, 0 °C, overnight, 80%; m) AcCl, $SnCl_2$, DMF, 0 °C, 2 h, 55%.

Scheme II.21 Synthesis of 3-carboxycephems

The synthesis of the corresponding 3-carboxyceph-3-em **58** was conducted similarly. After protection of the amino group, first the acetate at C10 was hydrolyzed with immobilized cephalosporin acetyl hydrolase. The thus obtained alcohol **49** has to be handled with great care to prevent lactonization under acid conditions. The esterification of the C4-carboxylic acid required subtle conditions, i.e. treatment with the isourea derived from DCC and *tert*-butyl alcohol, to avoid isomerization to Δ^2 by-products. Oxidation of the alcohol function at C10 was achieved again with IBX. Disappointingly, the seemingly simple oxidation of the formyl group in **55** to the corresponding acid failed. Therefore, a detour via sulfoxide **59** was necessary.
The preparation of 3-carboxyceph-3-em S-oxide **59** starts with the oxidation of the sulfur atom in **46** with *m*CPBA. Aldehyde **56** was obtained using the same reagents as described for the conversion of **46** into **55**. Gratifyingly, this formyl compound **56** was readily oxidized to the carboxylic acid stage using sodium chlorite giving 3-carboxycephem **59**. Subsequent reduction of this sulfoxide to sulfide **58** was readily accomplished by treatment with acetyl chloride in DMF and a catalytic amount of tin(II) chloride.
The synthetic sequences depicted in Scheme II.21 clearly demonstrate that each conversion has to be conducted carefully. Precise reaction conditions are highly essential.

4.2 Synthesis of 3-halocephems from 3-carboxycephems

The 3-carboxycephems, the preparation of which has been described in the preceding section, can be used for the replacement of the carboxyl group at C3 by halogen or hydrogen. For this purpose, Barton's radical decarboxylation procedure served as the key step.
Carboxylic acid **57** (Scheme II.21) was converted into the so-called Barton ester **60** using a DCC coupling with *N*-hydroxy-4-methyl-2-thiazolinethione (yield 90%). Subsequently, the Barton ester was irradiated in refluxing bromotrichloromethane in the presence of the radical initiator AIBN, affording the 3-bromo compound **61b** in a yield of 69% (Scheme II.22). The corresponding 3-chloro and 3-iodo compounds were obtained using the radical scavengers shown in Table II.8. Unfortunately, the yields are less satisfactory. Decarboxylation can also be accomplished, using *tert*-butyl mercaptane as the trapping agent. It should be noted that the β-lactam nucleus of the product with X = Cl is present in Cefaclor. The Barton reaction is rather capricious. The 3-carboxycephems with a Δ^3 double bond do not show a clean behavior in this radical decarboxylation process. The chemistry shown above allows a quick access to 3-halocephems, but the yields vary considerably for the respective halogens.

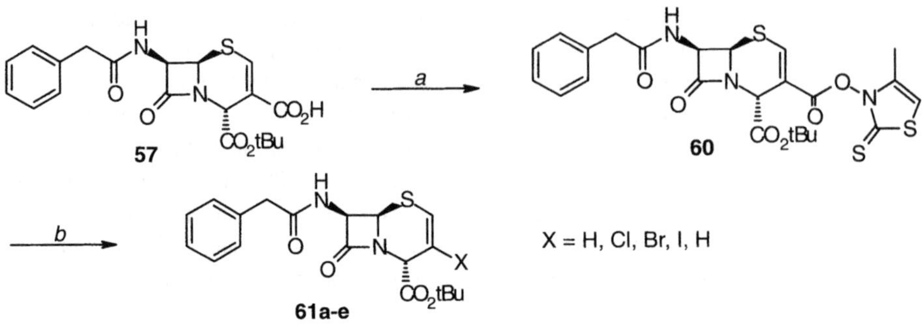

a) DCC, Barton reagent (OH compound), DMAP (cat.), -30°C-rt., overnight, 90%;
b) radical trap (solvent), reflux, 30% AIBN, 15-30', column chromatography.

Scheme II.22 The Barton decarboxylation reaction

Table II.8 Results of the Barton decarboxylation

entry:	trapping agent:	product (X =):	number:	yield (%):
1	CCl_4	Cl	**61a**	10
2	$CBrCl_3$	Br	**61b**	69
3	CHI_3	I	**61c**	28
4	*tert*-butylSH	H	**61d**	15

4.3 Wittig reactions with 3-formylcephalosporins

The search for cephalosporins with improved therapeutic properties receives considerable current attention. An important group of β-lactam antibiotics with an improved biological profile are the 3-alkenylcephalosporins. Examples are Cefdinir, Cefixime and Cefprozil.

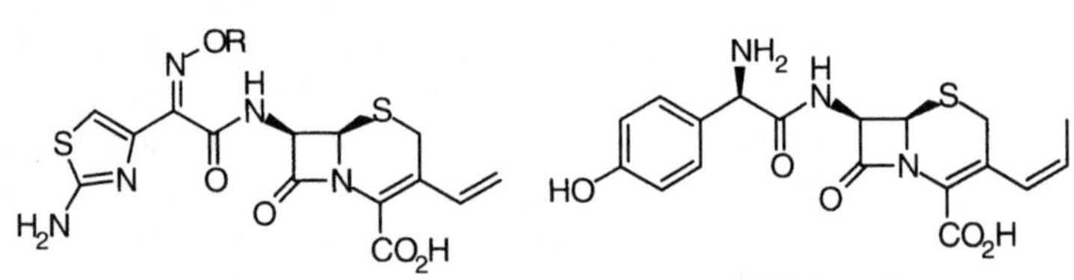

63 R = H (Cefdinir)
64 R = CH_2CO_2H (Cefixime)

65 Cefprozil

The synthesis of 3-alkenylcephalosporins most frequently proceeds via a Wittig reaction using phosphonium salts derived from cephalosporins with a halomethyl substituent at C3. An alternative approach makes use of 3-formylcephalosporins in an olefination reaction with phosphorus ylides. So far, the latter method has only been applied in a few cases. Therefore, the 3-formylcephalosporin **55** prepared according to Scheme II.21 was subjected to olefination reactions. Stabilized ylids derived from ethyl chloroacetate and chloroacetonitrile indeed gave olefination products with the *E*-geometry in moderate yields (Scheme II.23). Reaction of the formyl compound **55** with non-stabilized ylids, such as methylenephosphorane, only produced decomposition products, probably due to the too basic conditions. Thus, the reversed Wittig olefination mentioned above is here the method of choice.

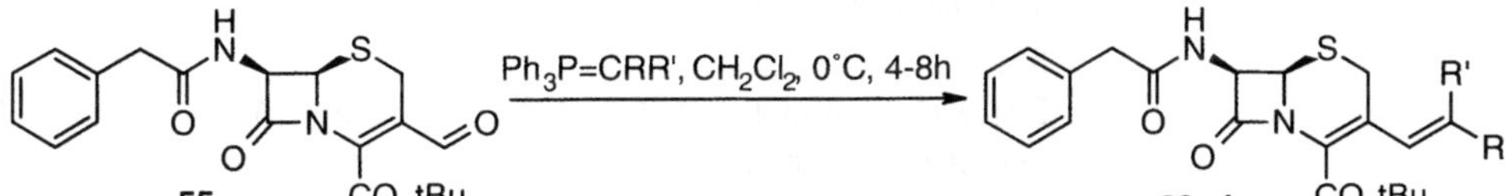

55 **62a-b**

entry:	*ylide:*	*R =*	*R' =*	*product (%):*
1	$Ph_3P{=}CHCO_2Et$	CO_2Et	H	**62a** (51)
2	$Ph_3P{=}CHCN$	CN	H	**62b** (53)

Scheme II.23

In spite of the limited scope in the Wittig reactions the 3-formylcephalosporins deserve further synthetic elaboration.

4.4 Concluding remarks

The chemistry presented in this chapter clearly shows that molecular precision plays a role in all cases. In the clathrate formation precise fitting of complexants in the cavities formed by the antibiotic molecules takes place with a high degree of selectivity. Even diastereoselective complexation of diastereomeric cephalosporins was observed and used in the asymmetric synthesis of ceplalosporins from racemic amino acid side chains. In the Mannich condensation of phenol with glyoxylic acid derivatives and urea fine-tuning of reaction conditions is essential to obtain predominant formation of the *p*-hydroxyphenylglycine. The structural modification of 7-ACA at C3 requires a careful selection of reagents to accomplish the desired conversion in high yields, because the cephalosporins are sensitive to acid as well as basic reagents due to the presence of the reactive β-lactam moiety.

In the topics discussed above, innovative results could only be obtained by using a delicate balance of all aspects of molecular precision.

§5 References

1. V.K. Svedas, A.L. Margolin, and I.V. Berezin, *Enzyme Microb. Technol.*, **1980**, *2*, 138-144; A.M. Blinkovsky, A.N. Markaryan, *Enzyme Microb. Technol.*, **1993**, *15*, 965-973; C.G.P.H. Schroën, V.A. Nierstrasz, P.J. Kroon, R. Bosma, A.E.M. Janssen, H.H. Beeftink, J.Tramper, *Enzyme Microb. Technol.*, **1999**, *24*, 498-506
2. D.H. Nam, C. Kim, D.D.Y. Ryu, *Biotechnol. Bioeng.*, **1985**, *27*, 953; C.G.P.H. Schroën, V.A. Nierstrasz, H.M. Moody, M.J. Hoogschagen, P.J. Kroon, R. Bosma, H.H. Beeftink, A.E.M. Janssen, J. Tramper, Modelling of the enzymatic kinetic synthesis of Cephalexin-Influence of substrate concentration, submitted.
3. US 4003896, **1977**, [Chem. Abs. **1977**, *86*, 171480m]
4. WO 93/12250, EP618979, US5470717, [Chem. Abstr. **1993**, *119*, 137533w]; A. Bruggink, E.C. Roos, E. de Vroom, *Org. Proc. Res. Dev.*, **1998**, *2*, 128-133; A. Bruggink, *Chimia*, **1996**, *50*, 431-432
5. J. Clark, *Chem. Brit.*, October, **1998**, 43-45
6. G.J. Kemperman, R. de Gelder, F.J. Dommerholt, P.C. Raemakers-Franken, A.J.H. Klunder, and B. Zwanenburg, *Chemistry, Eur. J.*, **1999**, *5*, 2163-2168.
7. K. Takahashi, *J. Chem. Soc., Chem. Commun.*, **1991**, 929-930
8. Z. Bocskei, K. Simon, G. Ambrus, E. Ilkoy, *Acta Crystallogr., Sect. C*, **1995**, *55*, 1319
9. T.M. Krygowski, S.J. Grabowski, J. Konarski, *Tetrahedron,* **1998**, *54*, 11311-11316
10. G.J. Kemperman, R. de Gelder, F.J. Dommerholt, P.C. Raemakers-Franken, A.J.H. Klunder, B. Zwanenburg, J. *Chem. Soc. Perkin Trans. 2*, **2000**, 1425-1429
11. (a) G.J. Kemperman, R. de Gelder, F.J. Dommerholt, P.C. Raemakers-Franken, A.J.H. Klunder, B. Zwanenburg, Creating cavities, layers and channels in the hosting framework of Cephradine, *submitted.* (b) G.J. Kemperman, *Thesis, University of Nijmegen, The Netherlands*, **2001**, Chapter 4.
12 (a) G.J. Kemperman, R. de Gelder, R. Wehrens, F.J. Dommerholt, A.J.H. Klunder, L.C.M. Buijdens, B. Zwanenburg, J. *Chem. Soc., Perkin Trans 2*, in press. (b) R. Wehrens, R. de Gelder, G.J. Kemperman, B. Zwanenburg, L.C.M. Buijdens, *Anal. Chim. Acta* **1999**, *400*, 413-424.
13 (a) G.J. Kemperman, R. de Gelder, F.J. Dommerholt, C.G.P.H. Schroën, R. Bosma, A.J.H. Klunder, B. Zwanenburg, *Green Chem.*, submitted. (b) G.J. Kemperman, *Thesis, University of Nijmegen, The Netherlands*, **2001**, Chapter 6. (c) C.G.P.H. Schroën, V.A. Nierstrasz, R. Bosma, G.J. Kemperman, M. Strubel, L.P. Ooijkaas, H.H. Beeftink, J. Tramper, *Enzyme Microb. Technol.*, submitted.
14. The author is indebted to the group of Prof.Dr. J. Tramper (Wageningen University, The Netherlands) for the collaboration during the enzyme inhibition studies.

15 (a) G.J. Kemperman, J. Zhu, A.J.H. Klunder, B. Zwanenburg, *Org. Lett.*, **2000**, *2*, 2829-2831. (b) G.J. Kemperman, J. Zhu, A.J.H. Klunder, B. Zwanenburg, *Eur. J. Org. Chem.*, in press. (c) G.J. Kemperman, J. Zhu, B. Zwanenburg, Process for the preparation of a β-lactam antibiotic, NL1013402, **1999**; PCT/NL00/0063, **2000**.
16. A. Neuberger, "Advances in Protein Chemistry"; M.L. Anson, J.T. Edsall, Eds.; Academic press: New York, **1948**; Vol. 4, p399; b) K. Toi, Y. Izumi, S. Akabori, *Bull. Chem. Soc. Jpn.*, **1962**, *35*, 1422; c) I. Sakieki, M. Mitsuno, *J. Chem. Soc. Jpn.*, **1959**, *80* 1035; d) H. Matsuo, Y. Kawazoe, M. Sato, M. Ohnishi, T. Tatsuno, *Chem. Pharm. Bull.*, **1970**, *18*, 1788
17. J. Olivard, D.E. Metzler, and E.E. Snell, *J. Biol. Chem.*, **1952**, *199*, 699; D.E. Metzler, J.B. Longenecker, and E.E. Snell, *J. Amer. Chem. Soc.*, **1954**, *76*, 639
18. J.C. Clark, G.H. Phillipps, and M.R. Steer, *J. Chem. Soc. Perkin I*, **1976**, 471-474; J.C. Clark, G.H. Phillipps, and M.R. Steer, *J. Chem. Soc. Perkin I*, **1976**, 475-481
19. S. Yamada, C. Hongo, R. Yoshioka, and I. Chibata, *J. Org. Chem.*, **1983**, *48*, 843-846
20. G.J. Kemperman, R. de Gelder, F.J. Dommerholt, P.C. Raemakers, A.J.H. Klunder, and B. Zwanenburg, *J. Chem. Soc., Perkin Trans. 2.*, **2001**, 633-638. (b) G.J. Kemperman, *Thesis, University of Nijmegen, The Netherlands*, **2001**, Chapter 5.
21. E. Dane, and T. Dockner, *Angew. Chem.*, **1964**, *76*, 342; b) E. Dane, and T. Dockner, *Chem. Ber.*, **1965**, *98*, 789-796
22. Bucherer, H. T.; Steiner, *W. J. Prakt. Chem.*, **1934**, *140*, 291-316.
23. Hoechst Celanese Corp., *European patent*, **1991**, EP0437058B1.
24. (a) Beecham Group Ltd., *European Patent*, **1979**, EP0001319. (b) Kanegafuchi Chemical Ind., *UK Patent*, **1980**, GB1572316.
25. Kanegafuchi Chemical Ind., *European Patent*, **1992**, EP0474112.
26. Hoechst France, *UK Patent*, **1981**, GB2053206.
27. Chemie Linz GMBH, *US Patent*, **1995**, US5436256.
28. K.K. Ajinomoto, *UK Patent*, **1979**, GB2012756.
29. (a) K. Yokozeki, S. Nakamori, S. Yamanaka, C. Eguchi, K. Mitsugi, F. Yoshinaga, *Agric. Biol. Chem.*, **1987**, *51*(3), 715-719. (b) O. Keil, M. P. Schneider, J. P. Rasor, *Tetrahedron: Asymmetry*, **1995**, *6*(6), 1257-1260. (c) US Patent 4094741.
30. G.T.M. Titulaer, J. Zhu, A.J.H. Klunder, B. Zwanenburg, *Org. Lett.*, **2000**, *2*, 473-475.
31. A. Fisher, G.N. Henderson, *Tetrahedron Lett.* **1983**, *24*, 131.
32. R. Keltjens, S.K. Vadivel, E. de Vroom, A.J.H. Klunder, B. Zwanenburg, *Eur. J. Org. Chem.*, in press.

Chapter III Biocatalysts and Biocatalysis in the Synthesis of β-Lactam Antibiotics ***103***

§1 Introduction ***103***

§2 Development of processes for side chain donors ***104***
2.1 Existing routes *104*
2.2 Dynamic kinetic resolution of phenylglycine esters via lipase-catalyzed ammonolysis *106*
2.3 Enzymatic hydrolysis of racemic phenylglycine nitrile *108*
2.4 Development of alternative side chain donors: thermodynamic coupling *110*

§3 Integration of activation, coupling and recycling of side chains ***112***
3.1 Salt-free esterification of α-amino acids *112*
3.2 Mixed coupling of PGA and PGM to 6-APA *115*
3.3 A two-step, one-pot synthesis of Cephalexin from D-phenylglycine nitrile *117*

§4 Development of improved catalysts for the synthesis of β-lactam antibiotics ***119***
4.1 Introduction *119*
4.2 Synthesis of β-lactam antibiotics in organic solvents *120*
4.3 Cross-linked aggregates of penicillin acylase as efficient coupling catalysts in aqueous medium *122*
4.4 Efficient synthesis of β-lactam antibiotics in the presence of cross-linked Starburst Dendrimer penicillin acylase *123*
4.5 Penicillin acylase immobilized on cross-linked p-xylylene diamine: a heterogeneous catalyst with the efficiency of free enzyme *126*
4.6 Adjuvants increase the synthesis/hydrolysis ratio *129*
4.7 Active site titration of penicillin acylase *130*
4.8 LC-MS of deactivated penicillin acylase *134*

§5 Peptide synthesis and amine resolutions catalyzed by penicilline acylase ***137***
5.1 Pen acylase-catalyzed synthesis of dipeptides: a chemo-enzymatic route to optically pure diketopiperazines *137*
5.2 Pen acylase-catalyzed resolution of amines in aqueous organic solvents *138*

§6 Concluding remarks & future prospects ***143***

§7 Acknowledgement ***145***

§8 References ***145***

Chapter III Biocatalysts and Biocatalysis in the Synthesis of β-Lactam Antibiotics

R.A. SHELDON, F. VAN RANTWIJK, L.M. VAN LANGEN, M.A. WEGMAN, L. CAO AND M.H.A. JANSSEN
(DELFT UNIVERSITY OF TECHNOLOGY, BIOCATALYSIS AND ORGANIC CHEMISTRY)

§1 Introduction

The commercially most important penicillins and cephalosporins are depicted in Figure III.1. They all contain either D-phenylglycine or D-*p*-hydroxyphenylglycine as the side chain.

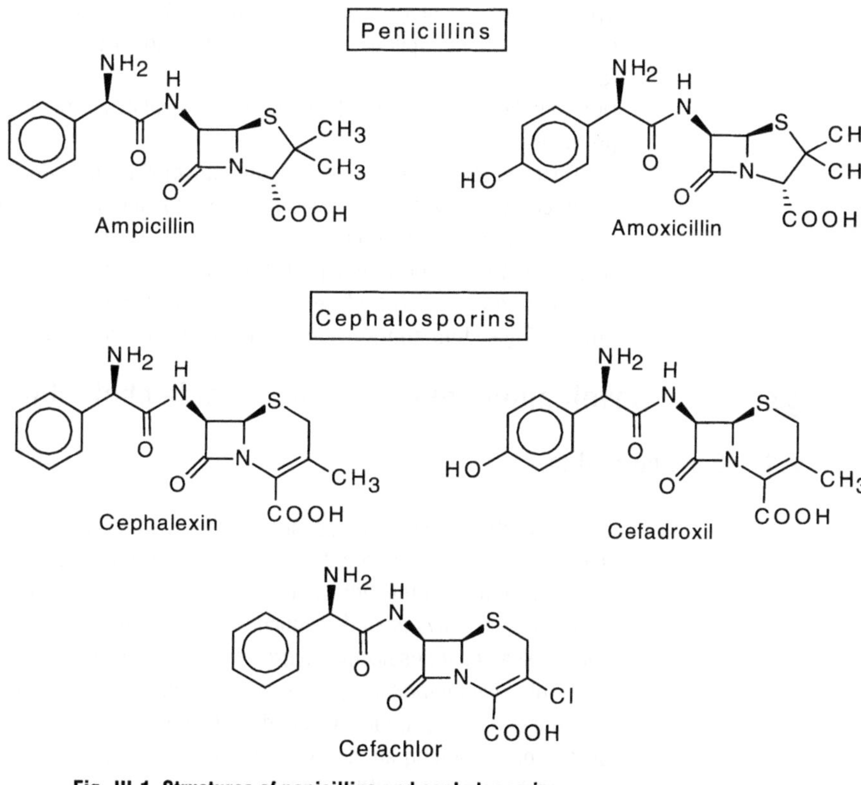

Fig. III.1 Structures of penicillins and cephalosporins.

The substitution of classical chemical syntheses of penicillins and cephalosporins by shorter, catalytic alternatives is expected to provide substantial environmental and economic benefits. This is readily apparent from a comparison of the classical chemical synthesis of Cephalexin with the biocatalytic alternative as already outlined in Chapter 1.The former is circuitous owing to the need for protection, deprotection and activation steps and involves the use of several undesirable reagents and solvents (*e.g.* CH_2Cl_2) leading to substantial waste generation. The biocatalytic route, in contrast, comprises fewer steps (6 instead of 10) and involves the use of enzymes in aqueous medium at ambient temperature.

Analogous chemical vs biocatalytic routes can be envisaged for penicillins, e.g. Ampicillin, in which the 6-aminopenicillanic acid (6-APA) nucleus is derived from penicillin G hydrolysis. The first example was the substitution of chemical deacylation by enzymatic deacylation in the synthesis of 6-APA and 7-ADCA[1,2]. This was relatively easy since in aqueous media the equilibrium lies far to the right. The next step was to substitute chemical coupling of the side chain to the 6-APA or 7-ADCA nucleus, which involves protection and activation of the side chain, by enzymatic coupling[3]. This is a much more difficult task since in aqueous media the equilibrium lies far to the left. In order to achieve a reasonable yield it is necessary to perform the coupling under kinetic control, using an ester or an amide. Competing hydrolysis of the side chain and/or the product is inevitable, however, and the efficiency of the coupling process is determined by the synthesis/hydrolysis ratio (S/H).

Our mandate for the development of (bio)catalytic methods comprised three protocols:

1. Development of processes for (alternative) side chain donors;
2. Integration of activation, coupling and recycling of side chains;
3. Development of the optimum biocatalyst for the coupling reaction.

In addition to the above three protocols we have also studied, as a spin-off activity, the application of pen-acylase in peptide synthesis and the resolution of chiral amines. These four sub-projects will be treated in the ensuing discussion.

§2 Development of processes for side chain donors

2.1 Existing routes

In this protocol we have focused on the synthesis of activated phenylglycine derivatives and precursors thereof. In the current process for Cephalexin via enzymatic coupling the side chain donor used is D-phenylglycine amide (D-PGA). Two processes have been developed by DSM for the synthesis of D-PGA. The first one involves enantioselective hydrolysis of the racemic amide catalyzed by whole cells of *Pseudomonas putida* ATCC 12633 which contain an L-specific aminopeptidase[4]. The racemic amide substrate is prepared by selective hydrolysis of the corresponding amino nitrile which is derived from the Strecker reaction of benzaldehyde with HCN/NH_3. A serious shortcoming of this kinetic resolution is that the desired D-PGA is obtained in a maximum yield of 50%. Recycling of the L-phenylglycine (L-PG) coproduct is circuitous - it requires three steps (racemization, esterification and ammonolysis) to regenerate the racemic amide - and is, hence, economically unattractive. A short-term solution to this problem is to racemize the L-PG and add the racemate to the existing process for resolution of phenylglycine by diastereomeric salt crystallization. However, this option may not be available in a longer term scenario where it is envisaged that the classical resolution process will be phased out.

In the second, more elegant approach an asymmetric transformation of the diastereomeric salts of D,L-phenylglycine amide with L-mandelic acid, in the presence of a catalytic amount of benzaldehyde, affords the L-mandelic acid salt of D-PGA in almost quantitative yield and >99% ee[5]. The benzaldehyde promotes *in situ* racemization via Schiff's base formation with the phenylglycine amide. The L-mandelic acid can be readily recycled after cleavage of the salt but the need for recycling of stoïchiometric quantities of resolving agent remains a limitation of this process.

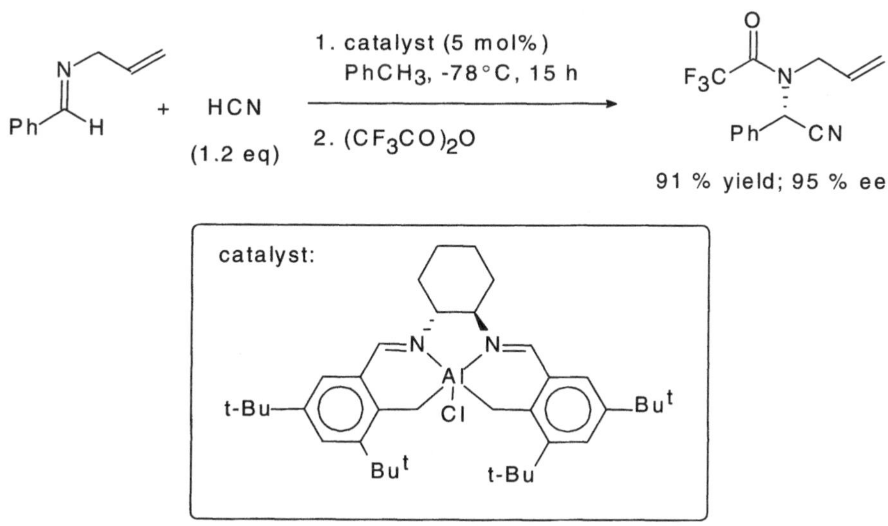

Fig. III.2 Catalytic asymmetric Strecker synthesis.

Hence, there remains an incentive to develop an effective catalytic method for the synthesis of D-PGA *via* a dynamic kinetic resolution of a chiral precursor or an asymmetric Strecker synthesis. Kanegafuchi (nowadays Kaneka Corporation) developed a process for the synthesis of D-phenylglycine via enzyme-catalyzed dynamic kinetic resolution of the corresponding racemic hydantoin[6]. The latter is prepared by a Bucherer-Berg reaction of benzaldehyde with HCN/H_2NCONH_2. Unfortunately, this methodology cannot be adapted for the synthesis of D-PGA. In principle, an enzyme catalyzed ammonolysis of the racemic hyantoin, with *in situ* racemization, would afford an economically attractive route to D-PGA but such a reaction has yet to be discovered.

Jacobsen and coworkers have described a method for performing asymmetric Strecker syntheses (see Figure III.2) using catalytic amounts (2-5 mol %) of an aluminium Schiff's base complex[7]. However, a very low temperature (-78°C) is required (the uncatalyzed reaction is rapid at ambient temperature) and the amino nitrile product is doubly derivatized (the amino group is substituted with trifluoroacetyl and allyl or benzyl). Moreover, the starting material is a Schiff's base which has to be synthesized from the aldehyde. Hence, this method requires three extra chemical steps - to prepare the substrate and to remove the two groups from the product - and is therefore economically unattractive.

2.2 Dynamic kinetic resolution of phenylglycine esters via lipase-catalyzed ammonolysis

We previously showed[8] that the C. *antarctica* lipase (Novozyme 435) - catalyzed ammonolysis of racemic phenylglycine ester affords D-PGA. We also observed a slow racemization of the ester substrate (5% in 24 h) during the course of the reaction, which suggested that a dynamic kinetic resolution could, in principle, be achieved if this racemization could be sufficiently accelerated (Figure III.3). For such a scheme to be effective the product amide should racemize much slower than the ester substrate. The fact that the amide product precipitates from the reaction medium (*tert*-butyl alcohol solvent) favors such a situation. The amino ester substrate can be prepared by a Pinner reaction of the corresponding amino nitrile, thus providing for a two-step process for the synthesis of PGA from the Strecker adduct in a theoretical yield of 100%.

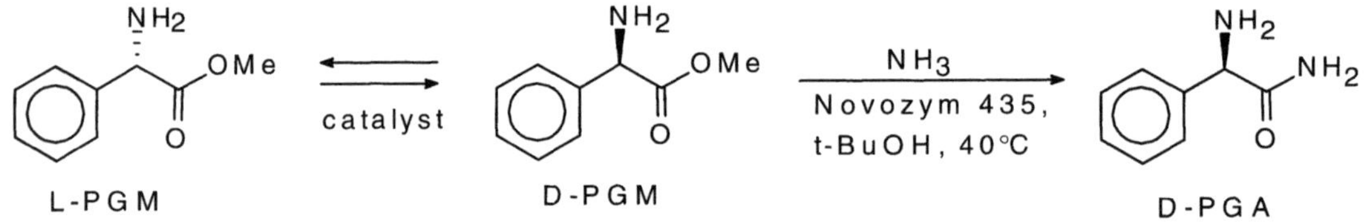

Fig. III.3 D-PGA via dynamic resolution of the ester.

The racemization of amino acid esters is known to be catalyzed by aromatic aldehydes such as salicylaldehyde and 3-hydroxy-2-methyl-5-hydroxymethylpyridine-4-carboxaldehyde (pyridoxal) under basic conditions[9]. Indeed, we previously observed that pyridoxal and salicylaldehyde readily racemize phenylglycine methyl ester in ammonia-saturated *tert*-butyl alcohol, thus providing the basis for a dynamic kinetic resolution process[10].

In order to establish the optimum conditions we carried out an extensive investigation[11] of the effect of reaction parameters - solvent, nature and amount of racemization catalyst and temperature - on the conversion and product ee. *A priori* one might expect the solvent to influence both the enantioselectivity and the rate of racemization. However, no improvement in enantioselectivity was observed (see Table III.1) compared to *tert*-butyl alcohol. Similarly, no improvement in racemization rate was observed in the solvents tested compared to *tert*-butyl alcohol.

Table III.1 Ammonolysis of phenylglycine methyl ester in various solvents.

Solvent	t (h)	Conversion (%)	ee_{amide} (% D)	E	Initial rate (mM min^{-1} mg^{-1})
tert-Butyl alcohol	4	46	78	16	16.7
tert-Amyl alcohol[b]	3	48	76	17	14.2
1,2-Dimethoxyethane	6	49	85	21	8.80
tert-Butyl methyl ether[b]	5	43	73	12	7.7
Dioxane	21	43	83	20	3.7
Acetonitrile	66	40	82	18	2.0
Hexane[b]	15	<5	–	–	–

[a] Reaction conditions: 1 mmol D,L-phenylglycine methyl ester was shaken with 50 mg Novozym 435 in 5 ml ammonia saturated solvent at 40°C.
[b] Novo SP 611, *Candida antarctica* lipase B on Accurel EP 100 (50 mg), was used instead of Novozym 435 to minimize hydrolysis of the phenylglycine methyl ester due to traces of water in the carrier material.

The enantioselectivity improved, as expected, with decreasing reaction temperature. This occurred, however, at the expense of reaction rate. The highest enantioselectivity - an E value of 52 - was observed when ammonolysis was performed at -20°C in *tert*-butyl alcohol/*tert*-butyl methyl ether (70:30 v/v) as solvent (*tert*-butyl alcohol alone would solidify at this temperature). When the reaction was performed under these conditions in the presence of pyridoxal as a racemization catalyst D-PGA was obtained in 85% yield and 88% ee after 66 hours (see Table III.2).

Table III.2 Dynamic kinetic resolution of PG methyl ester (PGM).

Racemisation catalyst	Ester/cat	Time (h)	Conv. (%)	ee_{amide} (%)
None	–	4	46	78
Pyridoxal	100	17	90	56
Salicylaldehyde	100	17	86	56
Pyridoxal[a]	50	66	85	88

[a] In tert-butyl alcohol/tert-butyl methyl ether (70:30, v/v) at -20°C.

In short, we have demonstrated the feasibility of the synthesis of D-PGA in a dynamic kinetic resolution via lipase-catalyzed ester ammonolysis. However, in order to become economically attractive further improvement of both the enantioselectivity and the reaction rate is required.

2.3 Enzymatic hydrolysis of racemic phenylglycine nitrile

Many species of bacteria, e.g. *Rhodococcus sp.*, are able to convert racemic amino nitriles to the corresponding L-amino acid[12,13]. This is achieved in two enzymatic steps involving an aselective nitrile hydratase and an L-specific amidase. When racemic phenylglycine nitrile is the substrate this affords a synthesis of D-PGA (see Figure III.4). The second step is analogous to the DSM aminopeptidase process for D-PGA described above. Consequently, this method suffers from the same shortcoming, *i.e.* a theoretical yield of 50% and L-phenylglycine as the co-product.

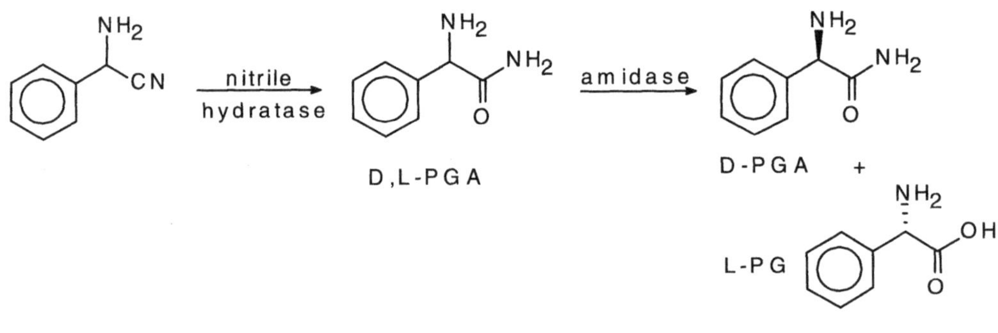

Fig. III.4 Hydration and subsequent hydrolysis of DL-PG nitrile.

However, we reasoned that this methodology could provide an interesting alternative to the above described existing technology. Moreover, if an enantioselective nitrile hydratase could be found this would provide the possibility of conducting a dynamic kinetic resolution of the racemic amino nitrile (aminonitriles are known to undergo facile racemization), affording D-PGA in a theoretical yield of 100% directly from the Strecker adduct (Figure III.5).

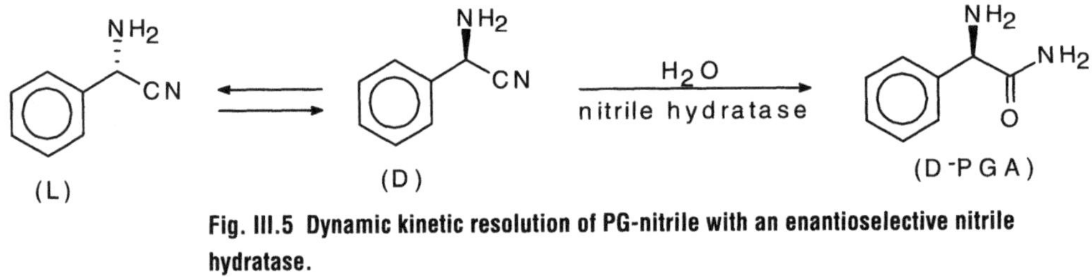

Fig. III.5 Dynamic kinetic resolution of PG-nitrile with an enantioselective nitrile hydratase.

This would clearly be superior to all existing methodologies for D-PGA synthesis. Alternatively, a less attractive but still interesting proposition, could involve the stereoretentive nitrile hydratase-catalyzed hydration of D-PG nitrile. The latter is readily accessible via asymmetric transformation of the L-tartaric acid salt of racemic PG nitrile. These considerations prompted us to screen bacterial strains for (enantioselective) nitrile hydratase/amidase activity (in collaboration with A. Stolz/ U. Heinemann of the University of Stuttgart).
We tested ca. 60 cultures and found 5 strains that were able to convert DL-PG nitrile

(see Table III.3)[14]. Owing to the propensity of the aminonitrile for decomposition into benzaldehyde *via* a retro-Strecker reaction we focused our attention on strains with the ability to catalyze rapid hydration of the nitrile moiety. Two strains, identified as *R. globerulus* and *R. rhodochrous* species fulfilled this criterion.

Table III.3 Hydration of DL-PG nitrile mediated by several bacterial strains.

Strain	D-PGA		L-PG		
	Yield (%)	ee (%)	Yield (%)	ee (%)	E
R. globerulus MAWA	48	>99	52	97	>100
R. rhodochrous MAWB	43	85	38	>99	>100
MAWC[b]	39	93	41	84	20
MAWD[b]	46	96	14	>99	>100
R. rhodochrous MAWE	40	35	24	88	20

[a] Reaction conditions: 0.01 mmol D,L-phenylglycine nitrile and resting cells ($A_{546\ nm}$ 20) in 1 ml 50 mM phosphate buffer pH 7 were shaken at 30°C for 4 h.
[b] Strain is not identified.

Both of these strains (*R. globerulus* MAWA and *R. rhodochrous* MAWB) converted DL-PG nitrile to DL-PGA within 5 minutes. The latter was subsequently hydrolyzed to a mixture of L-PG and D-PGA by an L-specific amidase, in essentially quantitative enantioselectivity (E > 100). Careful HPLC analysis revealed that the nitrile hydratase activity was aselective; the observed enantioselectivity was completely due to the amidase, as is generally the case[12,13]. The optimum result was obtained with *R. globerulus* MAWA cells grown on 2-methyl-3-butenenitrile: 48% yield of D-PGA with >99% ee and 52% yield of L-PG with 97% ee.

In short, we have demonstrated the feasibility of a nitrile hydratase/amidase-mediated conversion of DL-PG nitrile to a mixture of D-PGA and L-PG but we did not succeed in finding the hoped-for enantioselective nitrile hydratase activity. This could possibly be achieved in the future by subjecting an appropriate nitrile hydratase to *in vitro* evolution[15].

We next turned our attention to the stereoretentive nitrile hydratase-catalyzed hydration of D-PG nitrile[16]. To this end we used an immobilized *Rhodococcus sp.* from NOVO (NOVO SP361) and four bacterial strains from the above mentioned screening. Our objective was to achieve high chemical and volumetric yields of D-PGA with minimum racemization and degradation of the D-PG nitrile. As noted above, the bioconversion of α-amino nitriles is hampered by their facile degradation via a retro Strecker reaction. The latter takes place readily at pH ~ 7, conditions that are optimum for the enzymatic reaction.

We found that SP361 catalyzed the hydration of D-PG nitrile at pH 5, with minimum racemization and/or degradation. Under these conditions the amidase activity was low and highly L-specific and no formation of phenylglycine was observed. When the cells were recycled their performance decreased considerably, particularly when the D-PG nitrile L-tartaric acid salt was used as the substrate. We tentatively concluded that tartaric acid acts as an inhibitor and further experiments were performed with the free base.

Subsequent investigations revealed that competing retro Strecker reaction could be largely circumvented by conducting the reaction in a fed-batch operation (to maintain a relatively low concentration of substrate) at pH 7 and decreasing the reaction temperature to 5°C. It is worth noting, in this context, that degradation to benzaldehyde not only results in loss of substrate but also in a dramatic decrease in reaction rate. Apparently, benzaldehyde is also an effective inhibitor of the nitrile hydratase.

Under the above mentioned optimum conditions D-PGA was obtained in 96% yield and 95% ee after 7 hours using SP361. Similarly, other bacterial strains that had previously been shown to contain high nitrile hydratase activity with DL-PG nitrile (see above) were tested under the same conditions. As shown in Table III.4 they gave comparable results to SP361.

Table III.4 Fed-batch hydration of D-PG-nitrile catalyzed by bacterial cells.

Entry	Enzyme (U)	Time (h)	Yield (%)	ee (%)
1	SP 361 (-)	7	96.0	95.0
2	R. globerulus MAWA (69)	4	98.7	94.6
3	R. erythropolis MAWF (44)	7	95.6	97.4
4	R. rhodochrous MAWE (56)	7	98.6	96.8
5	R. rhodochrous MAWB (38)	6	95.4	98.3
		22	95.1	99.5

[a] Reaction conditions: 1 mmol D-phenylglycine nitrile in 5 ml 0.2 N H_2SO_4 was added in 3 h to the microorganism in 5 ml 10 mM phosphate buffer. The reaction mixture was shaken at 5°C.

Interestingly, we found that the enantiomeric purity of the product increased with prolonged reaction time. The amidase present in the microbial cells is completely L-specific and slowly converts the L-amide impurity in the product to the L-acid. With *R. rhodochrous* MAWB (entry 5 in Table III.4), D-PGA was formed in 95.4% yield and 98.3% ee after 6 hours which became 95.1% yield and 99.5% ee after 22 hours. In conclusion, we have shown that D-PG nitrile can be effectively converted to D-PG amide using various nitrile hydratase-containing bacterial strains: in ca. 95% yield, > 99% ee and a product concentration of 100 mM (15 g/ltr). We also showed that this reaction could be combined with pen acylase-catalyzed coupling of D-PGA to 7-ADCA to afford a one-pot, two-step synthesis of Cephalexin (see Section 3).

2.4 Development of alternative side chain donors: thermodynamic coupling

The (enzymatic) formation of an amide bond can be achieved by the acylation of an amine using either the free acid (thermodynamic control) or an activated derivative such as an ester or an amide (kinetic control). In an enzymatic coupling the reaction proceeds via an acylenzyme intermediate which subsequently transfers its acyl group to the amine nucleophile. Kinetically controlled synthesis is characterized by a relatively fast formation of the acylenzyme intermediate from the activated acyl donor. The product concentration passes through a maximum after which product hydrolysis dominates and the product concentration decreases eventually to zero. The efficiency of a kinetically controlled coupling is denoted by the synthesis/hydrolysis ratio (S/H) which is obviously dependent on, inter alia, the conversion. A serious shortcoming of kinetically controlled synthesis of penicillins and cephalosporins is the formation of substantial amounts of the free amino acid side chain by competing hydrolysis of the product and the acyl donor (amide or ester). This means that an excess of acyl donor is required and downstream processing necessitates recovery and recycling of the free amino acid.

Thermodynamically controlled coupling of the free amino acid (PG or HPG) with 6-APA or 7-ADCA would offer obvious economic benefits. Unfortunately, it was soon recognized that the synthesis of Ampicillin, Amoxicillin and Cephalexin by thermodynamic coupling is not viable[17,18]. Amino acids have a zwitterionic character over a broad pH range and the low energy of the zwitterion shifts the equilibrium nearly completely towards hydrolysis. Hence, thermodynamic coupling becomes feasible when the amino group is derivatized. The *N*-substituted derivatives can be regarded as 'normal' carboxylic acids which can undergo smooth thermodynamic coupling with 6-APA or 7-ADCA in high yield, by analogy with the known coupling with phenylacetic acid[19].

An obvious candidate for such a thermodynamic coupling is the D-*N*-carbamoyl derivative, readily prepared by hydantoinase-catalyzed enantioselective hydrolysis of the corresponding hydantoin (see earlier). The *N*-carbamoyl and other derivatives which have been considered are depicted in Figure III.6.

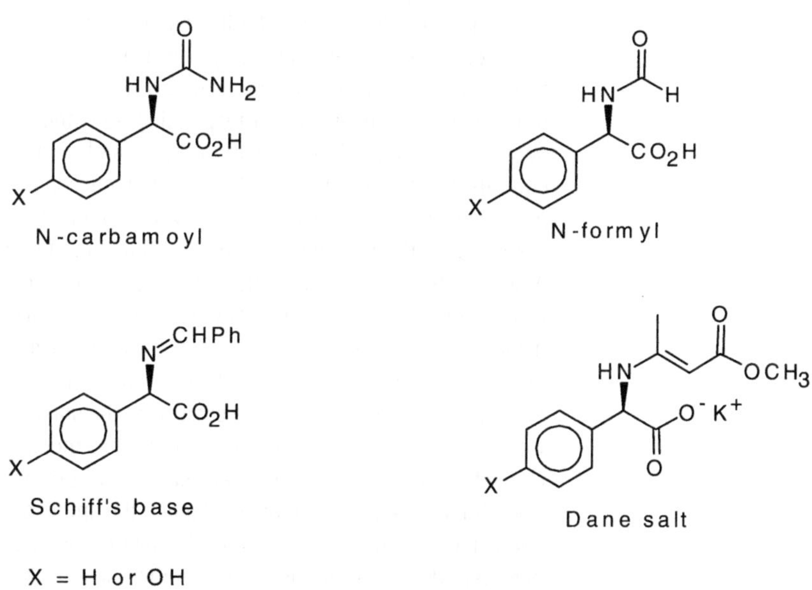

Fig. III.6 Alternative side-chain donors for thermodynamic coupling.

Obviously such a thermodynamic coupling procedure will also require the availability of an effective method for removal of the *N*-protecting group. Accordingly, we tested a variety of pen-acylases, lipases, esterases and an amino acylase in both the synthesis and hydrolysis of *N*-carbamoylamoxicillin but in no case was any activity observed.

A French patent application dating from 1969[20] described the successful thermodynamic coupling of Schiff's base derivatives of D-PG to 6-APA, mediated by whole cells of *E. coli* ATCC 9736. However, we were unable to reproduce these results using the Schiff's base derived from PG and benzaldehyde or the Dane salt (from PG and acetoacetic ester).

In conclusion, screening of a variety of existing enzymes has failed to identify an enzyme capable of mediating the coupling of *N*-derivatized side chains to 6-APA. We do not rule out, however, that this goal could be achieved in the future by, for example, *in vitro* evolution or bioimprinting.

§3 Integration of activation, coupling and recycling of side chains

3.1 Salt-free esterification of α-amino acids

As noted earlier amino acid esters can also be used as acyl donors in kinetically controlled coupling procedures or as precursors to the corresponding amino acid amides. This is particularly relevant in the case of D-*p*-hydroxyphenylglycine (D-HPG) which is not produced *via* a Strecker reaction (*p*-hydroxybenzaldehyde is too expensive). Consequently, the amino nitrile is not available as a starting material for the synthesis of D-HPG amide via analogous procedures to those used for D-PGA (see Section 2). This means that the amide should be prepared from D-HPG, via the ester. Alternatively, D-HPG ester itself could be used as the acyl donor.

One problem associated with the synthesis of amino acid esters is the need for more than one equivalent of mineral acid, *e.g.* HCl or H_2SO_4. Consequently, the ester product is formed as the corresponding salt and generation of the free ester necessitates neutralization with a base and concomitant generation of one equivalent of salt, *e.g.* NaCl or Na_2SO_4. In practice, substantially larger amounts are produced owing to the use of a large excess of mineral acid. This leads to problems associated with the disposal of copious amounts of inorganic salts.

Amino acids exist as zwitterions and in order to generate a free carboxyl group, for esterification, a stoichiometric amount of a mineral acid is required. We surmised, however, that the use of a solid acid catalyst, *e.g.* a zeolite could, in principle, yield the ester as the free base in a salt-free, non-corrosive process. This was based on the notion that desorption of the ester from the solid catalyst into the bulk solution would enable the adsorption of more amino acid molecules. In this way a situation is created whereby an acid-catalyzed reaction occurs on the surface of a solid acid even though the bulk solution, containing the amino acid ester, is basic.

The esterification of anthranilic acid catalyzed by acidic ion exchange resins has been reported[21] but in this case the difference in pKa of the carboxyl and amino groups ($\Delta pKa = 2.4$) is much smaller than in α-amino acids ($\Delta pKa = 7$).

For our initial screening of solid acid catalysts (acidic clays, zeolites, heteropoly acids and ion-exchange resins) we studied the esterification of D-PG with methanol. Only zeolites H-Y and H-USY (a naphtha cracking catalyst) catalyzed the esterification. A plausible explanation is that these zeolites are highly hydrophilic, facilitating the adsorption of D-PG in the presence of the methyl ester (D-PGM). Moreover, these zeolites consist of a three-dimensional network of large pores (7.4Å) intersected by supercages (12Å) which allows for facile migration of the substrate into, and the product out of, the channels. Because of the superior results H-USY was used in all further experiments. The conventional and solid-acid catalyzed esterification of α-amino acids are compared in Figure III.7.

Initially, experiments were performed under reflux conditions (64°C) where equilibrium is reached at an ester concentration of 30 mM, which corresponded to a yield of 36%. As would be expected, the equilibrium yield is independent of the amount of catalyst used. The initial rate increases with increasing amounts of catalyst, also as would be expected. The H-USY was filtered off and reused without any post-treatment but its activity slowly decreased after each cycle, presumably owing to deactivation by the water formed in the reaction. This was confirmed by deliberate addition of water to the reaction mixture. Upon calcination the activity of the H-USY was completely recovered. Moreover, its Si/Al ratio was unchanged after three cycles, consistent with no leaching of Al having occurred during the reaction. In principle, complete conversion should be possible by removing the water formed during the esterification. However, attempts to remove water by the addition of K-A molecular sieve or dimethyl carbonate (which reacts with water to give methanol and CO_2) did not significantly improve the yield of ester. Apparently it is difficult to remove water from the hydrophilic H-USY but we do not rule out that a method for continuous removal of water can be found.

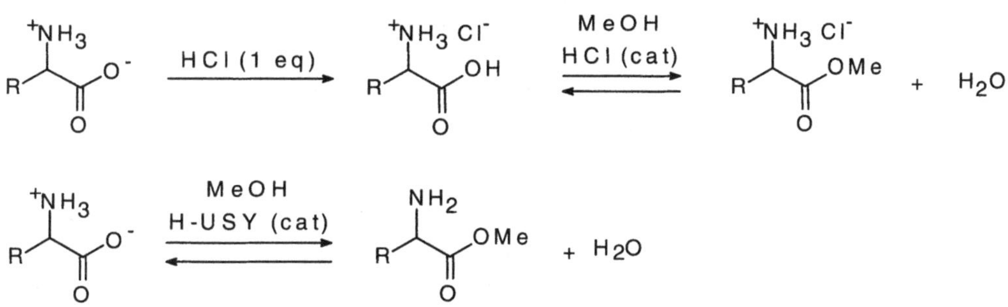

R = Ph, $PhCH_2$, $PhCH_2CH_2$, p-HO-C_6H_4

Fig. III.7 Conventional vs solid acid-catalyzed esterification of α-amino acids.

We also showed that the ester product inhibits the reaction, presumably by competing with the substrate for catalytic sites. We further observed that 1 g of H-USY adsorbed 0.5 mmol of ester which suggests that the amount of accessible catalytic sites is 0.5 mmol/g H-USY.

We also investigated the effect of temperature. As shown in Table III.5 the rate and equilibrium yield increased with temperature. At 130°C the ester was obtained in 86% yield after 30 minutes. We note, moreover, that the yield corresponds to the amount of ester in the bulk solution and when the 0.5 mmol of ester that is presumably adsorbed on the catalyst (see above) is taken into account the yield is 92%. Assuming that the number of active Al sites is 0.5 mmol/g H-USY the amount of ester product formed (ca. 6 mmol) corresponds to a turnover number of 12.

Table III.5 Effect of temperature on H-USY catalyzed esterification of D-PG.

T (°C)	p (bar)	t (h)	Yield (%)	*ee* (%)
65	1	20	36	4
100	15	2	73	29
130	20	0.5	86	51

A serious drawback of the method is, however, that the product is (partially) racemized (see Table III.5). The rate of racemization increased with temperature but this was offset by the much shorter reaction time (see Table III.5). A blank experiment showed that racemization occurred in the absence of the catalyst. It is known that amino acid esters undergo thermal racemization[9], which is presumably base-catalyzed (the amino ester is a base).

We next investigated the scope of the salt-free esterification. Phenylalanine and homophenylalanine were compared with PG as they comprise a series in which the phenyl group is progressively further separated from the stereogenic centre by an extra methylene group. As the chain length increased the equilibrium conversion decreased, possibly due to the increasing hydrophobic character of the amino acid. On the other hand, L-phenylalanine methyl ester was more stable than D-PGM towards racemization, consistent with the higher reactivity of the benzylic C-H bond at the stereogenic centre in the latter.

Because racemic phenylalanine methyl ester is a key intermediate in the DSM-Tosoh process for the artificial sweetener, Aspartame[22], we studied its synthesis in more detail. We found that as little as 10 mg H-USY sufficed to convert 2 g of L-phenylalanine (100°C, 20 h) to the corresponding methyl ester, which was partially racemized (52% ee), in 83% yield. This corresponds to a turnover number of 180 mol product/mol Al, assuming the unlikely participation of all the aluminum atoms present in the zeolite (5.6 mmol Al/g H-USY). Based on the more likely 0.5 mmol active sites per g H-USY (see above), the actual catalytic turnover would be > 1800. The salt-free esterification of D-HPG was also studied. The (partially racemized) methyl ester was obtained in 31% yield after 2.5 h at 100°C and 15 bar (N_2).

In conclusion, we have demonstrated the feasibility of salt-free esterification of α-amino acids with methanol over H-USY as a solid acid catalyst[23]. A drawback of the procedure is the concomitant racemization of the product. However, in the case of phenylalanine the desired product is the racemic phenylalanine methyl ester. Moreover, we also note that a solid acid catalyst could be used in conjunction with one equivalent of mineral acid to obviate the need for a large excess of the latter.

3.2 Mixed coupling of PGA and PGM to 6-APA

As noted above a major problem in the kinetically controlled coupling of side chain donors to 6-APA or 7-ADCA is competing hydrolysis which necessitates the use of a large excess of donor and the recycling of large amounts of, in the case of Ampicillin and Cephalexin, D-PG. Attempts to improve the synthetic efficiency by performing the reaction in a frozen medium at -20°C (enzymes "on the rocks") showed that the hydrolytic side reaction could be suppressed in such a system albeit at low substrate concentrations[24]. When the substrate concentrations were increased to more industrially relevant levels the effect largely disappeared. Consequently, we shifted our attention towards developing an effective and economic method for the recycling of D-PG.

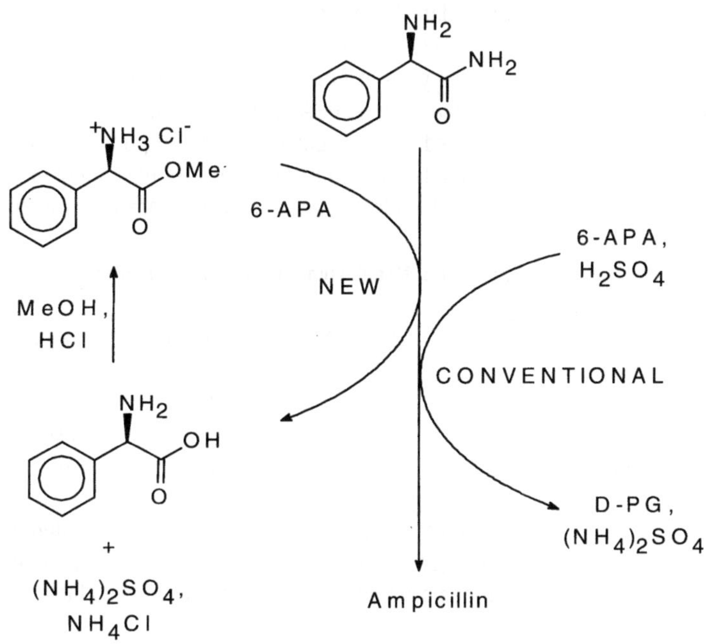

Per kg ampicillin:

Process	Input (kg)	Output (kg)	
	D-PGA	D-PG	salts
Conventional	0·62	0·26	0·25
Mixed donor	0·37	–	0·16

Fig. III.8 Ampicillin synthesis; conventional vs mixed donor coupling.

Both the amide (D-PGA) and the methyl ester (D-PGM) are suitable activated donors but the former is preferred as it is prepared from a precursor of D-PG (see above) while D-PGM is synthesized from D-PG and, hence, requires an extra step. However, when one considers the problem of recycling of D-PG one is confronted with a different picture. Recycling to D-PGA involves two steps, *via* D-PGM as an intermediate, and the generation of more than one equivalent of salt.

In the coupling of D-PGA to 6-APA or 7-ADCA one equivalent of ammonia is liberated and is neutralized by the addition of 1.5 equivalents (relative to Ampicillin) of acid (usually H_2SO_4). We reasoned, therefore, that the use of a mixture of D-PGA and the D-PGM salt in enzymatic coupling to 6-APA (or 7-ADCA) would be doubly beneficial. It would obviate both the need for a second step in the recycle loop and the need for extra acid in the coupling step, *i.e.* a substantial reduction in the amount of salt generated. This is illustrated in Figure III-8.

To validate the feasibility of such a 'mixed-donor' process, we studied the synthesis of Ampicillin using a mixture of D-PGA and D-PGM.HCl (60:40). As shown in Table III.6 the yield of Ampicillin in the mixed-donor process was comparable with that obtained with D-PGA alone. The amount of external acid consumed was, as predicted, ca. one equivalent less. Surprisingly, the rate was enhanced compared to the standard coupling with D-PGA. Subsequent experiments with D-PGM. $^1/_2\,H_2SO_4$ indicated that the rate enhancement was (partially) due to the chloride ion.

Table III.6 Mixed donor process for Ampicillin.

Acyl donor	Yield (%)	t50 (min.)	Acid consumption
D-PGA	97	380	1.53
D-PGA/D-PGM • HCl (3:2)	97	190	0.42
D-PGA/D-PGM • $^1/_2\,H_2SO_4$ (3:2)	91	250	0.47

Based on the assumption that 40% of the acyl donor used in the coupling process is hydrolyzed to D-PG the amount of waste produced per kg of Ampicillin in the various process options can be calculated (Table III.7).

Table III.7 Chemicals consumed and waste generated per kg of Ampicillin in various coupling procedures.

Recycling mode	Chemicals (g)				Waste (g)			
	D-PGA	HCl	NH_3	H_2SO_4	D-PG	NH_4Cl	$(NH_4)_2SO_4$	Total
None	620	--	--	186	260	--	250	510
Via amide	370	64	60	186	0	93	250	343
Via ester	370	64	0	51	0	93	70	163

It is clear from Table III.7 that a mixed donor coupling process reduces the amount of salt generated by a factor of two. Moreover, recycling *via* D-PGA produces an extra equivalent of salt compared to recycling *via* D-PGM.HCl. Hence, recycling via the ester in combination with mixed donor coupling reduces the amount of acid consumed and salt generated by a factor of 3 compared with recycling *via* D-PGA. A potential drawback of the mixed donor coupling is racemization of the D-PGM which is more facile than that of D-PGA. The presence of L-PGM in the reaction mixture could have deleterious consequences: possible formation of L-Ampicillin and phenylglycine dipeptides (see later) and deactivation of the pen acylase. We showed that, at 93% conversion of 6-APA to Ampicillin, the racemization of D-PGM in a mixed donor process corresponded to 0.9% of the amount of D-PGM added. The optical purity of both products, D-PG and Ampicillin, was excellent, however. Most (80%) of the L-PGM was precipitated as racemic PGM and, consequently, its negative effects were minimized. Thus, no loss in activity of the immobilized pen acylase was observed.
In short, we have demonstrated the feasibility of a mixed-donor process for Ampicillin synthesis[25]. The resulting process reduces the amount of acid consumed, and salts generated, by a factor of 3 compared to the standard coupling procedure.

3.3 A two-step, one-pot synthesis of Cephalexin from D-phenylglycine nitrile

Having demonstrated the nitrile hydratase-catalyzed stereoretentive hydration of D-PG nitrile into D-PGA (see Section 2) it occurred to us that it might be feasible to combine this transformation with the pen acylase-catalyzed coupling of D-PGA to 6-APA or 7-ADCA. This would afford a two-enzyme cascade, one-pot synthesis of *e.g.* Cephalexin from D-PG nitrile and 7-ADCA. Synthesis using enzymatic cascades are potentially more attractive than the conventional stepwise approach since they obviate the need for isolation and purification steps.
In order to identify the optimum system, the different nitrile hydratase-containing strains used in the stereoretentive hydration of D-PG nitrile were tested in a one-pot synthesis of Cephalexin (see Figure III.9). SP361 gave satisfactory results but it was not used in further experiments owing to uncertainty regarding its future availability. Apart from *R. globerulus* MAWA, which was strongly inhibited by 7-ADCA, all other strains gave good results. However, *R. rhodochrous* MAWE was used in further experiments since it exhibited the highest initial rate of nitrile hydration. Assemblase® 7500 (a preparation of *E. coli* pen acylase covalently attached to a gelatin-based carrier) was used for the *in situ* coupling of D-PGA to 7-ADCA.

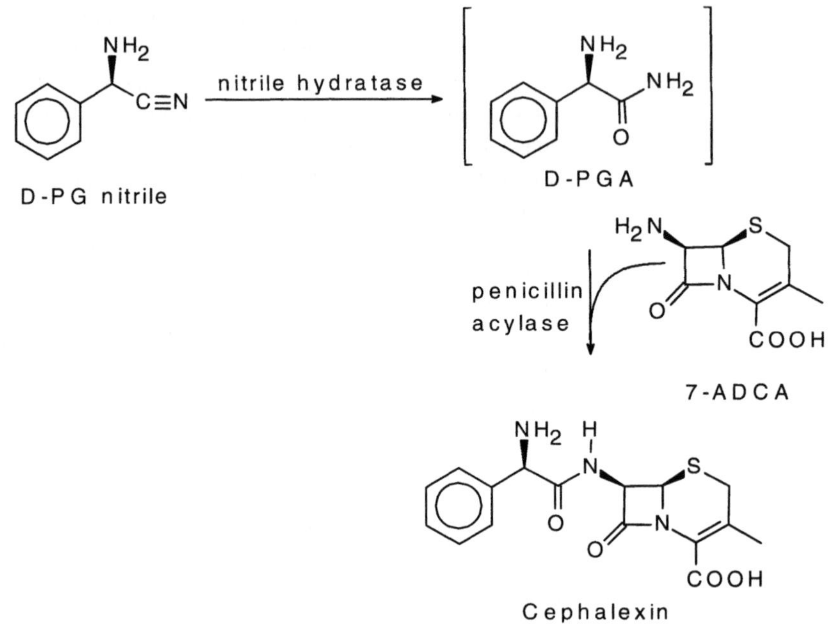

Fig. III.9 Two-step, one-pot synthesis of Cephalexin.

The D-PG nitrile was fed over time to the reaction mixture at 5°C and pH 7 in order to minimize its decomposition into benzaldehyde via a retro-Strecker reaction. A factorial experimental design was used to optimize the procedure with regard to reaction variables (amounts of Assemblase® and 7-ADCA and addition time of D-PG nitrile). The optimum result (48% yield of Cephalexin and S/H = 1.9) was observed when the D-PG nitrile was added all at once at the beginning. Presumably, slow addition of D-PG nitrile results in low steady state concentrations of D-PGA which favors competing secondary hydrolysis of the Cephalexin. Interestingly, we noticed that the hydration of D-PG nitrile was twice as fast in the one-pot protocol than when it was performed separately. This suggested that the nitrile hydratase-catalyzed hydration suffers from inhibition by the D-PGA product, which in the one-pot reaction is removed by reaction with 7-ADCA. This was confirmed in standard hydration experiments in which D-PGA was added at the beginning.

We subsequently found that the use of the T-CLEA pen acylase (tert-butyl alcohol used as precipitant, followed by crosslinking; see also section 4) gave better results in the one-pot procedure: 60% Cephalexin yield and S/H = 2.7. In order to improve the yield of Cephalexin and the S/H ratio even further we studied the effect of 1,5-dihydroxynaphthalene (1,5-DHN) which forms an insoluble complex with Cephalexin[26]. We found that Cephalexin was obtained in 79% yield with an S/H of 7.7, irrespective of whether Assemblase® or T-CLEA pen acylase was used, suggesting that diffusion limitations are less of an issue if the product is effectively removed from the reaction mixture. These results are summarized in Table III.8.

Table III.8 Two-enzyme cascade synthesis of Cephalexin.

Pen acylase	Complexing agent	Yield (%)	S/H
Assemblase® 7500	none	48	1.9
T-CLEA	none	60	2.7
Assemblase® 7500[b]	1,5-DHN	79	7.7

[a] 1 mmol of D-PG nitrile in 5ml 0.2 N H_2SO_4 added to 3 mmol 7-ADCA, 315 U pen acylase and 40 U *R. rhodochrous* MAWE in 5 ml 50 mM phosphate buffer (pH 7) at 5°C.
[b] The same result was obtained with T-CLEA pen acylase.

The S/H ratio of 7.7 was significantly higher than that observed (6.4) in the standard synthesis of Cephalexin from D-PGA under otherwise identical conditions. This prompted us to study the effect of D-PG nitrile on the S/H ratio in more detail. We found that the addition of D-PG nitrile effected a selective inhibition of (secondary) hydrolysis, resulting in the formation of Cephalexin in 80% yield with S/H = 19.3 (in the presence of 1,5-DHN), which is three times that observed in its absence. Further experiments suggested that the observed inhibition could be due to benzaldehyde, formed by degradation of the D-PG nitrile (see also Section 4).
In conclusion, we have shown that a one-pot synthesis of Cephalexin is possible using a combination of a nitrile hydratase and pen acylase. In the presence of a complexing agent Cephalexin was obtained in 79% yield and with an S/H of 7.7[27].

§4 Development of improved catalysts for the synthesis of β-lactam antibiotics

4.1 Introduction

It has been explained elsewhere (see earlier) that a practical enzymatic synthesis of β-lactam antibiotics necessarily involves a kinetically controlled reaction (aminolysis) of the β-lactam nucleus with an activated side chain donor. Water is the reaction medium of choice for such procedures, although, owing to competing hydrolysis of the side chain donor and secondary hydrolysis of the products, a large excess of the side chain donor is usually required for complete conversion of the β-lactam nucleus[28]. The generally accepted catalytic mechanism involves reaction of an acyl-enzyme intermediate with 6-APA (coupling) or water (hydrolysis)[29]. For this reason, an important parameter - the synthesis/hydrolysis ratio (S/H, the molar ratio of product and hydrolyzed side chain donor) - is used to evaluate the economic viability of the process[3].

4.2 Synthesis of β-lactam antibiotics in organic solvents

With the aim of achieving a high S/H ratio, the enzymatic coupling reaction can be alternatively performed in a hydrophilic organic solvent at low water content. Unfortunately, free penicillin G acylase is readily and irreversibly deactivated by organic solvents. Hence, we set out to develop a suitable methodology to maintain the activity of penicillin acylase in organic solvents[30].

It is well-known that the physical aggregation of protein molecules into super-molecular structures can be induced by the addition of salts, organic solvents or non-ionic polymers to protein solutions[31] without perturbation of the original three-dimensional structure of the protein. Indeed, aggregation induced by ammonium sulfate, polyethylene glycol, and some organic solvents such as alcohols, is a commonly used method of protein purification[32]. These solid aggregates are held together by non-covalent bonding and readily collapse and redissolve when dispersed in an aqueous medium.

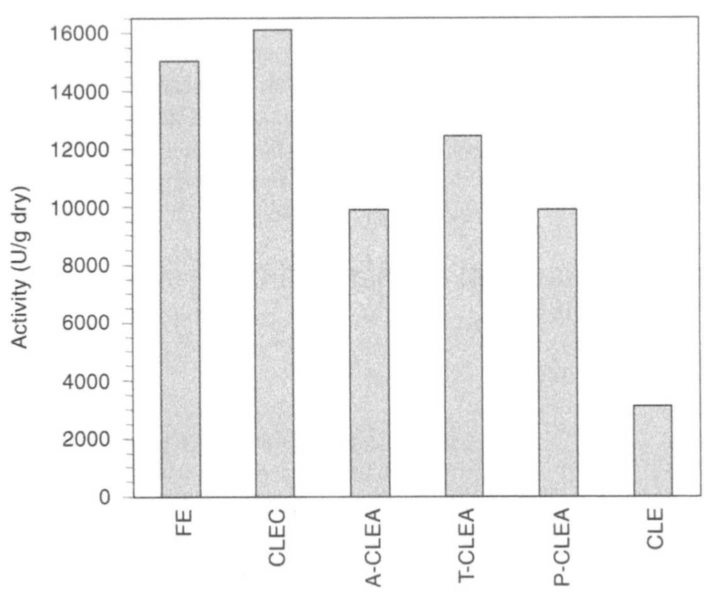

Figure III.10 The hydrolytic activity of free (FE) and crosslinked penicillin G acylases. A-CLEA, T-CLEA and P-CLEA represent the CLEAs prepared by using ammonium sulfate, tert-butyl alcohol and PEG 8000 as the precipitant, respectively.

We surmised that chemical cross-linking of these physical aggregates would produce crosslinked enzyme aggregates in which the pre-organized superstructure of the aggregates and, hence, their activity, would be maintained. Consequently, crosslinking of preformed physical aggregates of enzymes could constitute a simple method for the preparation of what we call cross-linked enzyme aggregates (CLEAs), with activities comparable to cross-linked enzyme crystals (CLECs).
To test our CLEA concept we selected three precipitants that are representative for the three types of precipitants mentioned above, ammonium sulfate, an organic solvent (tert-butyl alcohol) and a non-ionic polymer (poly(ethylene glycol), PEG). The activities of the recovered insoluble crosslinked enzyme increased with the precipitant concentration, reaching a maximum which corresponded to almost 100% retention of the activity compared with the enzyme in solution. The maximum activity was observed at the concentration of the precipitant required to precipitate all of the enzyme from the solution (ammonium sulfate: 30%; *tert*-butyl alcohol: 20%; PEG : 20%). Below this critical concentration the enzyme is present mainly in solution; upon cross-linking a crosslinked enzyme preparation (CLE) resulted in which only a small fraction of the original activity is retained (Figure III.10)[33,34].
In spite of a partial loss of activity upon isolation of the CLEAs, their specific activity in the hydrolysis of penicillin G (Figure III.10) is still of the same order as that of the penicillin G acylase CLEC and substantially higher than that of the CLE[33]. Presumably, in solid enzyme aggregates, as in enzyme crystals, the enzyme molecules can adopt an orderly arrangement in which their conformation is fixed in a rigid superstructure that is essential for their stability and activity and which is maintained upon crosslinking.
We found that CLEAs are active in the synthesis of Ampicillin in a broad range of organic solvents[33] (Table III.9). Log P, which is often used as an indicator of solvent behavior, does not show any obvious correlation with either the reaction rate or S/H. Very similar results were obtained with CLEAs of *Alcaligenes faecalis* penicillin acylase, which is an inefficient coupling catalyst in aqueous media. The conversion of 6-APA to Ampicillin could be improved to 84% at S/H 1.9 via optimization of the immobilization technique and the reaction conditions[35]. Attempts to lower the water concentration below 5% resulted, surprisingly, in a decrease in S/H. Hence, it was concluded that enzymatic synthesis of β-lactam antibiotics in low-water organic solvent is not a viable option.

Table III.9 Synthesis of Ampicillin catalyzed by T-CLEAs of *E. coli* penicillin acylase in organic solvents[a].

Solvent	Log P	Conv. (%)[b]	S/H
Triglyme	-1.8	5	2.0
2-Methoxyethyl ether	-1.3	11	2.9
1,2-Dimethoxyethane	-0.8	10	3.4
Acetonitrile	-0.4	17	4.8
2-Ethoxyethyl ether	-0.3	25	1.8
2,2-Dimethoxypropane	-0.2	56	1.3
tert-Butyl alcohol	0.8	18	1.3
tert-Amyl alcohol	1.4	33	1.5
tert-Butyl methyl ether	1.6	47	1.3

[a] Reaction conditions: 300 mM 6-APA, 500 mM D-phenylglycine amide in 10 ml solvent-water (95:5, v/v), 100 U enzyme at 0°C.
[b] Conversion was measured after 1 h.

4.3 Cross-linked aggregates of penicillin acylase as efficient coupling catalysts in aqueous medium

Recently, it was found that the S/H ratio depends not only on the nature of the side chain and the reaction conditions (pH, temperature, cosolvents)[28,36], but also on the enzyme immobilization method[37]. The latter effect was ascribed to diffusion limitations. The 6-APA nucleus is a much larger molecule than water and the diffusional limitation created in an immobilized enzyme is expected to lead to a lower S/H ratio.

The specific activity of the penicillin acylase CLEAs in the synthesis of Ampicillin from D-PGA and 6-APA was comparable to that of the free enzyme. We found that the CLEAs mediated the reaction with an S/H ratio that was close to that of the dissolved enzyme, whereas CLEC-EC afforded approximately half that value (Figure III.11). Remarkably, the S/H ratio of the T-CLEA declined less as the reaction proceeded, compared with the free enzyme, A-CLEA and P-CLEA. Apparently the structure of the enzyme is subtly changed by precipitation with tert-butyl alcohol[38].

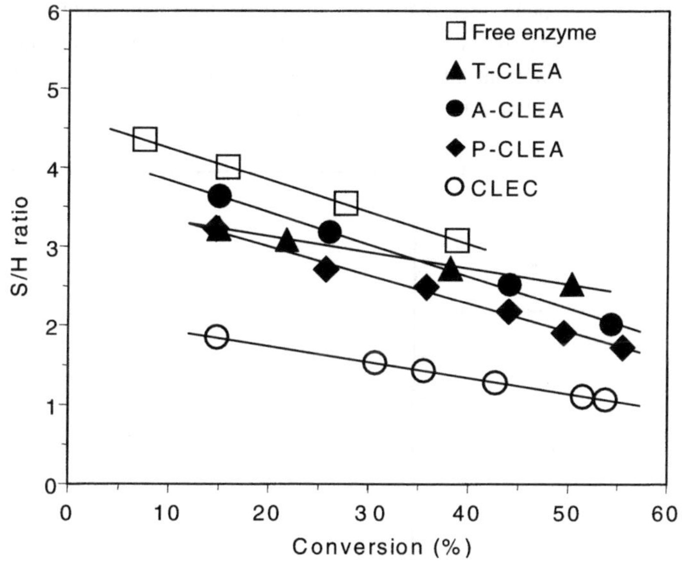

Fig. III.11 Relationship of S/H ratio with the conversion of 6-APA in the synthesis of Ampicillin catalyzed by free and crosslinked penicillin G acylases. Reaction conditions: 300 mM 6-APA, 500 mM D-phenylglycine amide and 20 U penicillin G acylase in 20 ml water were stirred at pH 7 and 20°C.

The chemical stability of the T-CLEA was checked by removing all solids from the reaction mixture at 60% conversion of 6-APA. The composition of the filtrate remained unchanged over time, demonstrating that no 'leakage' of enzyme into the solution had taken place. We also monitored the activity of the T-CLEA in the course of the reaction. Full activity was recovered up to the point where Ampicillin started to precipitate; at this point approx. 30% of the original activity was lost, presumably because the crystallization caused some breakup of the aggregates. We note that in similar experiments with commercially available immobilized penicillin G acylase preparations no active enzyme was recovered at all due to mechanical disintegration of the catalyst, presumably induced by product crystallization in the matrix.

4.4 Efficient synthesis of β-lactam antibiotics in the presence of cross-linked Starburst Dendrimer penicillin acylase

Dendrimers[39] are monodisperse and highly branched polymers with tailorable "sticky ends", which offers the opportunity of using dendritic molecules as building blocks. We reasoned that immobilization of enzymes via attachment to their "sticky ends" could result in a high loading of enzyme while maintaining high accessibility for the substrate molecules. Dendrimers have, until now, not been used as scaffolds for enzyme immobilization, to the best of our knowledge. We selected two types of dendrimer that both contain 64 terminal primary amino groups per molecule. Starburst® Generation 4 dendrimer has a backbone that contains tertiary amine as well as amide bonds, whereas the backbone of DAB-Am-64 is constructed of tertiary amine building blocks.

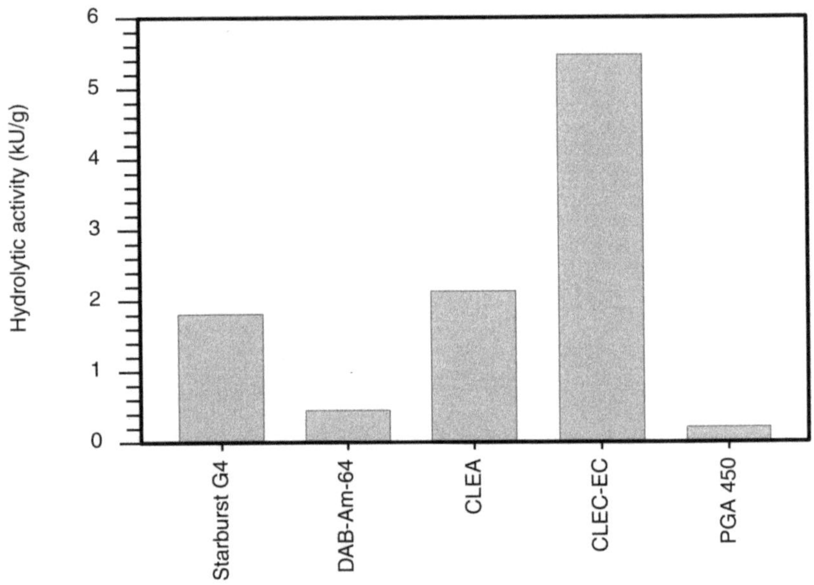

Fig. III.12 The hydrolytic activity of immobilized penicillin acylases, based on wet weight.

Generation 4 dendrimers are too small (approx. 46Å) to act as an enzyme scaffold. Hence, they were cross-linked, using excess glutaraldehyde, into particles of 5 - 10 μ size that contained approx. 90% water. Sufficient free aldehyde groups remained at the dendrimer surface to act as anchors for the covalent attachment of penicillin acylase. Cross-linked Starburst G4 dendrimer was particularly efficient; the resulting preparation had a much higher specific activity than other immobilized penicillin acylase preparations (see Figure III.12), with the exception of cross-linked crystals of penicillin acylase (ChiroCLEC-EC). Up to 2000 U (0.7 mmol[40,41]) of penicillin acylase were bound per g of cross-linked Starburst G4 dendrimer and approx. 90% of this activity was recovered in the immobilized preparation. The amount of penicillin acylase that could be attached to cross-linked DAB-Am-64 dendrimer was much less (see Figure III.13). Hence, we used the Starburst G4 preparation in further experiments.

4.4.1 Ampicillin

The kinetically controlled synthesis of Ampicillin from D-PGA and 6-APA was conducted in the presence of the Starburst G4 preparation as well as various other immobilized penicillin acylase preparations. A reaction with the free enzyme was included as a reference. The S/H of the Starburst G4 preparation was, in the initial phase, comparable with that of the free enzyme. The Starburst enzyme also performed significantly better than Assemblase® (S/H at 80% conversion 1.55 vs 1.22).

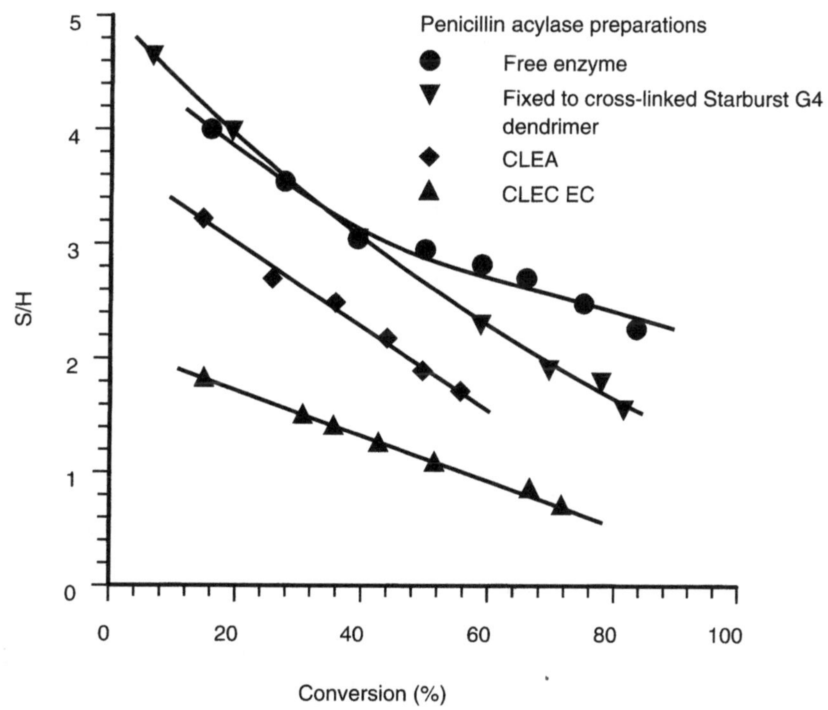

Fig. III.13 Synthesis of Ampicillin: effect of the penicillin acylase preparation on S/H vs conversion of 6-APA. Reaction conditions: 300 mM 6-APA, 500 mM D-phenylglycine amide and approx 200 U penicillin acylase in 50 ml aqueous buffer were stirred at pH 7 and 20°C.

4.4.2 Cephalexin

A standard synthesis of Cephalexin (see Fig. III.14) in the presence of the Starburst G4 penicillin acylase, as well as the free enzyme and its CLEC, demonstrated the synthetic efficiency of the Starburst preparation. Its S/H was close to that of the free enzyme up to high conversion. Moreover, it was a significantly more efficient synthesis catalyst than Assemblase® 7500, achieving the same final S/H at an approx. 10% higher conversion (data not shown).

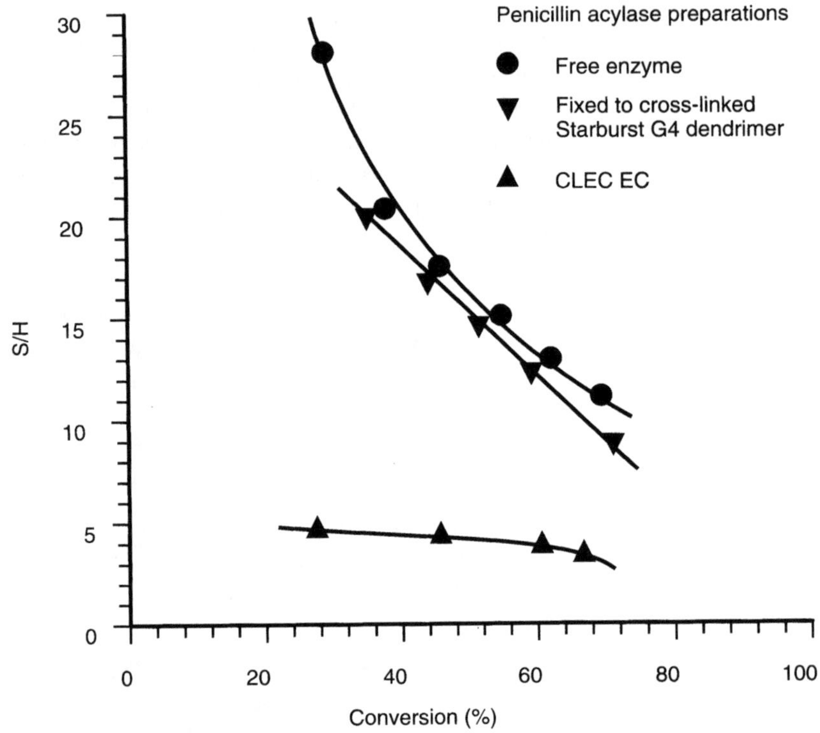

Fig. III.14 Synthesis of Cephalexin: effect of the penicillin acylase preparation on S/H vs conversion of 7-ADCA. Reaction conditions: 500 mM 7-ADCA, 550 mM D-phenylglycine amide and approx 300 U penicillin acylase in 50 ml aqueous buffer were stirred at pH 8 and 0°C.

4.5 Penicillin acylase immobilized on cross-linked p-xylylene diamine: a heterogeneous catalyst with the efficiency of free enzyme

Process economics dictate that enzymatic synthesis of β-lactam antibiotics should be performed with recycling of the biocatalyst. This has meant, in practice, that the catalyst should be a heterogeneous one. It has, however, become abundantly clear that heterogeneous (immobilized) penicillin acylase is a much less efficient catalyst in terms of S/H, than the free (dissolved) enzyme. This shortcoming of immobilization is mainly ascribed to diffusion limitations in the matrix of gel-type carriers or in the mesopores of solid, porous carriers, such as Eupergit C. The aggregation of solid microparticles to create a macroporous carrier could be a potential solution to the diffusion limitation problem.
We reasoned that the polymerization of 1,4-di(aminomethyl)benzene (*p*-xylylene diamine, XDA) with glutaraldehyde could result in solid microbeads which could be aggregated and crosslinked into macroporous particles of the desired size. Moreover, such a carrier would still bear a large number of underivatized amino groups. Some of these could serve as anchors for the immobilization of penicillin

acylase and the remainder would render the carrier material cationic under the reaction conditions. We note that the positive results obtained with penicillin acylase on cross-linked polyamine dendrimer (see above) had alerted us to the potential beneficial effect of cationic carrier materials[37].

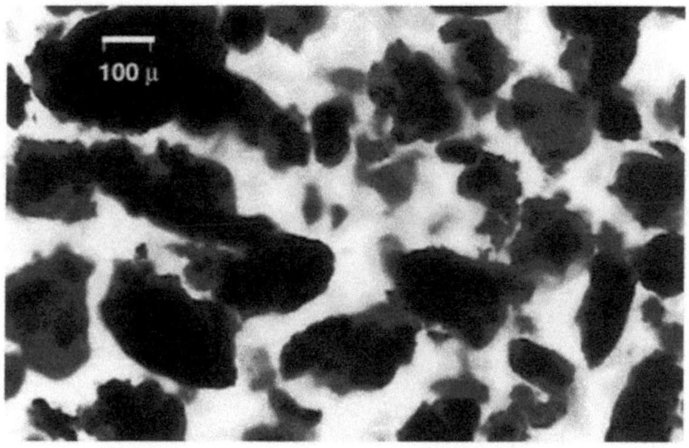

Fig. III.15 Microscopic image of poly-XDA-GA beads.

Beads of XDA-glutaraldehyde (poly-XDA-GA) were grown by mixing the constituents and repeatedly cycling the pH between 4 and 9 until all low-molecular weight amine had been converted. Irregular beads with approx 100 μ size resulted. Penicillin acylase was adsorbed by the beads and covalently immobilized via reaction with residual aldehyde groups on the surface. The specific activity was comparable with that of other immobilized penicillin acylases and the immobilisate was highly aquaphilic[42], which is thought to be advantageous for efficient synthesis (see figure III.15)

Table III.10 Properties of immobilized E. coli penicillin acylase preparations.

Preparation	Carrier	Hydrolytic activity U/g	Aquaphilicity
Poly-XDA-GA PA	Poly-XDA-GA	200	86
Assemblase(r) 7500	Gelatin-chitosan	260	65
PGA 450	Special polymer	210	n.d.
ChiroCLEC-EC		5660	16
PcA	Eupergit C	310	44

We subsequently showed that penicillin acylase on poly-XDA-GA is a remarkably efficient catalyst for the synthesis of β-lactam antibiotics. The course of S/H vs. conversion in the synthesis of Ampicillin catalyzed by penicillin acylase on poly-XDA-GA and other preparations are compared in Fig. III.16. The S/H at 85% conversion was 2.0, comparable with that of the free enzyme and much better than with Assemblase® (S/H 1.22 at 80% conversion).

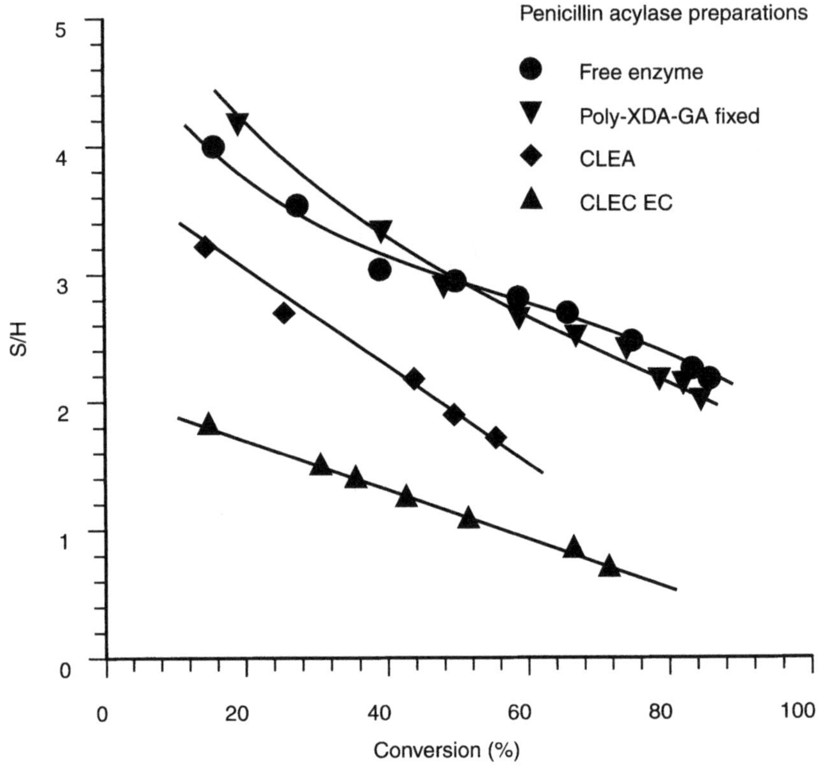

Fig. III.16 Synthesis of Ampicillin: effect of the penicillin acylase preparation on S/H *vs* conversion of 6-APA. For further details see Fig. III.13.

The poly-XDA-GA preparation also performed very well in the synthesis of Amoxicillin with S/H = 4.5 up to 75% conversion, which is slightly higher than free penicillin acylase (data not shown). Cephalexin was synthesized with S/H = 13 at 70% conversion, comparable with the free enzyme (see Figure III.17). For comparison, an S/H of 8.3 was observed with Assemblase® (data not shown) at 60% conversion.

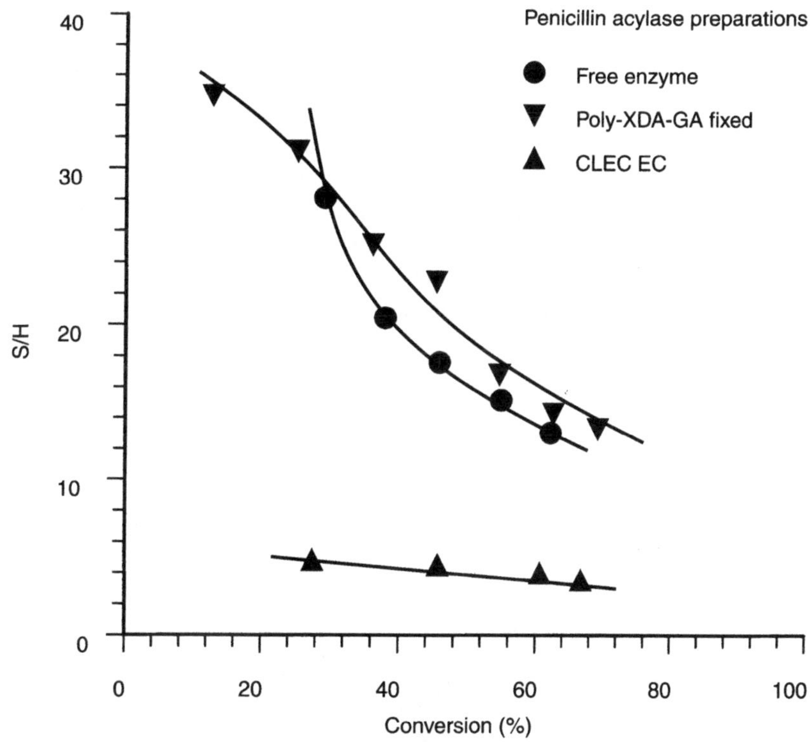

Fig. III.17 Synthesis of Cephalexin: effect of the penicillin acylase preparation on S/H *vs* conversion of 7-ADCA. For further details see Fig. III.14.

In conclusion, we have shown that a copolymer of 1,4-di(aminomethyl)benzene and glutaraldehyde is a new, promising material for the immobilization of penicillin acylase. The synthetic efficiency of the resulting biocatalyst is comparable with that of the free enzyme.

4.6 Adjuvants increase the synthesis/hydrolysis ratio

4.6.1 Effects of D-phenylglycine nitrile

It has often been noted that inhibitors tend to increase the synthesis/hydrolysis ratio of enzymatic coupling reactions[43]. D-PG-CN, which is a potent inhibitor of penicillin acylase, surprisingly also had such an effect (see Section 3.3): 0.1 M D-PG-CN increased the S/H of Assemblase® in the synthesis of Cephalexin by a factor of 3. Whether the effect is due to the cyano group, or to a decomposition product, such as benzaldehyde, is still open. It is known from literature that nitriles as well as aldehydes interact with the active serine in serine-dependent hydrolases.

4.6.2 Effects of hydrolyzed side chain donor

Starting the synthesis of Cephalexin in the presence of D-PG surprisingly increased the initial S/H by a factor of nearly 2 (see Fig. III.18A). The advantage disappeared, however, as the reaction proceeded to high conversion. The effect is specific for the D-enantiomer, because the initial S/H decreased when L-PG was added to the reaction mixture. A similar, much less pronounced, effect of D-PG was apparent in the synthesis of Ampicillin (Fig. III.18B); moreover, the initial S/H of the synthesis of Amoxicillin also increased in the presence of D-HPG (data not shown). All of these reactions have been performed with dissolved enzyme; hence, the increase in S/H cannot be due to, *e.g.*, salt effects on diffusion limitation.

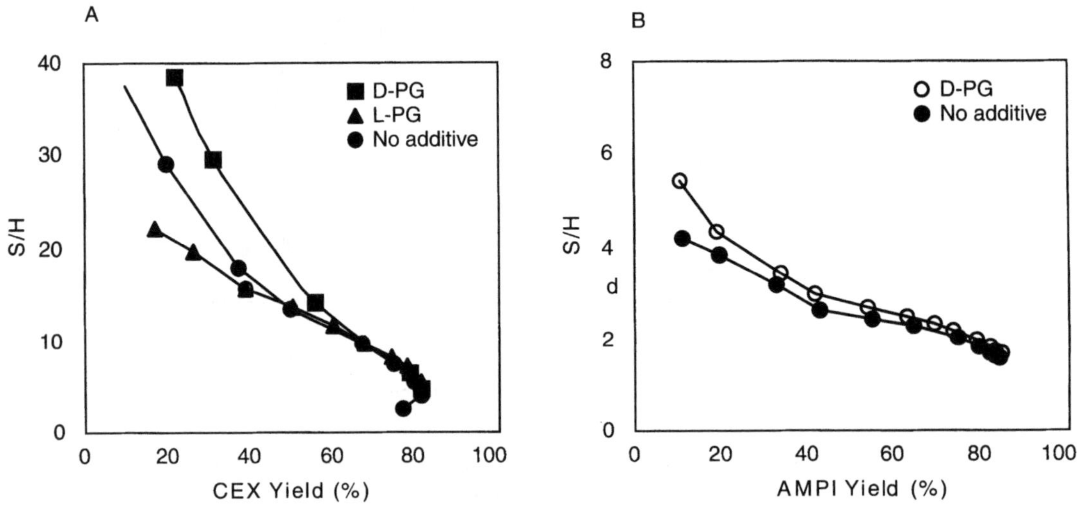

Fig. III.18 (A) Effect of saturation of the reaction medium at t = 0 with D-PG (■) or L-PG (▲) compared to the normal reaction with no additives at t = 0 (●) on the synthesis of Cephalexin with native enzyme. Reaction conditions: 0.50 M 7-ADCA, 0.55 M PGA, 50 ml reaction volume, pH 8.0, 0°C, 0.59 mg/ml enzyme. (B) Effect of saturation of the reaction medium at t = 0 with D-PG (O) compared to the normal reaction without additives at t = 0 (●) on the synthesis of Ampicillin with native enzyme. Reaction conditions: 0.30 M 6-APA, 0.50 M PGA, 50 ml reaction volume, pH 7.0, 21°C, 0.49 mg/ ml enzyme.

4.7 Active site titration of penicillin acylase

Enzyme immobilization usually results in a partial loss of activity that has, in general, defied further analysis. It is often obscure, for example, whether a 50% loss of activity is due to complete deactivation of half of the active sites, or a 50% decrease in turnover frequency of all of the active sites, or some combination of both. Active site titration - measuring the number of catalytically competent active sites - provides a basis for dissecting the activity loss. Moreover, active site titration makes it possible to compare the activity of enzyme preparations in any reaction in terms of turnover frequency (mol substrate converted per mol catalyst per second).

Two somewhat similar active site titration techniques have been described in the literature[44]. The first one, the "burst" method, involves a reaction of the enzyme sample with a chromogenic suicide substrate and measuring the burst of activity via UV spectrometry. The second, "incremental inhibition" is somewhat more complicated and involves the partial deactivation of the enzyme sample, using an inhibitor or suicide substrate, and assaying the residual activity of the enzyme sample. The procedure is repeated with increasing amounts of inhibitor until total deactivation is observed, the residual activity is plotted vs the amount of inhibitor and a line is fitted through the datapoints (see Figure III.19). The amount of inhibitor needed for total deactivation corresponds with the number of active sites.
The native penicillin acylases from *E. coli* (see Fig. III.19) and *A. faecalis* were treated with varying amounts of the suicide substrate phenylmethanesulfonyl fluoride (PMSF) and the residual activity in the hydrolysis of penicillin G was measured[40,45]. The active site concentrations and turnover frequencies (see Table III.11) were calculated on the basis of these measurements.

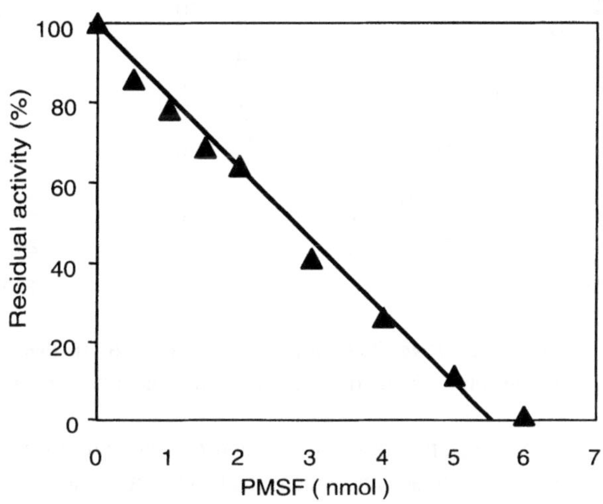

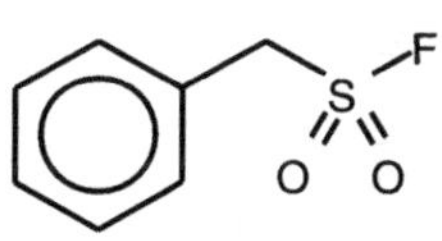

P M S F

Fig. III.19 Active site titration of native penicillin acylase from *E. coli* (20 μl, 33.2 U); the residual activity vs. the amount of added PMSF. The hydrolysis of penicillin G (2% penicillin G, pH 8, 34°C) was used to assay the residual activity. The axis intercept of the trendline corresponds with the number of active sites.

Preliminary to performing active site titrations of immobilized penicillin acylase preparations we checked for possible diffusion limitation effects of the inhibitor. The residual activity of partially inhibited immobilized penicillin acylases was monitored over time and we found that a stable response was obtained after < 10 min (data not shown).

The active site concentrations in the industrial penicillin acylase catalysts Assemblase® and Separase-G® were measured using the procedure described above. The datapoints (Fig. III.20) fitted a linear correlation of the residual activity and the amount of PMSF, indicating that all active sites contribute equally to the activity of the preparation. The turnover frequency of the immobilisates in the hydrolysis of penicillin G, which were calculated on the basis of these measurements, were much lower than those of the native enzymes (Table III.11). It is known that the hydrolysis of penicillin G generates a pH gradient in the carrier that can amount to 3 pH units[46]. Apparently, the *A. faecalis* acylase is more affected by such pH effects, as would be expected from the pKa values of the acidic group in its active site.

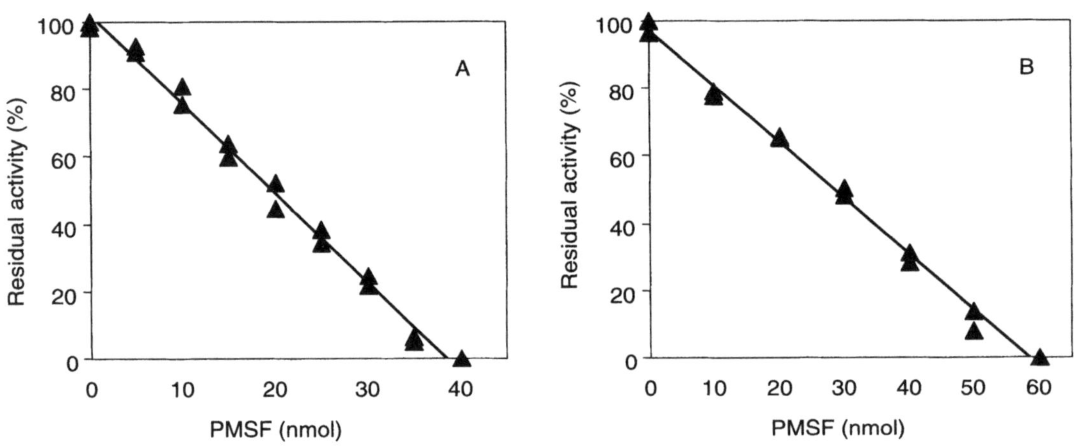

Fig. III.20 Active site titration of (A) Separase-G® (0.25 gram, 68.3 U) and (B) Assemblase® (0.20 gram, 75.6 U). Further details as in Fig. III.19.

The preparation of CLEAs via aggregation and cross-linking[33] is accompanied by some loss of activity, which was investigated using active site titration. The *E. coli* enzyme lost 27% of the catalytic sites in the course of the immobilization, and the turnover frequency of the remaining ones was 27% lower than that in the native enzyme. In contrast, the penicillin acylase from *A. faecalis* lost only 12% of the active sites, but the decrease in turnover frequency amounted to 36%.

Table III.11 Active site concentration and turnover frequency in penicillin acylase preparations.

Preparation	Activity (U)	Active sites	TOF [a] (s^{-1})	Recovery (%) Activity	Active sites
E. coli					
Native	1670 [b]	2.7×10^{17} [b]	61	-	-
Assemblase®	378 [c]	1.7×10^{17} [c]	22	d	d
CLEA	297 [e]	6.6×10^{16} [e]	45	53	73
PGA-300	990 [f]	3.3×10^{17} [f]	30	n.d.	n.d.
Eupergit C (1)	440 [c]	1.2×10^{17} [c]	37	55	94
Eupergit C (2)	590 [c]	1.8×10^{17} [c]	33	44	84
A. faecalis					
Native	1660 [b]	1.7×10^{17} [b]	98	-	-
Separase G [c]	273 [c]	9.4×10^{16} [c]	29	d	d
CLEA	485 [e]	7.5×10^{16} [e]	29	58	88
Eupergit C (1)	290 [c]	3.4×10^{16} [c]	86	70	80
Eupergit C (2)	660	8.1×10^{16} [c]	82	58	70

[a] 2% penicillin G, pH 8, 34℃.
[b] Per ml solution.
[c] Per g wet weight.
[d] No recovery can be given because the loading is not known.
[e] Per ml suspension.
[f] Per g dry weight.

We studied the immobilization of penicillin acylase on Eupergit C, an acrylic resin containing oxirane functionality, to assess the effects of the active site density (loading). Increasing the loading with the penicillin acylase from E. coli from 67% (preparation 1) to 100% occupation (preparation 2) reduced the active site recovery as well as the turnover frequency. We note that diffusion limitation seems to play a dominant role in the loss of turnover frequency, because it increased to a value close to the native one upon crushing the preparations (data not shown). The *A. faecalis* acylase behaved differently: the active site recovery decreased significantly when the coverage of the carrier with enzyme was increased, but the turnover frequency remained close to the native value.
The known active site concentrations make it possible to compare the turnover frequencies of different preparations in synthetic applications. *E. coli* penicillin acylase was immobilized on Eupergit C in a range of loadings, the resulting preparations were characterized and used in the synthesis of Cephalexin (Table III.12). The reactions were conducted at low concentration to avoid the necessity of vigorous stirring, which would destroy the catalyst particles.

Table III.12 Cephalexin synthesis in the presence of immobilised penicillin acylase [a].

Preparation	Active sites (10^{17}. g^{-1} wet)	TOF (s^{-1}) [b] Pen G	D-PGA	7-ADCA	S/H
Native	1670 [c]	56	13	13	30.0
Eupergit-1	0.60	50	13	12	9.3
Eupergit-2	1.23	26	12	10	4.9
Eupergit-3	1.75	26	11	9	4.2
Eupergit-4	1.80	23	12	8	3.5

[a] Reaction conditions: 0.1 M 7-ADCA, 0.15 M D-PGA in 0.1 M phosphate buffer pH 8.
[b] Initial values.
[c] Per ml solution.

The steep decrease of the turnover frequency in the hydrolysis of penicillin G could be due to accumulation of inhibiting phenylacetic acid in the carrier. This notion finds support in the absence of a large effect of the active site density on the turnover frequency of D-PGA. The turnover frequency of 7-ADCA also is affected much less than that of penicillin G; S/H decreased severely, however. The initial S/H could be increased to 10 - 17 by crushing the catalysts, indicating that slow diffusion of either 7-ADCA or Cephalexin causes the drop in S/H. In short, active site titration allowed us to analyze, to a certain extent, the activity loss of penicillin acylase upon immobilization.

4.8 LC-MS of deactivated penicillin acylase

4.8.1 Turnover-related deactivation of penicillin acylase

Immobilized penicillin acylase is known to be stable, with an operational half-life exceeding 100 cycles[47]. It is known, however, that some acyl donors, in particular L-amino acid derivatives, rapidly deactivate the enzyme in solution. Some examples are given in Table III.13.

Table III.13 Inactivation of *E. coli* penicillin acylase.

Formulation	Conditions of inactivation[a]	$t_{1/2}$ (min)
Native	0.5 M L-PGM, 0.5 M L-PheOMe, pH 7.5	5
Native	0.5 M L-PGM, pH 7.5	15
PGA 300	0.5 M L-PGM, pH 7.5	240
PGA 300	0.2 M L-PheOEt, pH 7.5	No inactivation
PGA 300	1 M L-PGM, 1 M L-PheOEt, pH 6.5	1 h

The irreversibility and rapidity of the deactivation of penicillin acylase point to a chemical modification of the enzyme, rather than to inhibition by a reaction product. Indeed, penicillin acylase could be used for the preparative synthesis of the dipeptide ester of L-phenylglycine from L-phenylglycine methyl ester[48], which shows that the enzyme can tolerate this product at saturated conditions (see Section 5.1). We surmised that deactivation could take place via reaction of the acyl-enzyme with the *N*-terminal amino group (Figure III.21). In order to provide support for the putative mechanism we studied samples of the native and deactivated pen acylase using LC-MS.

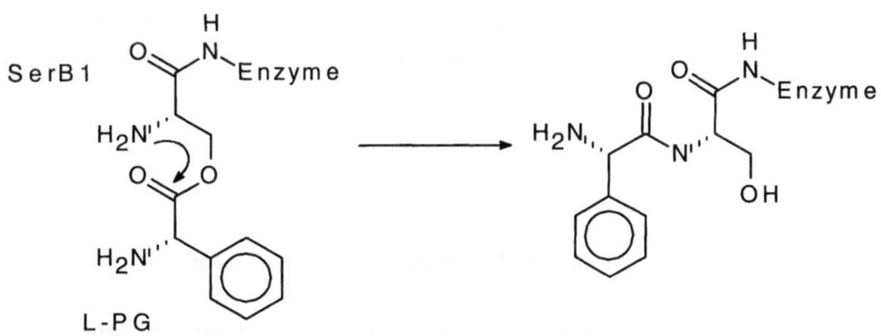

Fig. III.21 Proposed turnover related inactivation mechanism of penicillin acylase.

4.8.2 LC-MS of native and PMSF-deactivated penicillin acylase

LC-MS was validated by using it to characterize the native E. coli penicillin acylase as well as its PMSF-deactivated form. Penicillin acylase is heterodimeric; the α- and β-chains were separated in the LC step and separately subjected to MS. The masses could be established accurately as shown in Table III.14.

Table III.14 Mass-spectral analysis of native and PMSF-inactivated E. coli penicillin acylase[a].

Preparation	α-Chain		β-Chain	
	Theory	Found	Theory	Found
Native	23817	23817	62366	62367
PMSF inact		23817	62520	62520
PMSF inact (24 h)		23817		62345-62348

[a] LC/MS Qtof-2 in electrospray mode.

The mass spectrum of the β-chain (the one that bears the catalytic serine at its *N*-terminus) of partially deactivated enzyme, when measured a few hours after deactivation, shows the native β-chain as well as the one with the expected mass 62520. The mass of the α-chain of deactivated penicillin acylase remained identical to that of the native enzyme. Surprisingly, the deactivated form of penicillin acylase, PA-PMSF, was not stable, and its β-chain was partially converted after 24 h into a protein with a slightly lower mass (19-22 D) than that of the native one. We propose that the mass loss involves the formation of an aziridine (see Figure III.22) via an intramolecular nucleophilic substitution[49].

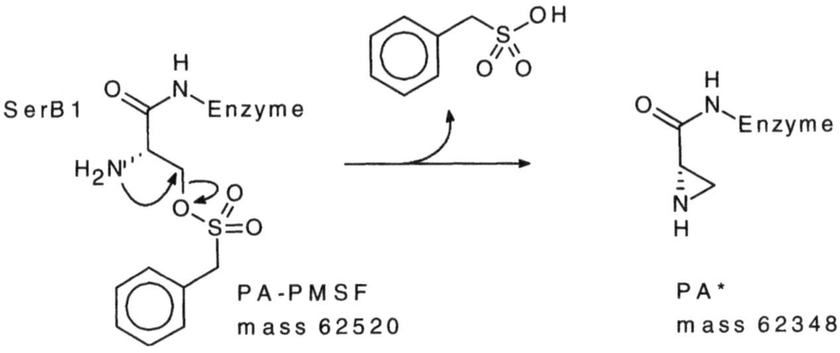

Fig. III.22 Possible transformation of PMSF-inactivated penicillin acylase.

4.8.3 Mass spectroscopy of penicillin acylase deactivated with L-PGM

L-Phenylglycine-deactivated penicillin acylase was subjected to LC-MS in a subsequent experiment. Penicillin acylase was completely deactivated in a mixture of 0.5 M L-phenylglycine methyl ester and 0.5 M L-phenylalanine methyl ester at pH 7.5. Unfortunately the protein samples of the deactivated enzyme were of a poor quality and had a lower protein content than would be expected, which caused a considerable error in the mass determination (Table III.15). The deactivation resulted in a change in the mass of the β subunit, whereas the α-chain remained unchanged. Several signals were detected, as shown in Table III.15, but surprisingly the main peak corresponded with a protein with a 23 D lower mass than that of the native β-subunit under these experimental conditions. Another peak (M 62455) corresponded with a mass increase of 127 D compared with the native enzyme.

Table III.15 Mass-spectral analysis of native and L-PG-OMe/L-Phe-OMe-inactivated *E. coli* penicillin acylase.

Preparation	α-Chain		β-Chain	
	Theory	Found	Theory	Found
Native	23817	23803	62366	62328
Inactivated		23801		62305[a], 62455

[a] Main peak.

We conclude that the deactivation indeed affects the β-chain. The formation of a β-chain with decreased mass is very similar to what was observed for the PMSF deactivated enzyme. Further measurements are being carried out to resolve the ambiguities.

§5 Peptide synthesis and amine resolutions catalyzed by penicillin acylase

During the course of our studies of the synthesis of penicillins and cephalosporins by pen acylase-catalyzed coupling reactions it occurred to us that this readily available enzyme may be useful for other commercially interesting biotransformations. This led us to study its use in the synthesis of dipeptides and the kinetic resolution of chiral amines as spin-off projects.

5.1 Pen acylase-catalyzed synthesis of dipeptides: a chemo-enzymatic route to optically pure diketopiperazines

Peptides and their derivatives constitute a commercially interesting class of biologically active compounds. Classical chemical syntheses of peptides are circuitous in that they involve protection, activation and deprotection steps. Enzymatic syntheses tend to be less circuitous and have been extensively studied using proteases[50] which *in vivo* generally catalyze the hydrolysis of peptides. However, a shortcoming of proteases is that their use is restricted to natural, L-amino acids, and they do not convert unnatural amino acids such as phenylglycine. We reasoned that pen acylase might be a suitable catalyst for the synthesis of *e.g.* D-PG-containing dipeptides, based on the fact that it catalyzes the coupling of D-PG derivatives to 6-APA and 7-ADCA. Dipeptide esters are known to readily undergo ring closure to the corresponding diketopiperazines which are of potential commercial interest in their own right.
The acyl donor binding subsite of pen acylase has a high affinity for phenylacetic acid[51] and can accept derivatives of PG but with moderate stereoselectivity[52]. The acyl acceptor subsite, in contrast, shows an extremely high preference for the L-enantiomers of a broad range of amino acids[51,53]. This characteristic feature simplifies the synthesis of D-phenylglycyl-L-phenylglycine since any formation of products with D-phenylglycine at the C-terminus or tripeptides is not expected. The formation of L-phenylglycyl-L-phenylglycine derivatives could be obviated by using free L-phenylglycine which acts as an acceptor at alkaline pH but is an extremely poor donor[18].
Accordingly, D-phenylglycine-L-phenylglycine was obtained in 69% yield by reaction of D-PGA with L-PG at pH 9.7 in the presence of immobilized *E. coli* pen acylase (Figure III.23). Similarly, optically pure L-phenylglycyl-L-phenylglycine methyl ester was obtained in 63% yield by self-coupling of L-PG methyl ester, which acted as both an activated acyl donor and a nucleophile (acceptor)[48].

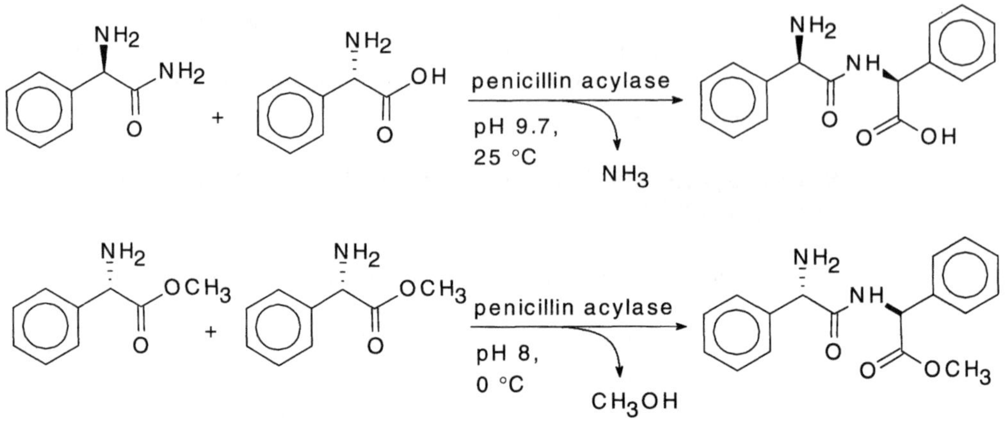

Fig. III.23 Synthesis of phenylglycyl dipeptides using *E. coli* pen acylase.

The corresponding diketopiperazines were readily prepared[48] by cyclization of the dipeptide esters in aqueous methanolic potassium hydroxide (Figure III.24). The *trans*-diketopiperazine formed from D-phenylglycyl-L-phenylglycine methyl ester is achiral. Cyclization of the L,L-dipeptide ester afforded the *cis*-L,L-diketopiperazine as the major product together with a small amount of the *trans*-product.

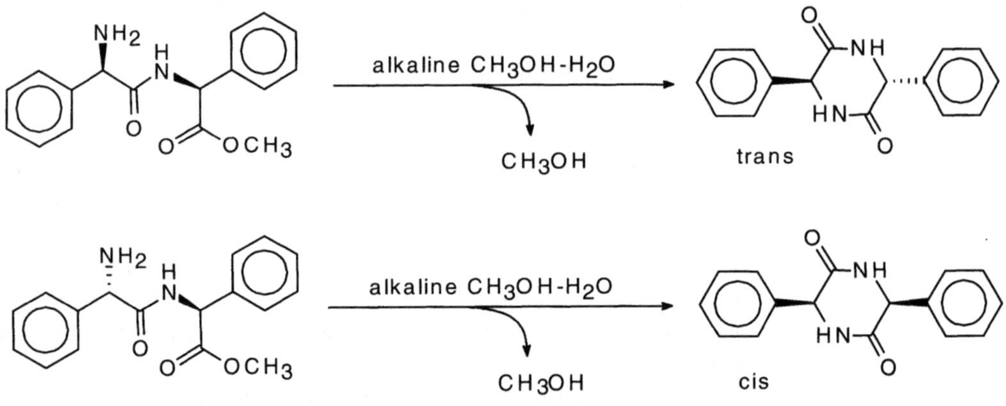

Fig. III.24 Diketopiperazines from ring closure of PG dipeptide esters.

5.2 Pen acylase-catalyzed resolution of amines in aqueous organic solvents

The resolution of chiral amines mediated by proteases and, particularly, lipases has been extensively studied[54]. A serious limitation of these reactions is, however, the low rate of reaction observed in organic media. More recently, Bayer[55] and BASF[56] scientists achieved dramatic rate improvements by employing an ester of methoxyacetic acid as the acyl donor, thus paving the way for commercialization, by BASF, of lipase-catalyzed resolution of amines.

It occurred to us that pen acylase might be a useful catalyst for amine resolutions via enantioselective reaction of the acylenzyme intermediate with the chiral amine nucleophile, analogous to its reaction with the amino group of 6-APA or 7-ADCA in the synthesis of penicillins and cephalosporins, respectively (Figure III.25).

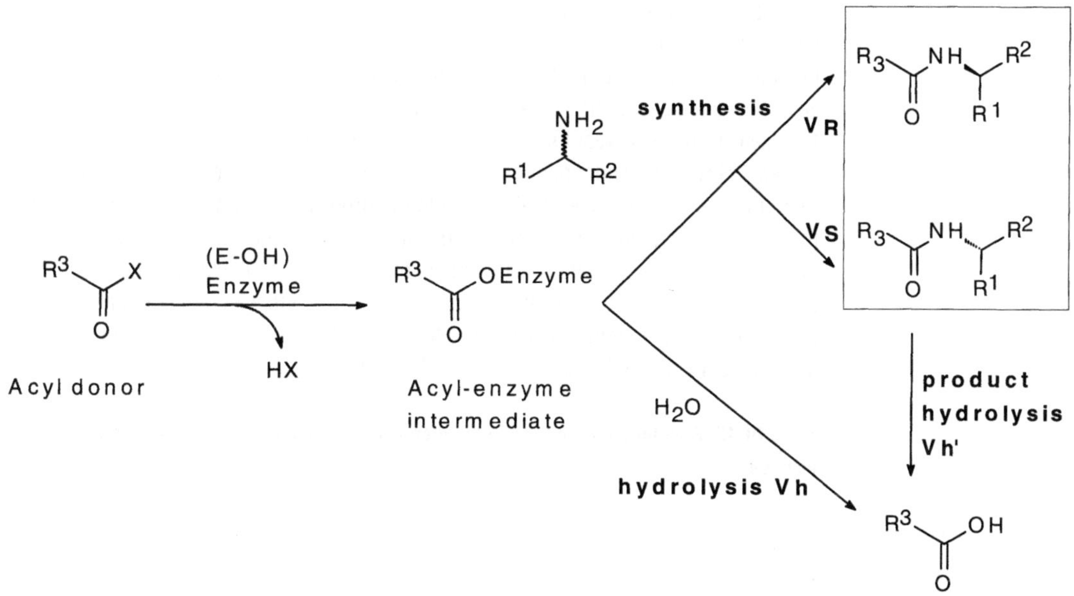

Fig. III.25 Enzymatic resolution of amines by kinetically controlled acylation.

In contrast with lipases and proteases, (free) pen acylase is very sensitive to deactivation in organic solvents and is inactive in anhydrous media, although CLEAs and CLECs of pen acylase have been used in organic media (see Section 4). Consequently, the use of pen acylase in amine resolution has been largely restricted to the hydrolysis of *N*-phenylacetylamines[57]. The alternative approach, kinetically controlled acylation in water, has received scant attention. This prompted us to study the acylation of a range of chiral amines, using phenylacetamide as the acyl donor, in the presence of two different pen acylases, from *E. coli* and *A. faecalis*, respectively[58]. The former is the enzyme used in penicillin and cephalosporin synthesis while the latter is a relatively unknown enzyme discovered in the late sixties[59]. It is more thermally stable than the *E. coli* enzyme and has a broad pH optimum[43,60]. Initial experiments demonstrated that the *A. faecalis* enzyme mediates efficient acylation of amines with phenylacetamide at high pH in aqueous medium and the acylation of 1-phenylethylamine was found to be highly enantioselective[61].

The reactions with the *A. faecalis* pen acylase were performed at pH 11 to fully deprotonate the amine. The stability of the *E. coli* enzyme is low at this pH, however, which necessitated performing the reactions at pH 10. The performance of the *A.*

faecalis and *E. coli* pen acylases were compared with regard to enantioselectivity, rate and synthesis/hydrolysis ratio (S/H). The results are shown in Table III.16.
E. coli pen acylase displayed poor enantioselectivities with all the amines tested (E values ranging from 2 to 12). In contrast, the *A. faecalis* enzyme was highly enantioselective (E > 100) with amines 1 and 4 which both contained a phenyl substituent. The rates and S/H ratios observed with the *A. faecalis* enzyme were generally an order of magnitude higher than those with *E. coli* pen acylase. Enantioselectivities could be further improved by adding a cosolvent such as acetonitrile (see Table III.16), ethanol or *tert*-butyl alcohol (data not shown).
It is clear from these results that the *A. faecalis* pen acylase is superior with regard to rate, S/H ratio and enantioselectivity. Reaction rates are high compared to resolution by pen acylase-catalyzed amide hydrolysis. The low hydrolysis rates in the present system are partially owing to the poor solubility of the N-phenylacetylamine, which is an advantage from a practical viewpoint (the product is easily isolated by filtration). On the basis of these results we conclude that our pen acylase catalyzed resolution procedure has commercial potential as an alternative to analogous lipase mediated resolutions of chiral amines.

Table III.16 Resolution of amines by kinetically controlled acylation using penicillin acylase.

	E. coli (pH 10)			A. faecalis (pH 11)								
	water			**water**			**10% ACN**			**25% ACN**		
Amine	E[a]	S/H[b]	Vi[c]	E[a]	S/H[b]	Vi[c]	E[a]	S/H[b]	Vi[c]	E[a]	S/H[b]	Vic
1	4 (R)	0.3	0.3	>100 (R)	2	23	N.D.			N.D.		
2	3 (S)	0.2	0.8	9 (R)	2.4	12	14 (R)	3	38	10 (R)	3.5	17
3	7 (S)	0.6	1.2	1.7 (S)	2	24	1.6 (R)	2	23	2.6 (R)	2	25
4	2 (R)	0.5	0.14	110 (R)	4	19	220 (R)	4	21	400 (R)	4	13
5	12 (R)	0.06	0.03	2.8 (R)	0.3	1.8	3.5 (R)	0.2	1.3	4.8 (R)	0.1	0.6

[a] The E value has been calculated based on the enantiomeric purity of the product at 10-30% conversion; in case of compound 5, which was slowly converted, the E value was calculated at 2-6% conversion. The absolute configuration of the product was determined by HPLC; standards of pure R-phenylacetylated amines were prepared by lipase catalyzed aminolysis of methyl phenylacetate.
[b] S/H is defined as the initial ratio between amide formation and formation of phenylacetic acid.
[c] The initial rate of amide formation in mmol per unit per hour.

Easy-on/easy-off technology for amine resolution

In order to be effective an amine resolution process should not only comprise an efficient enzymatic acylation step but also provide for facile hydrolysis of the acylated product and racemization of the unwanted isomer (Figure III.26). However, amide hydrolysis and amine racemization are anything but facile processes. Chemical hydrolysis of amides requires harsh conditions (alkaline pH and elevated temperatures) that are incompatible with many reactive functional groups. Enzymatic hydrolysis at close to neutral pH and ambient temperature offers obvious benefits. Hence, we focused our attention on the development of an 'easy-on/easy - off' amine resolution methodology whereby an enzyme is used for enantioselective acylation and another (or the same) enzyme is used to remove the acyl group.

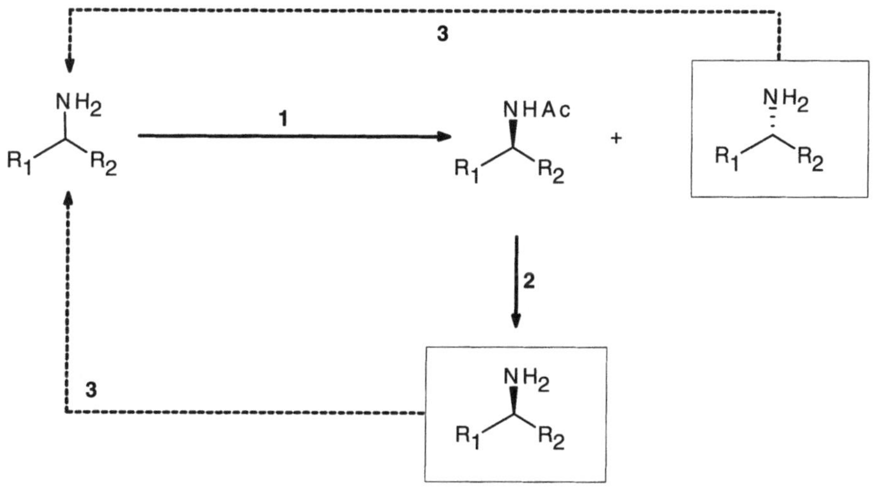

Fig. III.26 Resolution of a chiral amine. *Step 1:* enantioselective acylation of the amine and subsequent separation of the amide and unreacted amine. *Step 2:* hydrolysis of the optically active amide. *Step 3:* recycling of the unwanted enantiomer by racemization of the optically active amine.

To demonstrate the feasibility of this approach we performed the *C. antarctica* lipase B (Novozym 435)-catalyzed acylation of 2-aminopentane in di-isopropyl ether using D-phenylglycine propyl ester as the acyl donor (Figure III.27). The subsequent removal of the phenylglycyl group was, owing to its zwitterionic nature at neutral pH, easily and quantitatively achieved using pen acylase. The liberated (*R*)-amine was formed in 99% ee.

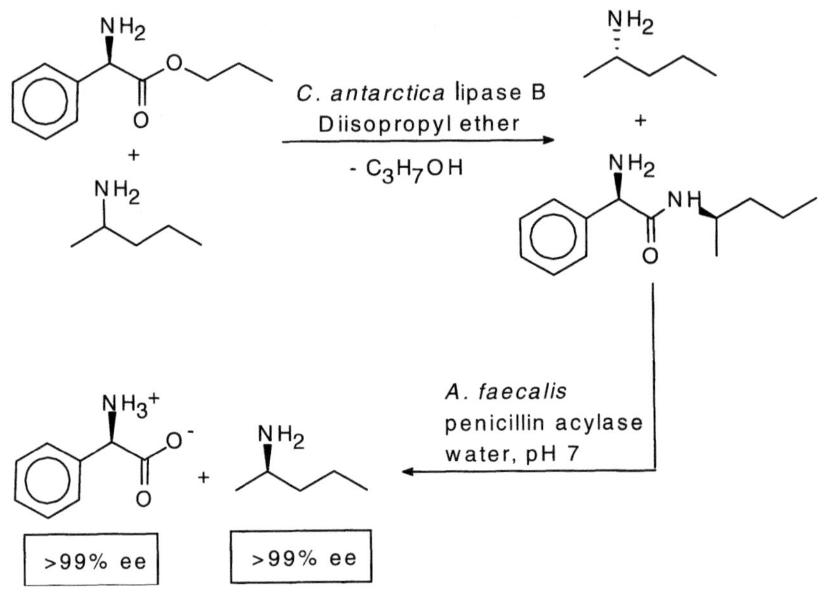

Fig. III.27 'Easy-on/easy-off' resolution of 1-aminopentane.

This 'easy-on/easy-off' resolution strategy has potential for the resolution of a broad variety of amines. Lipases accept a wide range of amines as nucleophile, often with high enantioselectivity[54]. Penicillin acylase also accepts a broad range of nucleophiles (see earlier)[51]. As Table III.17 shows the enantiopreference of both enzymes is similar.

Table III.17 Compatibility of *C. antarctica* lipase and penicillin acylase in amine resolution.

Enantiomeric ratio (E)

Amine	*C. antarctica* lipase B[a]	*E. coli* penicillin acylase	*A. faecalis* penicillin acylase
1a	55 (*R*)	5 (*S*)	ND
1b	>200 (*R*)	3 (*S*)	9 (*R*)
1c	68 (*R*)	7 (*S*)	1.7 (*S*)
2	>300 (*R*)	4 (*R*)	>100 (*R*)
3	650 (*R*)	*S*	ND
4	90 (*R*)	2 (*R*)	110 (*R*)
5	>200 (*R*)	12 (*R*)	2.8 (*R*)

[a] Data from reference 3.

The D-phenylglycyl group enhances the water solubility of the *N*-acylamine as a result of protonation of the amino function. Moreover, hydrolytic removal of this group liberates the zwitterionic D-PG, which is thermodynamically very stable and does not inhibit the pen acylase. Owing to its low solubility in water the D-PG is readily recovered by filtration and can, in principle, be recycled.

In short, we have developed an attractive 'easy-on/easy-off' procedure for the resolution of chiral amines[62]. Investigations of the scope of this methodology are currently underway.

§6 Concluding remarks & future prospects

Bearing in mind our original mandate, what have we accomplished? With regard to the development of new processes for side chain donors (protocol 1) we have shown that conversion of racemic PGM to D-PGA, via lipase-catalyzed ammonolysis, can be performed as a dynamic kinetic resolution by employing an aldehyde as a catalyst for *in situ* racemization. Unfortunately, the optimum result - 85% yield and 88% ee in 66 h - is not sufficient for economic viability. We have demonstrated the feasibility of a stereoretentive nitrile hydratase-catalyzed hydrolysis of D-phenylglycine nitrile to D-PGA, in 95% yield and > 99% ee. In combination with the synthesis of D-PG nitrile, by asymmetric transformation of the diastereomeric salts with L-tartaric acid, this provides an attractive, two-step synthesis of D-PGA from racemic D-PG nitrile (the product of the Strecker reaction of benzaldehyde with HCN/NH_3). Attempts to develop thermodynamic coupling procedures based on N-substituted phenylglycine derivatives were unsuccessful. No enzyme could be found that was able to catalyze the coupling of 6-APA to *N*-substituted phenylglycine derivatives.

With regard to the integration of activation, coupling and recycling of side chain donors, we have demonstrated that salt-free esterification of α-amino acids over a solid catalyst (H-USY, a naphtha cracking catalyst) is possible. Unfortunately, it leads to (partially) racemized product. However, this is not necessarily a problem in the case of phenylalanine since racemic phenylalanine methyl ester is the intermediate in the DSM-Tosoh process for the production of aspartame.

We have demonstrated the feasibility of a mixed-donor (D-PGM + D-PGA) coupling process for Ampicillin synthesis which reduces the acid consumption and concomitant salt formation by a factor three (65% less salt formation and 40% less D-PGA consumption) compared to the conventional enzymatic coupling. We have also developed a novel one-pot synthesis of Cephalexin by coupling the stereoretentive hydration of D-PG nitrile (see above) to the pen acylase-catalyzed coupling of the D-PGA product with 7-ADCA.

With regard to the development of the optimum biocatalyst for the coupling of side chain donors to the β-lactam nucleus we have developed novel methodologies for the immobilization of pen acylase. The immobilization of pen acylase as cross-linked enzyme aggregates (CLEAs), for example, endows it with stability in organic solvents containing 5% or more water. Unfortunately, this did not lead to a viable coupling procedure in organic media owing to the limited stability of the β-lactam nucleus under these conditions. Nonetheless, pen acylase CLEAs proved to be promising catalysts for coupling procedures in aqueous media. Moreover, the excellent stability of pen acylase CLEAs in aqueous/organic media provides a sound basis for their broad application in organic synthesis.

Very promising results were also obtained with pen acylase immobilized *via* covalent attachment to dendrimers or to the polymer produced from p-xylylene diamine and glutaraldehyde. As is the case with the CLEA technique these methodologies probably have broad potential for the immobilization of enzymes.

We also developed a method for active site titration of pen acylase preparations that gives useful insights into the efficiency of various immobilization techniques. Similarly, the use of LC-MS to study deactivated pen acylase is providing useful insights into the mechanism of catalytic turnover-related deactivation of the enzyme. Studies of the effect of various adjuvants on the S/H ratio are providing important clues to how the yields of coupling procedures can possibly be improved.

As spin-off activities we have shown that pen acylase can be used for the synthesis of dipeptides and the resolution of chiral amines in aqueous-organic media. The latter has commercial potential as an alternative to the currently used lipase-based procedures. As an extension of this study we developed an 'easy-on/easy-off' procedure for the resolution of chiral amines using a lipase in tandem with pen acylase. We expect that these procedures, and variants thereof, will have broad applicability and investigations of the scope are currently underway.

Another important benefit of the cluster project has been to the formation of collaborations with other groups, within and outside the project. In the latter context we note the collaboration with the group of Vytas Svedas (Moscow State University) and Andreas Stolz (University of Stuttgart). These collaborations and the many interesting developments discussed above provide an excellent basis for the further development of novel biotransformations in our group. For example, we are extending our studies of amine resolutions to a variety of commercially interesting substrates, including amino alcohols and β-amino acids. Our discovery of the novel CLEA-technique for immobilization of pen acylase is now being successfully applied to other enzymes in our group. Our studies of nitrile hydratases in collaboration with the group of Andreas Stolz have led to the formation of a European Network of companies and universities around the topic 'Nitrile Converting Enzymes' and an EU proposal has been submitted. We have also established collaborations with industrial and academic experts in the field of *in vitro* evolution of enzymes with a view to exploiting this methodology to improve the properties of, *e.g.* nitrile converting enzymes. Who knows? This may even lead to the development of an enantioselective nitrile hydratase for the hydration of PG nitrile. In short, we consider the cluster project to be an unqualified success.

§7 Acknowledgement

The project described in this chapter has greatly benefitted from the enthousiastic contributions of laboratory technicians and students. Ernst Neeleman, Lida Kock, Pedro Pereira, Marjolijne Luyt, Rute Madeira Lau, Natasja Oosthoek, Manuela de Leo, Carla Almeida, Jeoffrey Elzinga, Nick Lammers, Ewald Edink and Sílvia Pereira, as well as the invaluable Ernst Wurtz have all earned our gratitude for their efforts. Some of the described work could not have been performed without the contributions of our international collaborators. Vytas Svedas of Moscow State University and his students: Alexander Beltser, Xermes Chilov, Dorel Guranda, Andrej Khimiouk, Alexej Korennykh, Tatiana Shamolina and Maxim Youshko have earned our gratitude for their valuable contributions and hospitality. Thanks are due to Andreas Stolz and Ute Heinemann of Stuttgart University for making their nitrile hydratase culture collection as well as their expert knowledge available to us. The contribution by Lutz Fischer, now at the University of Stuttgart-Hohenheim, in teaching us how to grow *Rhodococci* at preparative scale, is gratefully acknowledged.
The supervisory team of DSM and Gist-brocades: Alle Bruggink, Erik de Vroom, Ben de Lange, Annemarie Schoevaars, Thom van der Does, Gerard Kwant and Sander Arendsen, has supported our efforts with their expert knowledge.

§8 References

1 E.J.A.X. van de Sandt and E. de Vroom, *Chim. Oggi* **18** (5) (2000) 72.
2 J. Verweij and E. de Vroom, *Recl. Trav. Chim. Pays-Bas* **112** (1993) 66.
3 A. Bruggink, E.C. Roos and E. de Vroom, *Org. Proc. Res. Develop.* **2** (1998) 128.
4 T. Sonke, B. Kaptein, W.H.J. Boesten, Q.B. Broxterman, H.E. Schoemaker, J. Kamphuis, F. Formaggio, C. Toniolo and F.P.J.T. Rutjes, in *'Stereoselective Biocatalysis'*, R.N. Patel, Ed., Marcel Dekker, New York, 2000, pp. 23-58.
5 W.H.J. Boesten, Eur. Pat. Appl. EP 442584 and EP 442585 (1991) to Stamicarbon.
6 J. Ogawa and S. Shimizu, in "*Stereoselective Biocatalysis*", R.N. Patel, Ed., Marcel Dekker, New York, 2000, pp.1-21.
7 M.S. Sigman and E.N. Jacobsen, *J. Am. Chem. Soc.* **120** (1998) 5315.
8 M.C. de Zoete, A.A. Ouwehand, F. van Rantwijk and R.A. Sheldon, *Recl. Trav. Chim. Pays-Bas* **114** (1995) 171.
9 E.J. Ebbers, G.J.A. Ariaans, J.P.M. Houbiers, A. Bruggink and B. Zwanenburg, *Tetrahedron* **53** (1997) 9417 and references cited therein.
10 M.A.P.J. Hacking, M.A. Wegman, J. Rops, F. van Rantwijk and R.A. Sheldon, *J. Mol. Catal. B: Enzymatic* **5** (1998) 155; see also F. van Rantwijk, M.A.P.J. Hacking and R.A. Sheldon, *Monatsh. Chem.* **131** (2000) 549.
11 M.A. Wegman, M.A.P.J. Hacking, J. Rops, R. Pereira, F. van Rantwijk and R.A. Sheldon, *Tetrahedron Asymm.* **10** (1999) 1739.

12 T. Sugai, T. Yamazaki, M. Yokoyama and H. Ohta, *Biosci. Biotech. Biochem.* **61** (1997) 1419.
13 M. Wieser and T. Nagasawa, in '*Stereoselective Biocatalysis*', R.N. Patel, Ed., Marcel Dekker, New York, 2000, pp. 461-486.
14 M.A. Wegman, U. Heinemann, F. van Rantwijk, A. Stolz and R.A. Sheldon, *J. Mol. Catal. B: Enzymatic*, in press.
15 F.H. Arnold and A.A. Volkov, *Curr. Opin. Chem. Biol.* **3** (1999) 54; J. Minshull and W.P.C. Stemmer, *Curr. Opin. Chem. Biol.* **3** (1999) 284.
16 M.A. Wegman, U. Heinemann, A. Stolz, F. van Rantwijk and R.A. Sheldon, Org. Proc. Res. Dev. 4 (2000) 318.
17 M.B. Diender, A.J.J. Straathof, L.A.M. van der Wielen, C. Ras and J.J. Heijnen, *J. Mol. Catal. B: Enzymatic* **5** (1998) 249.
18 C.P.G.H. Schroën, V.A. Nierstrasz, P.J. Kroon, R. Bosma, A.E.M. Janssen, H.H. Beeftink and J. Tramper, *Enzym. Microb. Technol.* **24** (1999) 498.
19 J.M. Guisan, R. Fernandez-Lafuenta, G. Alvaro, R.M. Blanco and C.M. Rosell, WO 91/09136 (1991).
20 French Pat. Appl. 2014689 (1969).
21 G.D. Yadav and M.S. Krishnan, *Org. Proc. Res. Dev.* **2** (1998) 86.
22 K. Oyama, in *'Chirality in Industry'*, A.N. Collins, G.N. Sheldrake and J. Crosby, Eds., Wiley, New York, 1992, p. 237.
23 M.A. Wegman, J.M. Elzinga, E. Neeleman, F. van Rantwijk and R.A. Sheldon, *Green Chemistry*, in press.
24 L.M. van Langen, E. de Vroom, F. van Rantwijk and R.A. Sheldon, *FEBS Lett.* **456** (1999) 89-92.
25 L.M. van Langen, E. de Vroom, F. van Rantwijk and R.A. Sheldon, manuscript in preparation.
26 G.J. Kemperman, R. de Gelder, F.J. Dommerholt, P.C. Raemakers-Franken, A.J.H. Klunder and B. Zwanenburg, *Chem. Eur. J.* **5** (1999) 2163-2168.
27 M.A. Wegman, L.M. van Langen, F. van Rantwijk and R.A. Sheldon, *Biotechnol. Bioeng.*, submitted for publication.
28 S. Ospina, E. Barzana, O.T. Ramirez and A. Lopez-Munguia, *Enzyme Microb.Technol.* **19** (1996) 462-469.
29 D.H. Nam, C. Kim and D.D.Y. Ryu, *Biotechnol. Bioeng.* **27** (1985) 953-960.
30 For earlier work on penicillin acylase in low-water organic media see: C. Ebert, L. Gardossi and P. Linda, *Tetrahedron Lett.* **37**, (1996) 9377-9380. C. Ebert, L. Gardossi and P. Linda, *J. Mol. Catal. B: Enzym.* **5**, 241-244 (1998). L. de Martin, C. Ebert, G. Garau, L. Gardossi and P. Linda, *J. Mol. Catal. B: Enzym.* **6**, 437-445.
31 D.L. Brown and C.E. Glatz, Chem. *Eng. Sci.* **47** (1986) 1831-1839.
32 F. Rothstein, in *'Differential precipitation of Proteins: Science and Technology'*, G.H. Roger, Ed. 1996, 115-208.
33 L. Cao, F. van Rantwijk and R.A. Sheldon, *Org. Lett.* **2** (2000)1361-1364.

34 A somewhat related system, involving the crosslinking of spray-dried enzymes, has been claimed in a patent application (S. Amotz, (Novo Industri A/S) US 4,665,028 (1987)). However, text, discussion and examples do not cover the system invented by our group, which has greater robustness and can also be used in organic solvents. When we applied this method to penicillin acylase only a minute fraction of the native activity was recovered.
35 L.M. van Langen, N.H.P. Oosthoek, L. Cao, F. van Rantwijk and R.A. Sheldon, in preparation.
36 R. Fernandez-Lafuente, C.M. Rosell and J.-M. Guisan, *Biotechnol. Appl. Biochem.* **24** (1996) 139-143.
37 V. Kasche and B. Galunski, *Biochem. Biophys. Res.Commun.* **104** (1982) 1215-1222.
38 Similar to the change in enantioselectivity of C. rugosa lipase upon treatment with isopropyl alcohol (I.J. Colton, S.N. Ahmed and R.J. Kazlauskas, *J. Org. Chem.* **60** (1995) 212-217).
39 D.A. Tomalia, D.J. Baker, M. Hall, G. Kallos, S. Martin, J.R.J. Roeck and P. Smith, *P. Polymer J.* **17** (1985) 117-132.
40 I.V. Berezin, A.A. Klyosov, P.S. Nys, E.M. Savitskaya and V.K. Svedas, *Antibiotiki (Rus.)* **19** (1974) 880-887.
41 V. Svedas, D. Guranda, L. van Langen, F. van Rantwijk and R. Sheldon, *FEBS Lett.* **417** (1997) 414-418.
42 M. Reslow, P. Adlercreutz and B. Mattiasson, *Eur. J. Biochem.* **172** (1992) 573-578.
43 K. Clausen, A. Nielsen, N. Petersen and A. Nikolov, US Pat. 5753458 (1998) to Gist-Brocades.
44 K. Brocklehurst in: P.C. Engel (Ed.) Enzymology labfax, Bios Scientific, Oxford (U.K.), 1996, p 59.
45 V.K. Svedas, A.L. Margolin, S.F. Sherstiuk, A.A. Klyosov, I.V. Berezin, *Bioorg. Khim.* **3** (1977) 546-553.
46 A. Spieß, R.-C. Schlothauer, J. Hinrichs, B. Scheidat, V. Kasche, *Biotechnol. Bioeng.* **62** (1999) 267-277.
47 R. Fernández-Lafuente, C.M. Rosell, B. Piatkowska J.M. and Guisán, *Enzyme Microb. Technol.* **19** (1996) 9-14.
48 L.M. van Langen, F. van Rantwijk, V.K. Svedas and R.A. Sheldon, *Tetrahedron: Asymm.* **11** (2000) 1077-1083.
49 O.C.Dermer and G.E. Ham, *Ethyleneimine and other aziridines*, Academic Press, New York, 1969, USA.
50 J.M. Frère, J.M. Ghuysen, H.R. Perkins and M. Nieto, *Biochem. J.* **135** (1973) 483-492. V. Kasche, U. Haufer and L. Riechmann, *Ann. N.Y. Acad. Sci.* **434** (1984) 99-105. B.Y. Golobolov, I.L. Borisov, V.M. Belikov and V.K. Svedas, *Biotechnol. Bioeng.* **32** (1987) 866-872.

51 V.K. Svedas, M.V. Savchenko, A.I. Beltser and D.F. Guranda, *Ann. N.Y. Acad. Sci.* **779** (1996) 659.

52 C. Fuganti, C.M. Rosell, S. Servi, A. Tagliani and M. Terreni, Tetrahedron Asymm. **3** (1992) 383.

53 V. Kasche, G. Michaelis and T. Wiesemann, *Biomed. Biochim. Acta* **50** (1991) 38; R.J. Didziapetris and V.K. Svedas, Biomed. Biochim. Acta 50 (1991) 237.

54 For a recent review see: F. van Rantwijk, M.A.P.J. Hacking and R.A. Sheldon, *Monatsh. Chem.* **131** (2000) 549.

55 U. Stelzer and C. Dreisbach, German Pat. DE 19637336 (1996) to Bayer.

56 K. Ditrich, F. Balkenbohl and W. Ladner, PCT Intern. Appl. WO 97/10201 (1997) to BASF; F. Balkenbohl, K. Ditrich, B. Hauer and W. Ladner, *J. Prakt. Chem.* **339** (1997) 381.

57 See, for example: V.A. Soloshonok, A.G. Kirilenko, N.A. Fokina, I.P. Shishkina, S.V. Galushko, V.P. Kukhar, V.K. Svedas and E.V. Kozlova, *Tetrahedron Asymm.* **5** (1994) 1119.

58 L.M. van Langen, N.H.P. Oosthoek, D.T. Guranda, F. van Rantwijk and R.A. Sheldon, *Tetrahedron Asymm.* **11** (2000) 4593.

59 M. Kulhanek and M. Tadra, *Folia Microbiol.* **13** (1968) 340.

60 R.M.D. Verhaert, A.M. Riemens, J.M. van der Laan, J. van Duin and W.J. Quax, *Appl. Environ. Microbiol.* **63** (1997) 3412.

61 D. Guranda, L.M. van Langen, F. van Rantwijk, R.A. Sheldon and V.K. Svedas, *Tetrahedron Asymm.*, submitted for publication.

62 L.M. van Langen, R. Madeira Lau, V.K. Svedas, F. van Rantwijk and R.A. Sheldon, manuscript in preparation.

Biocatalysts and Biocatalysis in the Synthesis of β-Lactam Antibiotics
Sheldon, van Rantwijk, van Langen, Wegman, Cao & Janssen

Chapter IV Process Technology and Process Integration in the Preparation of Penicillins *153*

§1 Introduction *153*
1.1 Production of semisynthetic penicillins *154*
1.2 Enzymatic production of Amoxicillin *154*
1.3 Shortcut routes from Penicillin G to Amoxicillin *157*
1.4 Outline of this chapter *159*

§2 Generalized description of the phase behavior of antibiotics *159*
2.1 Correlative model *160*
2.2 Solubilities in Mixed Solvents *161*
2.3 Partitioning in Mixed Solvents *164*
2.4 Relating different extraction systems *166*
2.5 Generalized Polarity Scales *167*
2.6 Notation *169*

§3 Thermodynamically controlled synthesis of Amoxicillin *169*
3.1 Background *169*
3.2 Feasibility of a thermodynamically controlled suspension-to-suspension process *171*
3.3 Cosolvent addition *171*

§4 Kinetically controlled enzymatic coupling of APA and HPGM *172*
4.1 Kinetically controlled suspension-to-suspension reactions *172*
4.2 Simulations *173*

§5 Towards shortcut routes from penicillin G to to Amoxicillin *175*
5.1 Extractive enzymatic hydrolysis of Penicillin G *175*
5.2 Modeling the extractive catalysis *177*
5.3 Experimental conversions and model predictions *177*
5.4 Combining extractive hydrolysis of Penicillin G with kinetically controlled synthesis of Amoxicillin *180*
5.5 One-pot shortcut route in anhydrous organic solvent *183*
5.6 Notation *184*

§6 Multifunctional bioreactors *184*
6.1 A Model for Fractionating Reactors *186*
6.2 Simulation results for fractionating hydrolysis reactions *188*
6.3 Case study: hydrolysis of Penicillin G *190*
6.4 Fractionating synthesis reactor *191*

§7 Single and multi component crystallization of SSA *192*
7.1 Model *192*
7.2 Batch-wise experiments *194*
7.3 Crystallization: Morphology and Kinetics *195*
7.4 Simulation of batch-wise crystallization process *197*
7.5 Notation *199*

§8 Conclusions *199*

§9 Acknowledgements *200*

§10 References *200*

Chapter IV Process Technology and Process Integration in the Preparation of Penicillins.

L.A.M. VAN DER WIELEN, M. OTTENS AND A.J.J. STRAATHOF (KLUYVER LABORATORY FOR BIOTECHNOLOGY, DELFT UNIVERSITY OF TECHNOLOGY)

§1 Introduction

The antibiotic penicillin G (Pen G) is the most common raw material for semisynthetic β-lactam antibiotics. Key-intermediate for β-lactam antibiotics is 6-aminopenicillanic acid (APA), the β-lactam nucleus, which has a worldwide annual production volume of approximately 10,000 tons. Pen G is produced fermentatively by adding phenylacetic acid (PAA) to a crude fermentation broth. Pen G is converted, either chemically or enzymatically, to APA and PAA. Coupling chemically or enzymatically different side chains to APA can yield a wide range of semisynthetic penicillins with different specificities and stabilities. Examples of bulk semisynthetic antibiotics are Amoxicillin (Amox) and Ampicillin (Ampi) (Van de Sandt and De Vroom, 2000; Bruggink et al., 1998). These antibiotics have market sales worth of ca. $ 3 billion per year as bulk-formulated drug; further market information is given in chapter 1. Figure IV.1 shows a general overview of the route for synthesis of Amox.

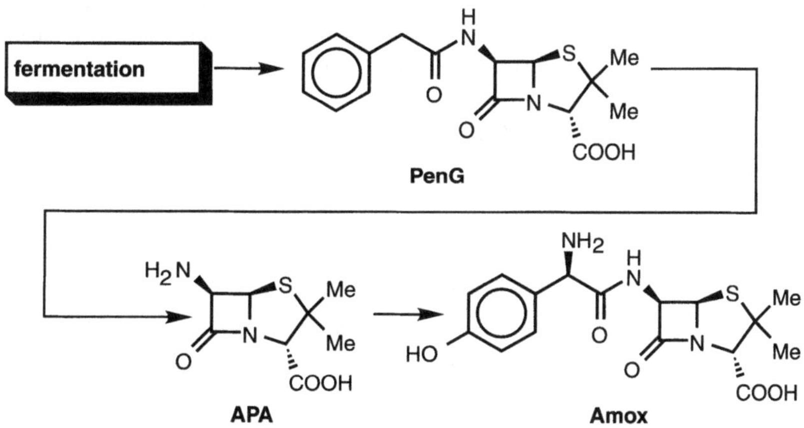

Fig. IV.1 General production route of the penicillin-derived antibiotic Amoxicillin.

1.1 Production of semisynthetic penicillins

Traditionally, the conversion of Pen G into semisynthetic antibiotics was performed chemically. In these chemical production methods, much waste is produced, i.e. 10-100 kg of waste per kg of product. These classical processes make use of halogenated organic solvents and need energy-consuming sub-zero temperatures (Van de Sandt and De Vroom, 2000). The biocatalytic processes, however, are performed in water at ambient or slightly elevated or decreased temperatures and use only a titrant as auxiliary chemical. The amount of waste using the biocatalytic processes compared to the chemical processes is reduced by a factor of 5 for the hydrolysis reaction of Pen G to APA and a factor of 3 for the coupling reaction of APA and side chain (Van de Sandt and De Vroom, 2000).

In the present downstream processing of Pen G, Pen G is extracted from the fermentation broth with an organic solvent like butyl acetate or amyl acetate. Subsequently, Pen G is either back-extracted to an aqueous phase and then crystallized, or crystallized directly from the organic phase by adding solid potassium acetate or sodium acetate. The Pen G crystals are used for the production of APA by enzymatic hydrolysis. APA is recovered by crystallization, dissolved again and may be coupled enzymatically with an activated side chain to form a semisynthetic antibiotic. A schematic overview of an enzymatic process of Pen G purification and biotransformation to Amox is given in Figure IV.2. As can be seen in this overview, this enzymatic process comprises many process steps. The next section focuses on the enzymatic coupling reaction of the β-lactam nucleus and the side chain. The succeeding section discusses a potentially more efficient antibiotic production process resulting from a combination of several process steps (shortcuts). This discussion originates from the work of Diender (2001).

1.2 Enzymatic production of Amoxicillin

The enzymatic coupling reaction between the β-lactam nucleus and the side chain can be performed using two distinctive strategies:

1. *thermodynamically-controlled synthesis* in which a direct condensation of the nucleus and a non-activated side chain, such as hydroxyphenylglycine (HPG) takes place (see Figure IV.3a). This strategy can only be applied successfully if conditions for a favorable equilibrium position can be found. The enzyme does not influence the position of the equilibrium, but only the rate at which the equilibrium is established (Kasche, 1986). To shift unfavorable reaction equilibria towards synthesis, moderate to high concentrations of water-miscible organic solvents may be added (Homandberg et al., 1978; Fernandez-Lafuente *et al.*, 1991a).

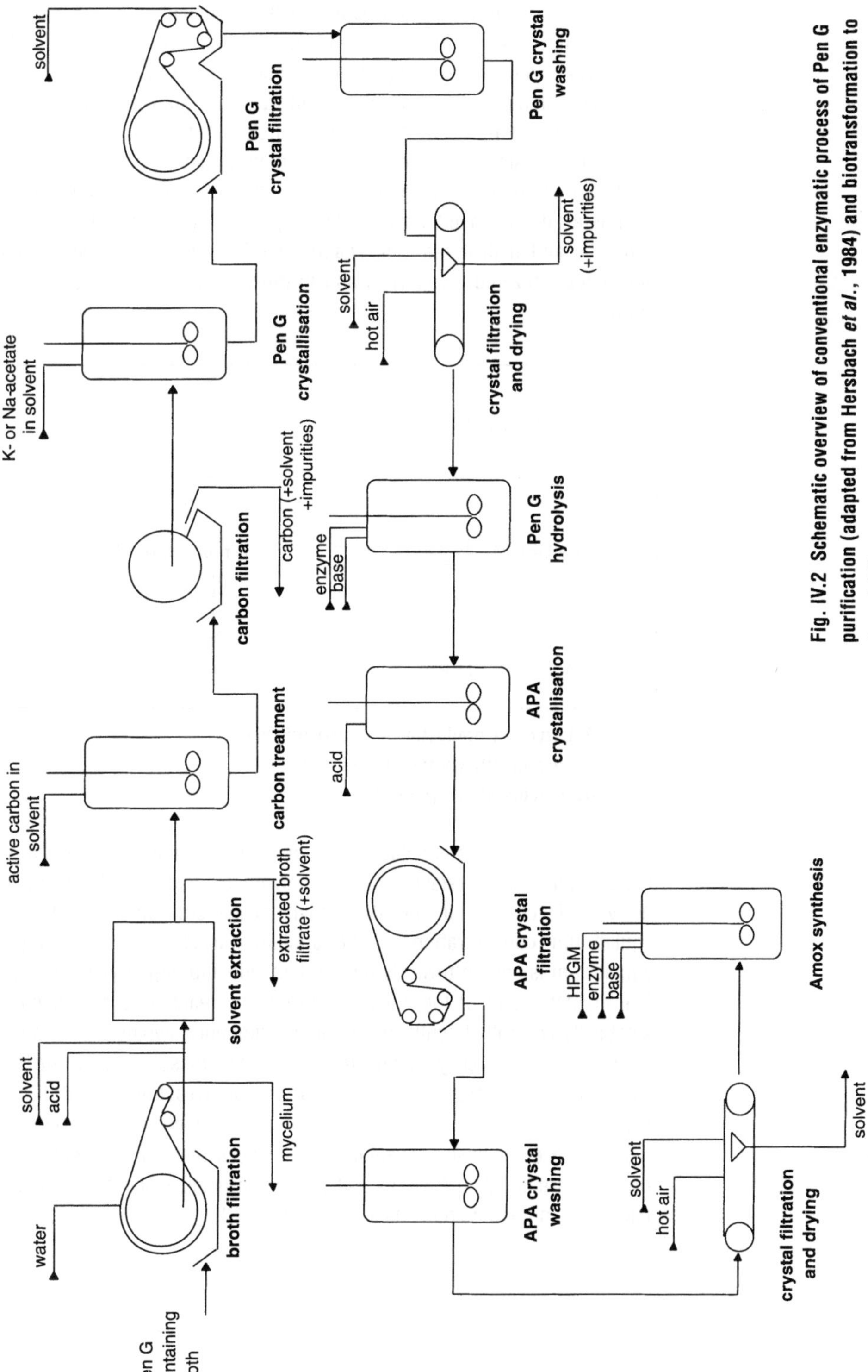

Fig. IV.2 Schematic overview of conventional enzymatic process of Pen G purification (adapted from Hersbach *et al.*, 1984) and biotransformation to Amoxicillin.

2. *kinetically-controlled synthesis* in which the nucleus is coupled to an activated side chain, such as hydroxyphenylglycine methyl ester (HPGM) or hydroxyphenylglycine amide (HPGA) (reaction I, Figure IV.3b). Using this strategy, high yields on nucleus are achievable, but part of the activated side chain hydrolyzes (reaction II, Figure IV.3b) and is not coupled to the nucleus, so recycling and reactivation of side chain are necessary or extra waste is produced. Also part of the produced antibiotic may hydrolyze (reaction III, Figure IV.3b). An important parameter -the so-called synthesis/hydrolysis ratio (i.e. the molar ratio of produced antibiotic and hydrolyzed side chain)- is influenced by reaction conditions and biocatalyst (enzyme) properties and formulation (Kasche, 1986; Bruggink et al., 1998).

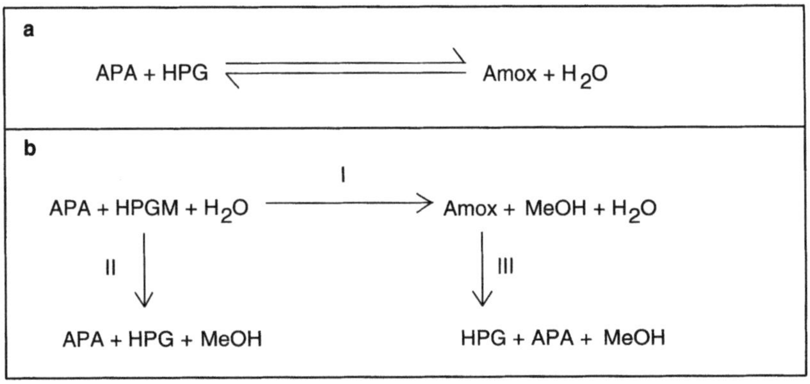

Fig. IV.3 Enzymatic production of Amoxicillin via
a. thermodynamically-controlled synthesis,
b. kinetically-controlled synthesis.

Because of the low water solubility of semisynthetic antibiotics like Amox and Ampi, the antibiotic may crystallize during the synthesis reaction. This crystallization can be very favorable for the thermodynamically-controlled synthesis when the equilibrium concentration of the antibiotic exceeds its solubility. Then the equilibrium is shifted towards the product side and higher yields will be achieved. The crystallization of the antibiotic during the reaction is also favorable for the kinetically-controlled synthesis as this phenomenon avoids the unwanted product hydrolysis by the enzyme due to much lower dissolved concentrations. When hydrolytic side reactions of substrate or product (reactions II and III, Figure IV.3b) are to be reversed or suppressed, the so-called suspension-to-suspension reactions are especially advantageous. In these reactions the substrates and the products are mainly present as undissolved particles, while the enzymatic reactions take place in the aqueous phase (see Figure IV.4).

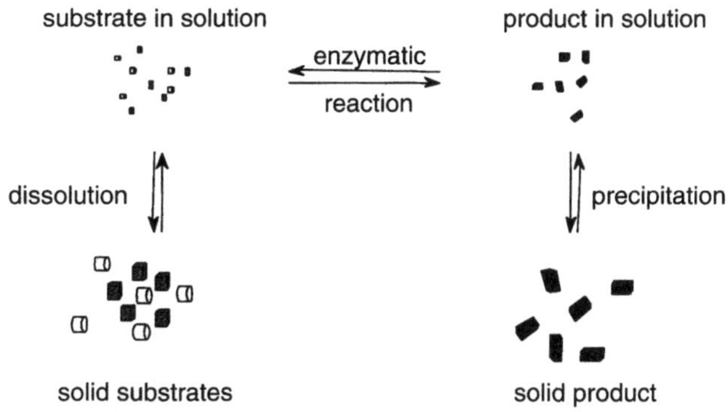

Fig. IV.4 Schematic overview of suspension-to-suspension reaction.

In this type of reactions, the product to reactor volume ratio is much higher than in the conventional systems in which all reactants are completely dissolved (Halling et al., 1995).

1.3 Shortcut routes from Penicillin G to Amoxicillin

As described earlier, a reduction in the number of process steps in the production of semisynthetic antibiotics by combining several steps could lead to more efficient production processes, while producing less waste. A possibility for a shortcut route is to perform the hydrolysis reaction of Pen G salt and the synthesis reaction of Amox simultaneously. In this way, the equilibrium of the hydrolysis reaction will be shifted towards the product side as the intermediate APA is consumed in the synthesis reaction. Both reactions can be performed using the same enzyme, penicillin acylase. By performing the reactions simultaneously, APA formed in the hydrolysis reaction can be used directly for the synthesis reaction without having to purify it in between the reactions. This would reduce 1. the number of process units 2. losses of APA during downstream processing and 3. salt formation as the number of pH shifts (for enzymatic steps and crystallization) can be decreased (see Figure IV.5).

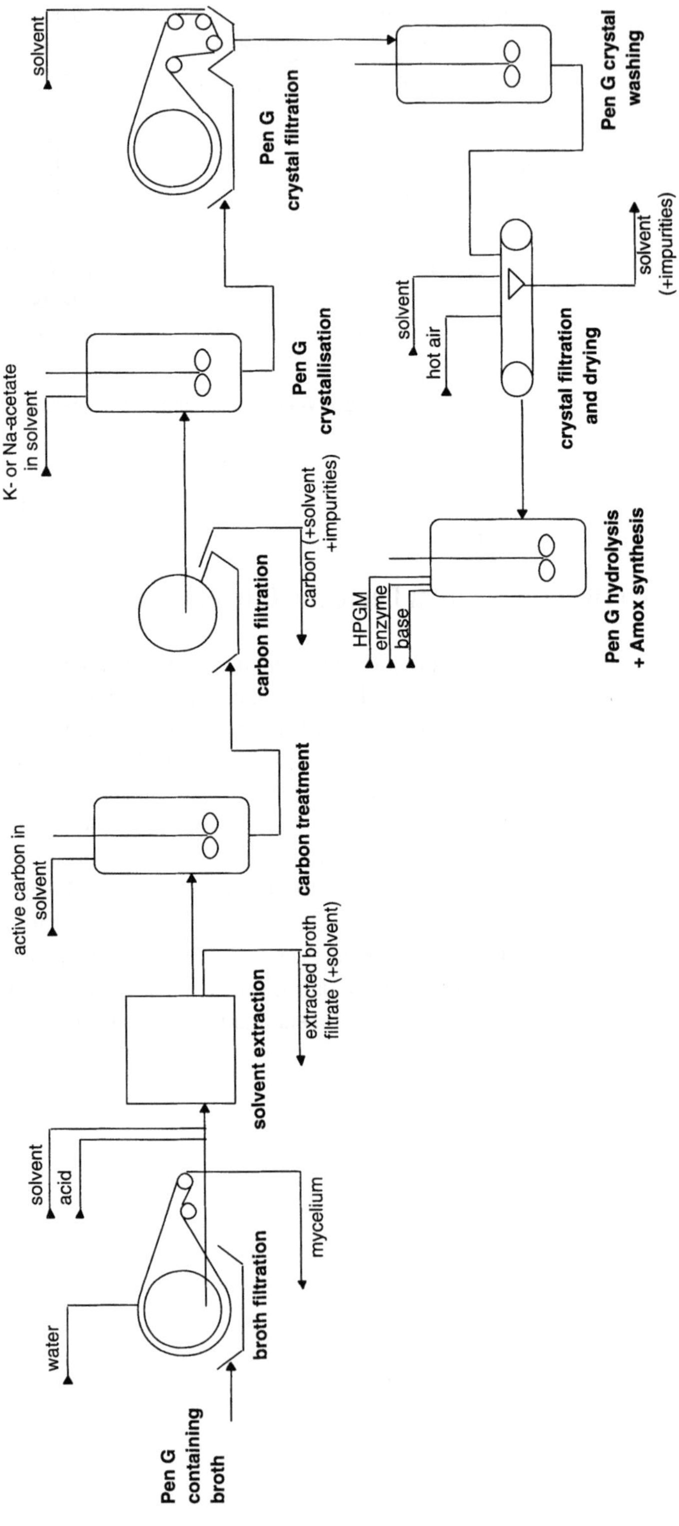

Fig. IV.5 Schematic overview of a process concept of Pen G purification and simultaneous enzymatic Pen G hydrolysis and Amoxicillin synthesis.

An even more advanced shortcut route would be to start with Pen G acid dissolved in the organic solvent directly after its extraction from the fermentation broth. The hydrolysis reaction of Pen G and the kinetically-controlled synthesis reaction of Amox could both take place in the organic solvent or in organic solvent/water mixtures. This shortcut route might reduce the salt production and the number of process steps even more than the previous shortcut, because in this case also Pen G does not have to be purified prior to the enzymatic reactions (see Figure IV.6).

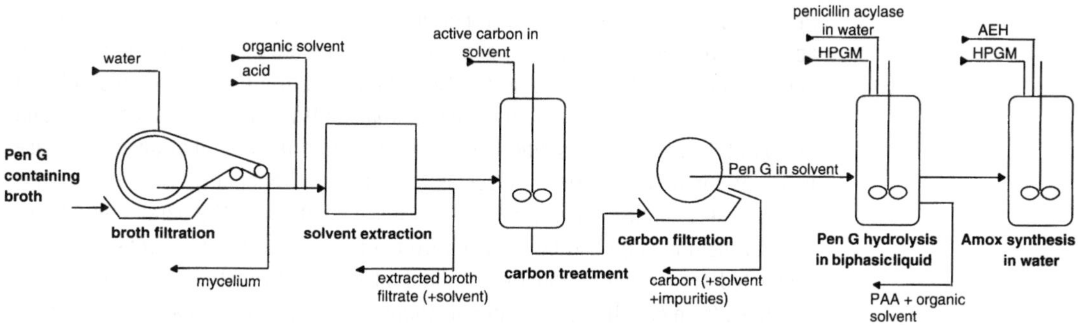

Fig. IV.6 Schematic overview of a process concept of partial Pen G purification and subsequent Pen G biotransformation to Amoxicillin in aqueous/organic solvent two-phase system.

1.4 Outline of this chapter

The particular aim of the work described in this chapter is to develop new enzymatic process concepts for the production of semisynthetic penicillins, with the emphasis on the enzymatic hydrolysis of penicillin G, the enzymatic coupling reaction of APA and a side chain, and the recovery of the products by crystallization.
A thermodynamical approach is required to evaluate the earlier-mentioned systems to produce Amox. In this thermodynamical approach reaction equilibria, dissociation constants, solubilities and partition behavior of the reactants play a role. The different process concepts for the production of Amox were described using the thermodynamical approach. Therefore the phase behavior of the compounds involved in this study is discussed first.

§2 Generalized description of the phase behavior of antibiotics

For a quantitative comparison of process alternatives for semi-synthetic antibiotics (SSAs) and in particular for selection of purification systems for mixtures of near-identical species, quantification of molecular properties and operating conditions is required. The approach for model development presented here is strongly related to 'chemical feel'. Intuitive qualitative chemical concepts such as 'hydrophobicity' or 'lipophilicity', essentially based on empirism, have been used to quantify and rank molecular properties and operating conditions of processes in order to facilitate the selection of process alternatives. Identifying generalizing quantitative principles to correlate thermodynamic properties can translate this valuable knowledge into quantitative tools. This is the so-called 'correlative' approach.

On a mole fraction basis, most industrial fine chemical processes operate under diluted and usually aqueous conditions with respect to the dissolved bioproducts and their contaminants. An aqueous Penicillin G (Pen G) solution of 100 g/L has a Pen G mole fraction of only 0.5%. For amino acids, SSAs and their precursors in aqueous solution, (asymmetric) activity coefficients are essentially unity, so even concentrated industrial systems of SSAs operate from a thermodynamic point of view at infinite dilution conditions. The approach introduced above recognizes and exploits this aspect as a sound basis for developing theories that describe phase behavior for a wide variety of systems realistically. Secondly, relatively simple models likely capture the deviations from these conditions. The parameters in this general methodology can be obtained from a limited number of experiments, and translated across the boundaries of different separation techniques. They can be predicted from data that are commonly found in the pharmaceutical or toxicological characterization of the final products.

2.1 Correlative model

The general thermodynamic conditions for equilibrium between two or more phases (α, β...) is the equality of chemical potentials of each transferable component i in each of these phases:

$$\mu_i^\alpha = \mu_i^\beta \tag{1}$$

The expression for the chemical potential of solute s in phase α reads as follows:

$$\mu_s^\alpha = \mu_s^{\Theta,\alpha} + RT\ln \gamma_s{}^\alpha + RT\ln x_s{}^\alpha \tag{2}$$

The pure liquid reference value for the chemical potential of the solute $\mu_s^{\Theta,\alpha}$ is hypothetical since most pure biochemicals are solids. A more realistic reference chemical potential is that where activity coefficients of solute s approach unity at infinite dilution of solute s for any -pure or mixed- solvent system. So instead of equation 2, we have:

$$\mu_s^\alpha = RT\ln H_s{}^\alpha + RT\ln \gamma_s^{*\alpha} + RT\ln x_s{}^\alpha \quad \text{with} \quad \lim_{x_S \to 0} \gamma_s^* = 0 \tag{3}$$

where H_s is the Henry coefficient of solute s in a single component or multicomponent solvent systems of interest. As shown in earlier work (Gude et al., 1996a; Van Ness and Abbott, 1982), a general expression for a Henry's constant of a solute in a multicomponent solvent system reads as follows:

$$\ln H_{sm} = \sum_j x'_j \ln H_{sj} + \ln \gamma_{sm}^\infty - \sum_j x'_j \ln \gamma_{sj}^\infty = \sum_j x'_j \ln H_{sj} + \ln H_{sm}^E \tag{4}$$

where H_{sm}^{E} is an excess Henry's constant which is a function of the activity coefficients of the solute in the mixture and in the single solvents, x'_j denotes the solute free solvent mole fraction and j runs over all solvent components. Hence, the Henry's constant in mixtures of solvents can, in case of ideal systems, be calculated on the basis of binary solute-solvents data only. An important advantage of the framework developed here is that in most situations of interest for bioseparation processes, the asymmetric activity coefficients are approximately unity. The excess Henry's constant is a function of the mixture and binary solute-solvent activity coefficients only and is usually a simple function of composition. For example, the corresponding excess Gibbs energy of a binary solvent system (1,2) with a single solute (s) is approximated with three binary and one ternary term:

$$g^E = g_{12}^E + g_{1s}^E + g_{2s}^E + g_{12s}^E \tag{5}$$

When this approximation holds, the terms that represent the interactions of the solute with the solvents as well as the ternary term vanish when the solute concentration decreases to zero. The excess contribution is then predominantly determined by the interactions among the solvent components (in which no solute contribution is incorporated). In analogy with Haynes and coworkers (Johansson et al., 1998), this term represents the "excess self-energy" of the solute-free phase system. This term reflects the extra energy that is required to disrupt interactions between solvent molecules to insert a solute in a 'cavity' in the phase system. Positive values represent unfavorable insertions, and negative values represent favorable processes.

2.2 Solubilities in Mixed Solvents

Most biomolecules at ambient conditions are solids when pure. Therefore crystallization is usually one of the steps in the purification process. In this work, it is the starting point for the application of the thermodynamic framework. The solubility of solute s in a single solvent is described with the following equation:

$$\frac{\mu_s^s}{RT} = \ln H_{sj} + \ln x_{sj}^{sat} + \ln \gamma_{sj}^{*\,sat} \quad \text{or} \quad \ln H_{sj} = \frac{\mu_s^s}{RT} - \ln x_{sj}^{sat} - \ln \gamma_{sj}^{*\,sat} \tag{6}$$

where μ_s^s is the chemical potential of the crystal phase and H_{sj} is the Henry's constant of solute s in solvent j. In the following, we assume that the crystal phase composition is constant and not affected by the solvent composition. Effects of varying hydration or solvatation, as well as polymorphism must be accounted for separately, which is beyond the aim of this text. Hence, μ_s^s remains constant. For the chemical potential of solute s in a mixture (m) of multiple solvents, we find at its solubility:

$$\frac{\mu_s^s}{RT} = \ln H_{sm} + \ln x_{sm}^{sat} + \ln \gamma_{sm}^{*\,sat} \tag{7}$$

Substitution of eq. 4 into eq. 7 results in the following expression:

$$\frac{\mu_s^s}{RT} = \sum_j x_j' \ln H_{sj} + \ln H_{sm}^E + \ln x_{sm}^{sat} + \ln \gamma_{sm}^{*\ sat} \tag{8}$$

The unknown chemical potential of the solid phase is eliminated by substituting eq. 6 in eq. 8, which leads to the expression below:

$$\ln x_{sm}^{sat} = \sum_j x_j' \ln x_{sj}^{sat} - \ln H_{sm}^E + \sum_j x_j' \ln \gamma_{sj}^{*\ sat} - \ln \gamma_{sm}^{*\ sat} \tag{9}$$

For liquid mixtures, that are considered ideal in this framework, eq. 9 reduces to a simple logarithmic average of the solubilities of the solute in each of the 'pure' solvents in the mixture, weighed by their respective solute-free mole fraction:

$$\ln x_{im}^{sat} = \sum_j x_j' \ln x_{ij}^{sat} \tag{10}$$

In real liquid phase mixtures of biomolecules, mixed solvent solubilities are usually very low and their asymmetric activity coefficients are close to unity. An expression for the excess Henry's constant based on a Flory-Huggins type combinatorial expression combined with a Margules residual term (Gude et al., 1996ab) seems a convenient model. The combinatorial part of this expression reads for a single solute (*s*) in a binary solvent system (1,2) as follows:

$$\ln H_{sm}^E\Big|_{comb} = \left(x_1'^m \ln r_1 + x_2'^m \ln r_2\right) - \ln\left(x_{1m}' r_1 + x_{2m}' r_2\right) + \left(x_1'^m \frac{r_s}{r_1} + x_2'^m \frac{r_s}{r_2}\right) - \frac{r_s}{x_{1m}' r_1 + x_{2m}' r_2} \tag{11}$$

and the residual part is:

$$\ln H_i^{E,m}\Big|_{res} = A_{12}\left(1 + C_{12S}\right) x_1'^m\, x_2'^m \tag{12}$$

The binary solvent-solvent interaction A_{12} can be obtained from conventional VLE data from literature (Gmehling and Oncken, 1977). This leaves the ternary parameter C_{12S} as the only adjustable parameter. This parameter is specific for a solute and fairly insensitive to the actual solvent system, in case of amino acids in binary (water + alkanol; Gude et al., 1996ab) and multicomponent (water + two alkanols; Van Berlo et al., 1997) solvent mixtures, as well as for semi-synthetic antibiotics in binary solvent systems (Rudolph et al., 1999; Rudolph et al., 2001a). Rudolph et al. (1999) showed that a UNIQUAC based combinatorial term, replacing the Flory-Huggins type combinatorial expression may be preferred in some cases. This has an important practical implication, namely that liquid phase interactions of these solutes are in general negligible. Exceptions may occur for (reversible) chemical reactions, such as dimerization or polymerization. Hence, solubilities of these solutes are hardly affected by the presence of other solutes at similarly low concentrations. This has been investigated by Rudolph et al. (2001b). Some characteristic results are presented in Figure IV.7. Figure IV.7 shows the aqueous solubility of pure Amox (A), of pure HPG (B) as well as the solubilities of both species (A+B) in mutually saturated aqueous solutions. More results are available and show that this trend is general. Only very soluble components can affect slightly (usually enhance) the solubility of poorly soluble species. In most cases, this 'salting-in' behavior seems related to the increased ionic strength of the solution by the well soluble species, and not to its specific chemical structure.

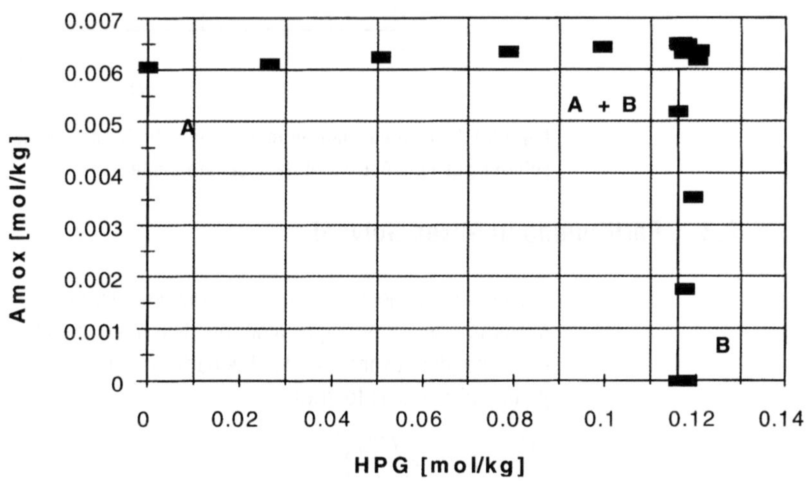

Fig. IV.7 Aqueous solubilities of Amox and HPG in pure Amox (A), pure HPG (B) and mutually saturated systems (A+B) at 298 K (adapted from Rudolph et al, 2001b).

The effect of added salt on the solubility of SSA and their precursors seems to follow a similar trend of salting-in, but salt concentrations are usually at least an order of magnitude larger than these of the best soluble SSAs. This seems mostly due to long range charge(salt)-dipole(SSA) interactions to decrease the SSA activity coefficients, and thereby reasonably independent of the salt used. At higher salt concentrations (molar range), the SSAs are salted-out, mostly due to short range, "ideal" hard sphere interactions and thereby relatively salt-dependent. This is illustrated by the calculated contributions to the activity coefficient of Ampi in aqueous NaCl solutions using a Perturbed Hard Sphere model (Rudolph et al., 1999; Rudolph et al., 2001b; model not shown here).

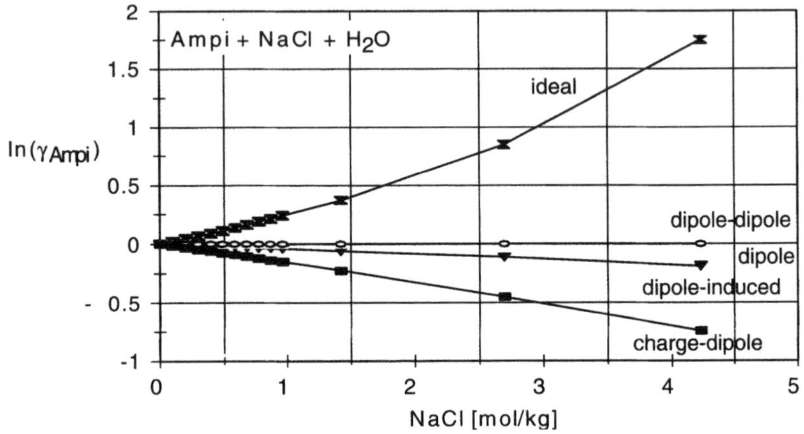

Fig. IV.8 Calculated contributions to the activity coefficient of Ampi in aqueous NaCl solutions using a Perturbed Hard Sphere model.

2.3 Partitioning in Mixed Solvents

Partition coefficients of solutes at low concentrations in multicomponent solvent systems can in principle also be described by the proposed framework. The corresponding expression for the partitioning of solute s over a top (t) and bottom phase (b) reads as follows:

$$\frac{x_s^t}{x_s^b} = K_s^{t/b} = \frac{H_s^b \gamma_s^{*b}}{H_s^t \gamma_s^{*t}} \tag{13}$$

with

$$\ln K_s^{t/b} = \sum_j \left(x_j'^t - x_j'^b \right) \ln H_{sj} + \ln\left(\frac{H_{sb}^E}{H_{st}^E} \right) + \ln\left(\frac{\gamma_{sb}^*}{\gamma_{st}^*} \right) \tag{14}$$

In real systems, the reasoning for partition coefficients of solutes is similar to that for the solubility in real mixed-solvent systems. The last term of the above expression usually vanishes because the solutes are mostly diluted, and their asymmetric activity coefficients approach the limiting value of unity at infinite dilution ($\gamma^*_{st(\text{or } sb)}$ =1). The term containing the excess Henry's constants of top and bottom phases, however, remains. This equation can be further fine-tuned for particular applications. When eq. 6 is substituted in eq. 14, the following result is obtained:

$$\ln K_s^{t/b} = \sum_j \left(x_j'^t - x_j'^b \right) \ln\left(x_{sj}^{sat} \gamma_{sj}^{*sat} \right) + \ln\left(\frac{H_s^{Eb}}{H_s^{Et}} \right) + \ln\left(\frac{\gamma_{sb}^{*\,b}}{\gamma_{st}^{*\,t}} \right) \tag{15}$$

The usefulness of eq. 15 is demonstrated in Figure IV.9 which shows the mole fraction based partition coefficients of amino acids in 1-butanol and water systems. Markers are experimental data from Gude et al. (1996b) and the curves are calculated with eq. 15.

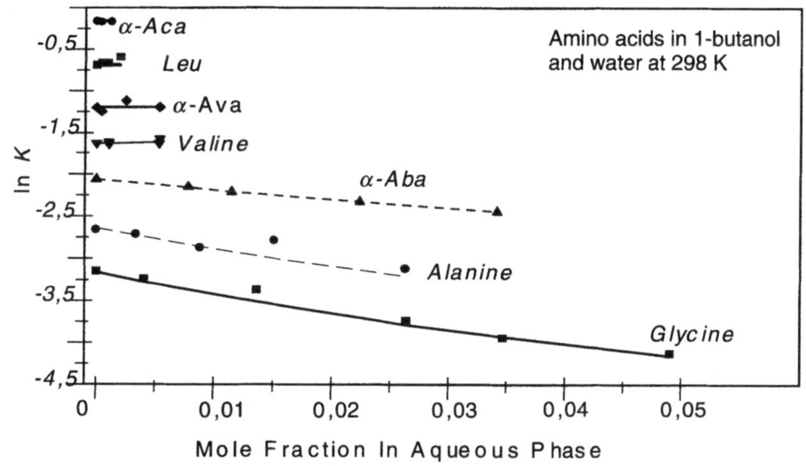

Fig. IV.9 Partition coefficients of amino acids in a biphasic system containing 1-butanol and water. Markers are experimental data from Gude et al. (1996b), curves are calculated with eq. 15.

In thermodynamically ideal systems, the logarithmic term containing the excess Henry's constants vanishes, and the following simple result is obtained:

$$\ln K_s^{t/b} = \sum_j \left(x_j'^t - x_j'^b \right) \ln x_{sj}^{sat} \tag{16}$$

In case of extraction systems with aqueous-mixed organic solvent systems, the solvents are usually selected such that the solubility of the solvents in the aqueous phase is small. Therefore, it can be assumed that the composition of the aqueous phase remains constant and the partition coefficient of the solute as a function of the solvent composition can roughly be estimated as follows:

$$\ln K_s^{m/a} = \sum_j x_j'^m \ln K_{sj,ideal}^{j/a} \quad \text{with } K_{s,ideal}^{j/a} = \frac{x_{sj}^{sat}}{x_{sw}^{sat}} \tag{17}$$

where K_{sj} is the partition coefficient of the solute in the water - single solvent j system and x_{sw}^{sat} is the solubility of solute s in water. Hence, a first estimate of partition coefficients in partially miscible solvents can be obtained already when only the solubilities in each of the solvent components are known. When the molar volumes of the solvents are not similar, incorporation of the term from eq. 15 that contains the excess Henry's constants, may be necessary to improve the overall description. As a matter of fact, a similar (combinatorial) correction for effects of unequal size of solvent molecules is often used in practice. Then, the solute free mole fractions x in equation 17 are replaced by the corresponding solute free *volume fractions* ϕ, and the mole fraction based partition coefficients are replaced by partition coefficients on a molar basis. The resulting relation reads as follows:

$$\ln K_s^{(c)m/a} = \sum_j \phi_j'^m \ln K_{sj}^{(c)j/a} \qquad (18)$$

Although these corrections seem equivalent, eq. 17 and 18 are not identical. Both eq. 17 and 18 are simple but powerful tools for optimization of multi-component organic solvent blends such as used for the extraction or liquid-liquid chromatography of biomolecules. Successful examples of using eq. 18 are given by Van Buel *et al.* (1997) and Van Halsema *et al.* (1998).

2.4 Relating different extraction systems

Partitioning data in 1-octanol/water ('Log P') are generally available, and often required for the toxicological or pharmaceutical characterization of the bioproduct. An overview of the data available until 1977 was given by Rekker (1977) who considered partition coefficients of numerous diluted solutes in aqueous-organic solvent mixtures. Collander (1951) was the first to investigate the relation between molar partition coefficients P_s in octanol/water of species *s* and those in iso-butanol/water systems. This resulted in a relation of the following general form:

$$\log P_i^{b/w} = a \log P_i^{o/w} + b \qquad (19)$$

when the *'solute* (and solvent) *sets are not too dissimilar'* (Rekker, 1977). This relation is of a general value and of tremendous use. It can be derived from eq. 17 (or eq 18) by subtracting eq. 17 (or 18) for any solvent system *b* (e.g. containing iso-butanol) from this equation for any solvent system o (e.g. containing 1-octanol). Appropriate corrections should be made for differences in the molar volumes of the solvents. Experimental values of *a* and *b* are approximately constant for solutes (and solvent) in a similar hydrogen bonding class (donor, acceptor, non-bonding). When solutes from different hydrogen bonding classes are included in the correlation, the standard deviation increases drastically (Grant and Higuchi, 1990). An example of the validity of the Collander relation is presented by Tsuji *et al.* (1977) for the partition coefficients of 11 ß-lactam antibiotics in octanol/water and iso-butanol/water mixtures. Their data, which are given in Figure IV.10, follow the Collander equation within experimental error.

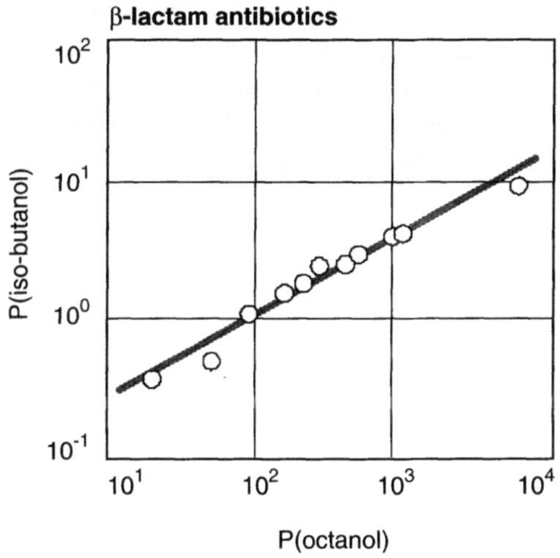

Fig. IV.10 $P^{o/w}$ of ß-lactam antibiotics in octanol/ water and iso-butanol/water. Curve is calculated using the Collander equation (eq. 19). Data: Tsuji et al. (1977).

2.5 Generalized Polarity Scales

Following the correlative framework, generalization towards other phase systems of use for separation processes can be given. An example relevant in the context of SSA is the ion exchange of antibiotics anions (Van der Wielen et al., 1995) or amino acid cations (Dye et al., 1990). On the basis of binary parameters (i.e. ion exchange selectivities), the uptake in multicomponent systems could be predicted fairly accurately. Binary selectivities at infinite dilution of the target species could also be correlated to their hydrophobicity. It is evident that for situations with weak electrolytes and a variation in pH, the behavior of the various ionic forms should be taken explicitly into account.

The tendencies in the phase behavior of biomolecules, which have a broad variety of interacting polar, nonpolar, H-bonding and charged groups can be scaled among like molecules. This has resulted in 'hydrophobicity'-scales, which can be used to obtain at least qualitative information, and in some cases also semi-quantitative information about their **relative** phase behavior. Because the parameter indicating the extent of 'hydrophobicity' increases regularly in a homologous series of solutes, such as amino acids, it is tempting to compare hydrophobicity scales from different origin quantitatively. Here, we have used, arbitrarily, the experimental log P values of glycine and phenylalanine given by Rekker (1977) and obtained from shake flask experiments, to establish a relative hydrophobicity Δg_x^{rel} of species *x* in the same system, which is calculated as follows:

$$\Delta g_x^{rel} = \frac{\Delta G_x^t - \Delta G_{Gly}^t}{\Delta G_{Phe}^t - \Delta G_{Gly}^t} \qquad (20)$$

where ΔG_x^t is a measure for the rank of species x on a particular hydrophobicity scale, which can be C_{12S}, ln H_{iS}, log P or log S. Some known hydrophobicity scales of amino acids in very different solvent (and sorbent) systems have been recalculated in terms of the relative hydrophobicities of the amino acids and plotted in Figure IV.11 versus the hydrophobicity scale based on octanol-water partitioning.

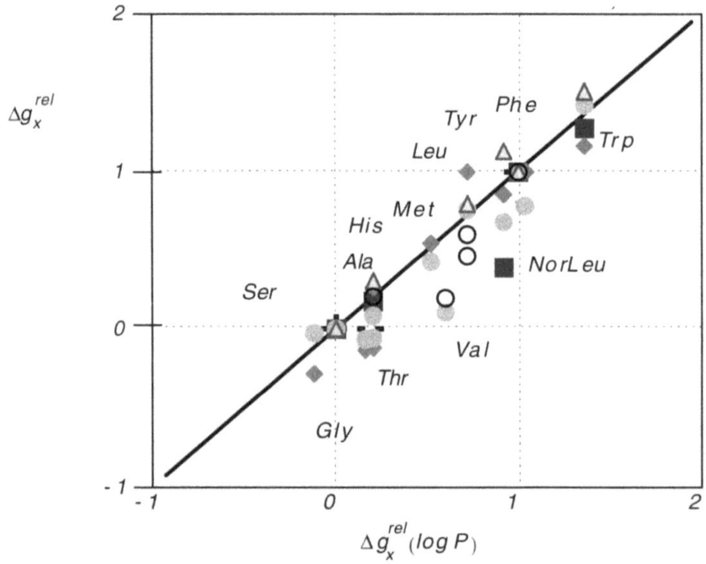

Fig. IV.11 Relative hydrophobicities of amino acids versus the relative hydrophobicity in 1-octanol-water systems. Reversed-phase HPLC (● Sasagawa et al, 1990), Aqueous two-phase systems (▲ Eiteman and Gainer, 1991-2), cation exchange resins (○ Dye *et al*, 1990), CPC (■ El-Tayar et al, 1993) and calculated with Rekker's method (◆).

We would like to stress that the relatively simple model describes solubilities and partition coefficients of solutes in single component as well as in multi-component, partially miscible solvent mixtures. The model requires the solubilities of a single solute in each of the pure solvents only and one single parameter to account for the excess effects. This parameter can be correlated to polarity scales, and thereby to data that are generated in completely different systems. It has been demonstrated elsewhere (Van der Wielen and Rudolph, 1999; Rudolph et al., 1999, 2001) that this approach can be straightforwardly extended to more complex solvent systems, sorption and ion exchange systems.

2.6 Notation

$g^{(E)}$	molar (excess) Gibbs energy	J mol^{-1}
Δg_x^{rel}	relative hydrophobicity	
G	Gibbs energy	J
H	Henry's law constant	Pa kmol^{-1}
$K_i^{2/1}$	partition coefficient of solute i over phase 2 and 1	
R	gas constant, 8.3144	J mol^{-1} K^{-1}
T	temperature	K
x,y	mole fraction	
Greek		
γ	activity coefficient	
μ	chemical potential	J mol-1
superscripts, subscripts		
*	asymmetric convention	
m	mixture	
∞	infinite dilution	
s	solute	
sat	saturated	
'	solute free	

§3 Thermodynamically controlled synthesis of Amoxicillin

3.1 Background

The most frequently applied strategy for enzymatic synthesis is a kinetically controlled reaction, in which an activated acyl donor is coupled to an antibiotic nucleus. High yields on nucleus are achievable, but part of the activated acyl donor hydrolyzes and is not coupled to the nucleus, so recycling and reactivation of the acyl donor is necessary, leading to costly extra process steps. In addition, hydrolysis of the product may occur (Fernandez-Lafuente et al., 1996). A simpler strategy for enzymatic synthesis would be a direct condensation of antibiotic nuclei and unactivated acyl donors in a thermodynamically controlled reaction, which avoids these recycles and reactivation. This strategy can only be applied successfully if conditions for a favorable equilibrium position can be found. The enzyme does not influence the position of the equilibrium, but only the rate at which the equilibrium is established. In water, however, hydrolysis of the antibiotic may be favored to synthesis, because of unfavorable thermodynamics. To shift the reaction equilibrium towards synthesis one may add moderate to high concentrations of water miscible organic solvents (Kim and Lee, 1996; Tewari et al., 1995) and/or increase the substrate concentration up to the solubility level or even higher, which gives a suspension of substrate. In favorable cases the concentration of the product that is formed exceeds its solubility. Then a suspension of solid substrates can be

converted into a suspension of solid products. This situation, to which we refer as the 'suspension-to-suspension' process, has been achieved for several enzymatic reactions (Kasche, 1986; Halling et al., 1995).
The purpose of this section is to describe the thermodynamics of the synthesis of Amoxicillin in order to study the feasibility of the 'suspension-to-suspension' concept. The thermodynamic condition for a 'suspension-to-suspension' conversion to occur is that the ratio of dissolved concentrations (i.e. the apparent equilibrium constant, K_{app}) is larger than the ratio of solubilities (R_S):

$$K_{app} = \frac{c_{Amox}}{c_{APA}\, c_{HPG}} > \frac{S_{Amox}}{S_{APA}\, S_{HPG}} = R_S \tag{21}$$

where c_i is the actual concentration of species i and S_i its solubility.
Using measured values of K_{app} in water, $K_{app,aq}$ was calculated as a function of pH (see Figure IV.12., model line)(Diender et al., 1998a). The measured values support the model, because experimental $K_{app,aq}$-values decrease with increasing pH according to the prediction. However, the $K_{app,aq}$-values observed by us were an order of magnitude smaller than the values for Ampicillin hydrolysis reported by Svedas et al. (1980) and Blinkovsky and Markaryan (1993). They gave values of $K_{app,aq}$ of 158 $mol^{-1} \bullet l$ at pH=4.5, 78 $mol^{-1} \bullet l$ at pH 5.0 (Svedas et al., 1980) and 5.5 $mol^{-1} \bullet l$ at pH 5.5 (Blinkovsky and Markaryan, 1993), all at 25°C. Ampicillin is very similar in structure to Amoxicillin. We had expected the apparent equilibrium constant of Amoxicillin and Ampicillin to be approximately equal. The low value of K_{app} observed for Amoxicillin has important implications for the feasibility of the 'suspension-to-suspension' reaction, as will be shown below.

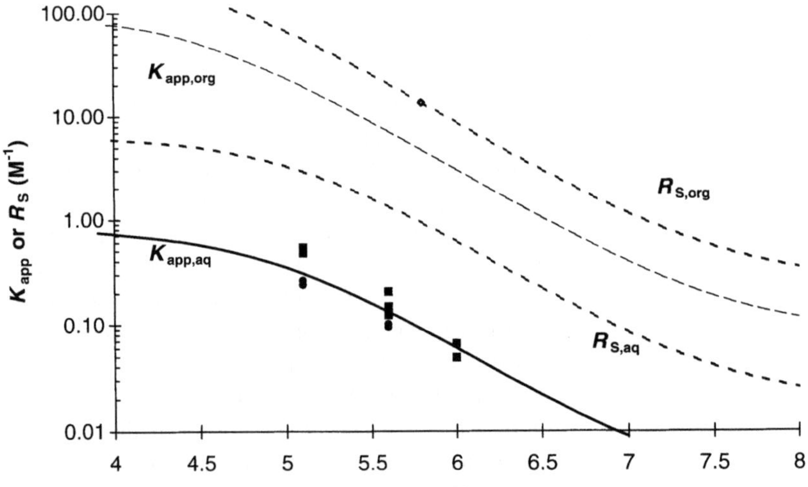

Fig. IV.12 Apparent equilibrium constants for the synthesis of Amoxicillin in water and a 50% (w/w) DMF mixture as a function of pH at 25°C. Experimental values for synthesis (dots) and hydrolysis (squares), and predicted values of $K_{app,aq}$ and $K_{app,org} \bullet 10^{-2}$ in 50% (w/w) DMF mixture. K_{app}=[Amox]/[APA][HPG]. Calculated ratio of solubilities $R_{S,aq}$, $R_{S,org}$. Experimental value of $R_{S,org}$.

3.2 Feasibility of a thermodynamically controlled suspension-to-suspension process

In order to study the feasibility of a 'suspension-to-suspension' conversion of Amoxicillin in water the solubilities in water of the reactants and product were determined as a function of pH. A study of the feasibility of the 'suspension-to-suspension' conversion of Amoxicillin based on the equilibrium constant reported for Ampicillin (Blinkovsky and Markaryan, 1993) and these solubilities may suggest that 'suspension-to-suspension' conversion could be successful. However, our experimental equilibrium concentrations of Amoxicillin in water are an order of magnitude lower than its solubility, so the 'suspension-to-suspension' conversion should not occur. Also, preliminary results by others indeed showed no 'suspension-to-suspension' conversion in water (L.M. van Langen and R. Sheldon, personal communication). This conclusion seems valid in the whole pH range from 4 to 8, because the calculated value of $R_{S,aq}$ is always larger than $K_{app,aq}$ (Fig. IV.12). Schroën et al. (1999) describe similar results for Cephalexin.

3.3 Cosolvent addition

One may add moderate to high concentrations of water-miscible organic solvents to shift the reaction equilibrium towards synthesis. The addition of organic cosolvents lowers the water activity, and stabilizes the neutral forms of dissociating groups of the substrates (mainly carboxylic acid groups).
It is possible to predict the synthetic yield of a reaction simply by measuring the equilibrium in water and the apparent pK_a values of the reactants in the cosolvent-water mixtures with the model of Diender et al. (1998b). This model has been used to predict the apparent reaction equilibrium constant for the synthesis of Amox in water and in a 50% (w/w) DMF mixture. We have chosen the following reference reaction (with the same functional charged groups in both reactants as in the product):

$$\mathrm{APA^- + HPG^+ \rightleftharpoons Amox^{+-} + H_2O}$$

With the results of equilibrium measurements of Amox in water and the apparent pK_a values of Amox, APA and HPG the apparent equilibrium constant for the synthesis of Amox in cosolvent-water mixtures can be predicted. Because Fernandez-Lafuente et al. (1991b) achieved the highest yields for Pen G synthesis in a 50% (w/w) DMF/water mixture, this solvent mixture was selected to predict the equilibrium of Amox synthesis.
As mentioned before, the equilibrium of Amox in water is heavily in favor of hydrolysis. However, when synthesis is performed in a 50% (w/w) DMF/water mixture, K_{app} is calculated to increase nearly a factor 100 in the DMF/water mixture compared to water. To check this prediction, an enzyme preparation that would be active and stable in the DMF/water mixture would have to be available. However, it should be noticed that the actual equilibrium concentrations of Amox that might be reached in the DMF/water mixture probably are lower than in water because the

solubilities of the reactants in the DMF/water mixture are much lower than in water, and $K_{app,org} < R_{S,org}$, as shown in Figure IV.12 (Diender *et al.*, 1998b).
Thus a 'suspension-to-suspension' conversion of Amoxicillin is expected not to be feasible in a water-cosolvent mixture, but this can be tested only if an enzyme is available that is and remains sufficiently active under such conditions.

§4 Kinetically controlled enzymatic coupling of APA and HPGM

4.1 Kinetically controlled suspension-to-suspension reactions

For enzymatic reactions, the focus is usually on media consisting only of aqueous or organic solvent, or monophasic or biphasic mixtures thereof. As mentioned before, recently another type of enzymatic reactions, involving solid substrate and/or product, is getting more attention. In the so-called suspension-to-suspension (or solid-to-solid, or solid phase) reactions the substrate(s) and product(s) are mainly present as undissolved particles, whereas the enzymatic reaction takes place in the liquid phase. The product to reactor volume ratio is much higher than in the conventional systems. This type of processes can also lead to yields similar to those in conventional enzymatic reactions. Suspension-to-suspension conversions are especially advantageous when hydrolytic reactions are to be reversed or suppressed (Erbeldinger et al., 1998; Halling et al., 1995; Straathof et al., in press; Lee and Kim, 1998; Kasche, 1986; Kasche and Galunsky, 1994).
During a suspension-to-suspension process different subprocesses take place, e.g. dissolution and crystallization of the substrate(s) and product(s), enzymatic synthesis of the product(s), and enzymatic hydrolysis of the substrate(s) and product(s). The enzymatic reaction will be influenced by pH and, for reactants that show acid-base dissociation in the pH range of interest, solubility and hence dissolution and crystallization will be influenced. Then the subprocesses themselves will produce or consume protons and hence influence the pH. A thorough understanding of these different subprocesses is necessary for optimizing the suspension-to-suspension process.
Wandrey and Flaschel (1979) calculated the pH in homogeneous aqueous solutions as a function of conversion for the hydrolysis of N-acetylmethionine. Later, Flaschel et al. (1992), Nakanishi and Matsuno (1986) and Nakanishi et al. (1986) predicted the pH profile in a water organic-solvent two-phase system. Different groups have investigated the kinetics of a suspension-to-suspension reaction (Lee and Kim, 1998; Erbeldinger et al., 1999). However, these studies do not allow a prediction of the extent of pH control that is required in an industrial suspension-to-suspension process.

In this section, we show predictions of the pH and concentrations of the reactants during the kinetically controlled synthesis of the semisynthetic antibiotic Amoxicillin without pH control. The kinetically controlled synthesis of Amoxicillin (Amox) from D-*p*-hydroxyphenylglycine methyl ester (HPGM) and 6-aminopenicillanic acid (APA) was chosen as a model system. The enzyme penicillin acylase catalyzes the synthesis (reaction I), by coupling HPGM and APA, but also the undesired substrate hydrolysis (of HPGM, reaction II) and product hydrolysis (of Amox, reaction III). Both side reactions lead to hydroxyphenylglycine (HPG).

I Synthesis: $\text{APA} + \text{HPGM} \longrightarrow \text{Amox} + \text{MeOH}$

II Substrate hydrolysis: $\text{HPGM} + \text{H}_2\text{O} \longrightarrow \text{HPG} + \text{MeOH}$

III Product hydrolysis: $\text{Amox} + \text{H}_2\text{O} \longrightarrow \text{HPG} + \text{APA}$

In a batch reactor, both substrates may initially be mostly undissolved, whereas most of the Amoxicillin will crystallize during its production (also, the byproduct HPG might crystallize during reaction) (Bruggink et al., 1998; Clausen and Dekker, 1996). The quantitative model that is used is based on mass and charge balances, pH-dependent solubilities of the substrates and products and finally enzyme kinetics.

4.2 Simulations

Using the results from an experiment starting with 0.1 M APA and 0.1 M HPGM, the ratio between the mass-action based rate constants k_{synth} and k_{subshydr} was found to have a value of 21.4. The predicted concentrations and pH visually gave the best results to match the experimental values when $k_{\text{synth}}/k_{\text{prodhydr}}$ was also 21.4.

In the model changes in solubility are supposed to be instantaneous. However, in the course of the reaction there could be a rate limitation in the crystallization or the dissolution of the crystals, so the actual concentration in the liquid phase may be higher or lower than calculated.

An insight in the subprocesses taking place during suspension-to-suspension conversions is obtained by discussing in detail the simulation of an experiment starting with 0.2 M APA and 0.4 M HPGM. In the experiment, the pH was increasing gradually from the start and then suddenly decreasing sharply. Analyzing only the overall concentrations, we would not be able to explain the pH profile. However, when the total amounts of the substrates are split into solid and dissolved amounts by performing a simulation with the model (Figure IV.13, for data of the substrates and IV.14, for data of the products), the causes of the pH changes become clear.

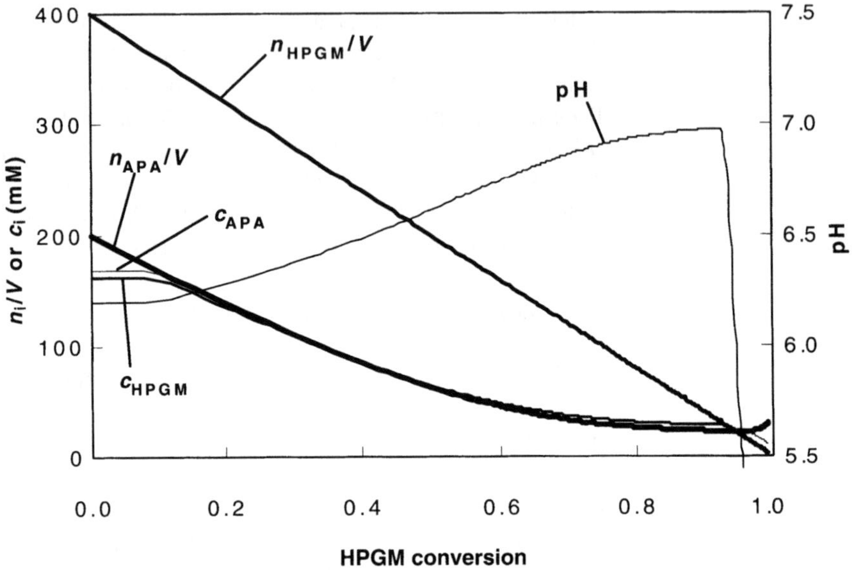

Fig. IV.13 Dissolved concentrations and total amounts per reactor volume of APA and HPGM and pH calculated as a function of HPGM conversion. Start conditions 0.2 M APA and 0.4 M HPGM.

In the initial stage of the experiment, the pH is constant. In this part both APA and HPGM are saturating the liquid phase. In the crystals APA and HPGM are present as APA^{+-} and $HPGM^{0}$ respectively. When the predominant species are shown the initial enzymatic reactions become:

Synthesis: $$HPGM^{+} + APA^{-} \longrightarrow Amox^{+-} + MeOH$$

Substrate hydrolysis $$HPGM^{+} + H_2O \longrightarrow HPG^{+-} + MeOH + H^{+}$$

During the dissolution of HPGM protons are consumed (in the liquid phase HPGM is present as cation), which are mostly provided by the dissolution of APA (in the liquid phase APA is predominantly present in the anionic form) while the rest is provided by HPGM hydrolysis. Hence the pH is constant. At the end of this stage, the pH starts to increase, because n_{APA}/V_{aq} drops below its solubility and becomes equal to c_{APA}. There is not sufficient proton production anymore, as the solid phase of APA is depleted, but the proton consumption continues as HPGM is still dissolving. Gradually product hydrolysis becomes more important. Like substrate hydrolysis, this leads to the production of protons. When the ratio between synthesis and hydrolysis decreases during the reaction, the pH increase declines. At the end of the second stage, the total amount of HPGM has decreased down to its solubility (n_{HPGM}/V_{aq} becomes equal to c_{HPGM}). At that moment, the pH starts to decrease sharply. In this third stage of the experiment almost no HPGM is used for Amox synthesis and most is hydrolyzed to HPG and MeOH, causing proton production.

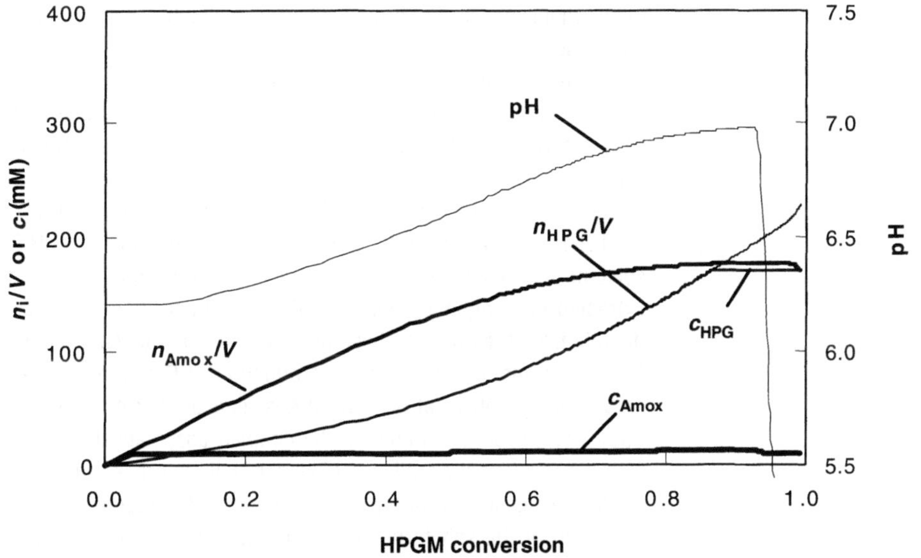

Figure IV.14 Dissolved concentrations and total amounts per reactor volume of Amox and HPG and pH calculated as a function of HPGM conversion. Start conditions 0.2 M APA and 0.4 M HPGM.

As this case shows, the model can explain pH shifts during the suspension-to-suspension reaction. The discontinuities in pH changes during suspension-to-suspension reactions are caused by the depletion of the solid phase of one of the substrates. Similar discontinuities will be found when a product starts to crystallize, if the predominant dissolved product species is an anion or cation. The model can be used to find the optimal conditions to produce Amox. For example, when the enzyme stability or activity is low in a certain pH range the model can predict whether or when the pH will be in that range and pH control is necessary. In this way no unnecessary buffers, acids or bases are used for pH control, which can simplify downstream processing. The model can also predict when to stop the reaction to achieve the highest yield of product.

§5 Towards shortcut routes from penicillin G to Amoxicillin

5.1 Extractive enzymatic hydrolysis of penicillin G

Enzymatic extractive catalysis, i.e. the coupling of an enzymatic reaction and an extraction of the product(s) in a biphasic system has been applied and described by many researchers (e.g. Martinek et al., 1981[a,b]; Andersson et al., 1984; Eggers et al., 1989; Woodley and Lilly, 1990; Janssen et al., 1995; Hernandez-Justiz et al., 1998). One major advantage of extractive catalysis is that the equilibrium of the reaction can be shifted towards the product side by extracting the product from an aqueous to an organic phase. The organic phase can also be a reservoir for substrates with a low solubility in water. Substrate or product inhibition can be avoided if the inhibitor prefers the organic phase to the aqueous phase.

In this section extractive enzymatic catalysis for the hydrolysis of penicillin G (Pen G) is described. This can be combined with the synthesis of β-lactam antibiotics, as will be discussed later.

In the industrial process (Figure IV.2), after fermentation at pH 6.2-6.8 (Hersbach et al., 1984), Pen G is extracted from the fermentation broth at pH 2.0-2.5 by organic solvents like butyl acetate or amyl acetate. Subsequently, Pen G is either back-extracted to an aqueous phase at pH 6.8-8.0 and then crystallized, or crystallized directly from the organic phase by adding solid potassium or sodium acetate. The Pen G crystals are used for the production of 6-aminopenicillanic acid (APA) by enzymatic hydrolysis at pH 7-8. APA is recovered at its isoelectric point (Mwangi, 1994), dissolved again and can be coupled enzymatically to a side chain base to form a semisynthetic antibiotic (Diender et al., in press). In each process step salts are formed as side products due to the pH shifts (Reschke and Schugerl, 1984[a]). A reduction of the salt formation would be achieved if the extracted Pen G in the organic phase could directly be added to the enzymatic reactor and could be hydrolyzed efficiently in a two-phase system by extractive catalysis. In such a case, Pen G, initially only present in the organic phase, partitions between both phases depending on the pH and is hydrolyzed in the aqueous phase. The side chain that is formed by hydrolysis, phenylacetic acid (PAA), also partitions between the aqueous and organic phase, while the second hydrolysis product, APA, having a zwitterionic nature, stays in the aqueous phase. By extracting PAA, the equilibrium position is shifted towards the product side and good yields can be achieved. This was shown by Barenschee et al. (1992) and Rindfleisch et al. (1997). They investigated several possibilities for integrated processes for the production and biotransformation of Pen G and in this way reduced the number of process steps. Further improvement of the Pen G conversion in those biphasic systems can be achieved by choosing the process conditions such that the equilibrium concentration of APA exceeds its solubility, causing APA to crystallize. This crystallization of APA could improve the yield considerably.

To adjust the pH from 2.0-2.5 of the extraction liquid to the pH at which the enzymatic hydrolysis is performed, a base (e.g. ammonia or potassium hydroxide) should be added. To avoid salt production because of this pH shift, the side chain base that will be coupled to APA in a successive reaction (see previous section) can be added instead. For example, hydroxyphenylglycine methyl ester (HPGM), the side chain for Amoxicillin formation, which would be added to APA after hydrolysis of Pen G, may also be added before hydrolysis.

To improve the existing processes in this manner, we studied the enzymatic hydrolysis of Pen G in water/butyl acetate two-phase systems. This could lead to a shortcut process as shown in Figure IV.6. For the process development a model is required that quantitatively describes the equilibrium position of Pen G hydrolysis in water/butyl acetate two-phase batch systems as a function of the pH, cation presence, initial Pen G concentration and organic/aqueous phase volume ratio.

5.2 Modeling the extractive catalysis

Figure IV.15 shows a schematic picture of the phase equilibria during the extractive hydrolysis of penicillin G in a water/butyl acetate biphasic system.

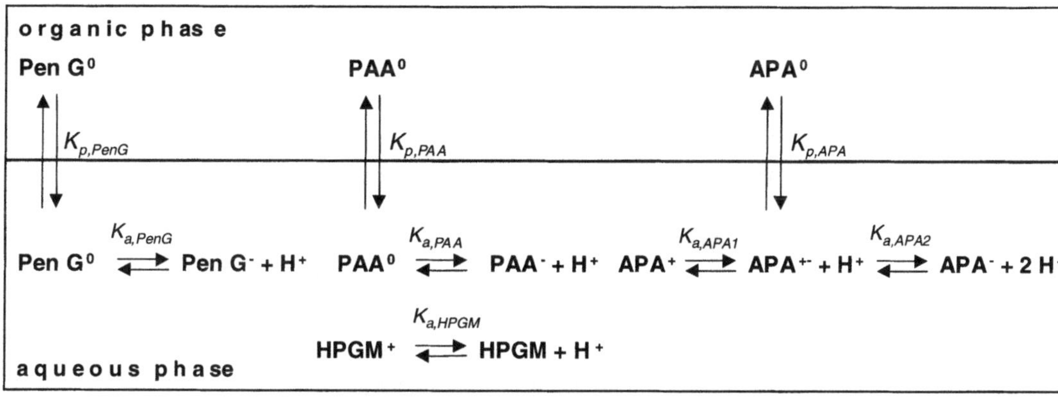

Fig. IV.15 Schematic overview of dissociation and partition equilibria in aqueous/organic two-phase systems.

The following equilibrium constants are required in an equilibrium model:

- partition coefficients for the neutral or zwitterionic species
- dissociation constants
- reaction equilibrium constant

In addition, crystallization of APA can occur if its concentration in the aqueous phase exceeds its equilibrium concentration. This crystallization phenomenon is beneficial for the yield as the enzymatic hydrolysis reaction is shifted towards the product side. The solubility of APA can be described as a function of pH with the following equation:

$$S_{\mathrm{APA}} = \frac{S_{\mathrm{APA}^{+-}}}{F_{\mathrm{APA}^{+-}}} \tag{22}$$

in which $S_{\mathrm{APA}^{+-}}$ is the solubility of the zwitterionic species of APA (0.008 mol•l^{-1}) and $F_{\mathrm{APA}^{+-}}$ the fraction of APA in the zwitterionic form. The equilibrium model and its parameters are described in detail by Diender et al. (2001a).

5.3 Experimental conversions and model predictions

Figure IV.16 shows an example of the course of pH and the concentrations of the reactants in both phases during a hydrolysis experiment. During some hydrolysis experiments (especially the experiments at low pH) chemical degradation of Pen G and APA occurred. The amounts of enzyme were maximized in order to minimize the chemical degradation in the experiments at low pH. In the experiment shown in Figure IV.16, the pH is decreasing from 7.0 at the start of the experiment to 5.4 at equilibrium.

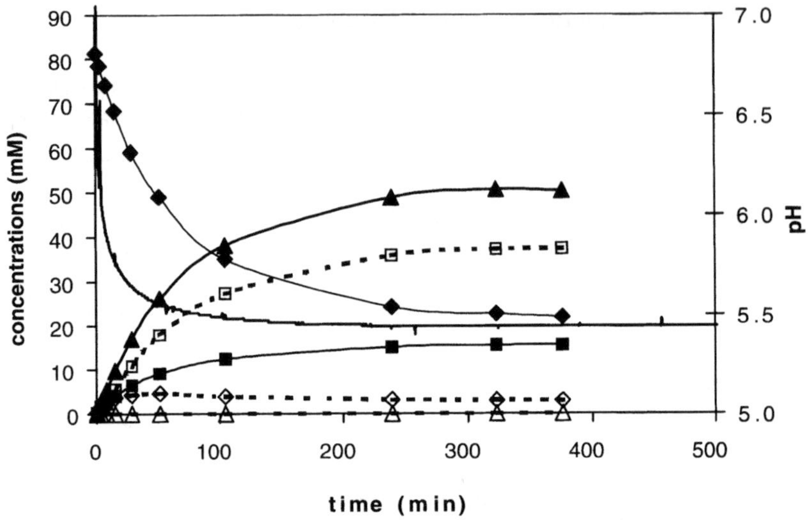

Fig. IV.16 Concentration profile of reactants and pH profile during hydrolysis of Pen G in a 1:1 (w/w) water/butyl acetate two-phase batch system at 25°C (experiment 1, Table IV.I). Symbols: closed markers, concentrations in aqueous phase; open markers, concentrations in organic phase; ◇: Pen G; Δ: APA; □: PAA; solid line, no marker: pH.

Protons are produced as Pen G is hydrolyzed. This is only partly compensated by extraction of PAA to the organic phase, and a steep decrease of pH occurs at the start of the experiment. The enzymatic reaction is finally reaching its thermodynamic equilibrium causing the pH to become constant. As shown by the results, equilibrium was reached during the time of the experiment. The enzyme was still active at low pH and not denatured.

Table IV.1 Initial and equilibrium conditions of Pen G hydrolysis experiments in two-phase water/butyl acetate batch system at 25°C.

Exp.	counter ion	$c^{bi}_{counter\ ion}$ (mM)	V_{org}/V_{aq}	$c^{bi}_{PenG^0}$ (mM)	pH initial	pH equilibrium	yield exper. (%)	yield pred. (%)
1	K^+	43	1	41	7.0	5.4	64	71
2	K^+	46	1.1	44	7.4	5.5	69	72
3	K^+	37	1.1	39	5.4	5.2	80	80
4	K^+	36	1.2	38	5.2	5.2	79	81
5	K^+	19	1.2	42	4.4	4.5	86	92
6	K^+	0	1.1	41	2.9	3.0	94	96
7	K^+	73	1.2	67	7.0	5.5	69	73
8	K^+	36	1.2	82	4.4	4.6	86	91
9	$HPGM^+$	22	1.2	54	4.0	4.4	93	93
10	$HPGM^+$	12	1.2	44	4.0	4.0	88	95

In some experiments at low pH (experiments 5, 6, 8, 9 and 10, see Table IV.1) APA crystallized during the experiment. This was observed by the appearance of crystals during the experiment. Also, the aqueous APA concentrations showed initially a maximum, decreased again and became constant at the solubility concentration. Apparently, at those pH values supersaturation of APA occurs.
Table IV.1 presents the measured and predicted yield (based on PAA) of each experiment. In general, the measured yields are a bit lower than the predicted yields. An explanation for this could be that the solubility and pK_a's used in the calculations (Diender et al., 2001[a]) have been measured in the absence of organic solvent and at a different ionic strength than in the present experiments. Another explanation could be that equilibrium was not reached during the experiment. This could be due to degradation of Pen G or APA or due to the crystallization of APA. In experiment 10, for example, the measured aqueous concentration of APA at which we assumed equilibrium was still higher than the solubility at that pH. This indicates that equilibrium was not reached yet. However, the model predicts trends in the yield as a function of pH and the initial Pen G concentration quite well.

In purely aqueous systems the yield is strongly dependent on pH, as reported before (e.g. Tewari and Goldberg, 1988). At pH 8 almost 99% conversion of 0.1 mol•l^{-1} Pen G can be achieved, while at pH 4.5 only 33% of Pen G is converted. Due to entropic effects, higher yields can be achieved in more dilute solutions (Martinek and Semenov, 1981[a]). At pH 8 in water/BuAc biphasic mixtures partitioning is negligible. Then the maximal yield achievable is lower than in aqueous systems (at similar initial amounts of Pen G per liquid volume). This is caused by the relatively high aqueous concentrations of PAA and APA in the biphasic systems, due to the smaller volume of the aqueous phase. Below pH 6, the produced PAA is extracted into the organic phase, shifting the equilibrium to the product side. Higher yields are achieved in the two-phase systems than in pure aqueous systems in that pH range. At lower pH values (depending on the initial aqueous Pen G concentration), APA crystallization is taking place, which improves the yield considerably for all systems. For example, at pH 5, yield is calculated to increase from 40 to 90% by APA crystallization in a 1:10 (w/w) water/BuAc system. The pH at which APA starts to crystallize is higher as the volume ratio of organic to aqueous phase is increasing. This is due to the high concentration of APA in the small remaining volume of aqueous phase. Figure IV.17 shows the influence of the initial Pen G concentration on the yield in a 1:1 (w/w) water/BuAc biphasic system (Diender et al., 2001[a]).

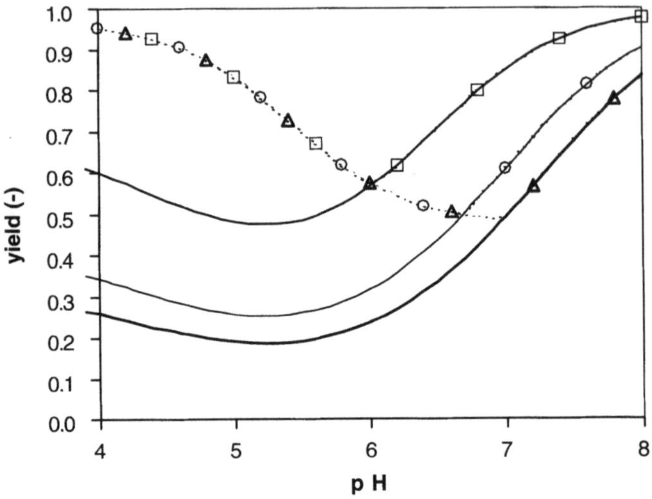

Fig. IV.17 Calculated yields as a function of pH for the enzymatic hydrolysis Pen G/ HPGM salt for different initial Pen G concentrations in 1:1 (w/w) water/BuAc at 25°C. Full lines: no APA crystallization; Dotted line: with APA crystallization; □ 0.1 mol Pen G • l^{-1} mixture; o 0.5 mol Pen G • l^{-1} mixture; Δ 1.0 mol Pen G • l^{-1} mixture.

Clearly, in more dilute solutions higher yields can be achieved, due to the entropic effects as mentioned before. However, due to the crystallization of APA high conversions of Pen G can be achieved even at high initial concentrations. This favorable situation allows the further development of processes for the conversion of Pen G into semisynthetic antibiotics with reduced salt formation as mentioned in the Introduction.

5.4 Combining extractive hydrolysis of penicillin G with kinetically controlled synthesis of Amoxicillin

One major problem in the shortcut route to Amoxicillin is that one of the products from the Pen G hydrolysis reaction, phenylacetic acid (PAA), essentially inhibits the catalytic activity of penicillin acylase for the Amox synthesis (Svedas et al., 1980). Therefore, the aqueous APA suspension that is obtained in the aforementioned extractive reaction should be converted into Amox using a different enzyme for the second reaction. A potentially useful enzyme is α-amino acid ester hydrolase (AEH) from *Acetobacter pasteurianus* ATCC 9325. This enzyme is known to be able to catalyze the synthesis reaction in the presence of PAA (Nara et al., 1977). However, the synthesis/hydrolysis ratio (S/H, the molar ratio of produced Amox and HPG, the

hydrolyzed side-chain donor), seemed to be negatively influenced by the presence of PAA (unpublished results). The extent of this influence is not known. This means that the shortcut route using AEH requires removal of PAA to an unknown level. The removal does not have to be absolute, as is required for any route using penicillin acylase for the Amox synthesis.
Therefore the kinetics of Amox synthesis catalyzed by AEH from *A. pasteurianus* ATCC 9325 should be known. Kato et al. (1980) and Ryu and Ryu (1987) have studied AEH from *X. citri*, which closely resembles AEH from *A. pasteurianus* ATCC 9325 with respect to molecular mass, the ratio of specific enzyme activities for Cephalexin synthesis and hydrolysis and the pH optima for these reactions. They found that AEH is influenced by the ionic strength of the reaction mixture. Higher synthetic yields could be obtained by decreasing the ionic strength.
We synthesized Amox from APA and HPGM with AEH from *A. pasteurianus* ATCC 9325. The influence of the ionic strength, due to the presence of reactants, on the performance of AEH was studied. From initial rates mass action-based rate constants were calculated. The rate constant for the synthesis reaction, k_{synth}, seemed to be affected by the ionic strength, whereas the rate constant for HPGM hydrolysis, $k_{subhydr}$, was not significantly influenced.

The influence of other molecules than the substrate and product, e.g. inorganic salts for pH adjustment or contaminants such as Pen G and PAA, on the performance of AEH was studied as well. The influence of Pen G and PAA was similar and correlated to the ionic strength. The concentrations of these compounds should be kept as low as possible (< 5 mM). The influence of the studied inorganic salts (potassium phosphate, NaCl and $(NH_4)_2SO_4$) gave more scattered results. A quantitative kinetic model was based on enzymatic conversion rates (described as a function of ionic strength), pH-dependent solubilities and stoïchiometric, macroscopic and charge balances (Diender et al., 2001b). The model predicted the concentrations during the course of the experiments in the absence and presence of PAA and Pen G reasonably well. At low ionic strength higher yields of Amox will be obtained, while at high ionic strength the unwanted hydrolysis of HPGM is favored over synthesis. Therefore, the synthesis reaction of Amox should be done preferably in the absence of PAA and Pen G and by feeding HPGM instead of an inorganic base for pH control. Figures IV.19 and IV.20 show the difference between feeding with HPGM and feeding with KOH. In the latter case the ionic strength during the reaction is much higher, although in both simulations the same total amount of HPGM was added (Diender et al., 2001b).

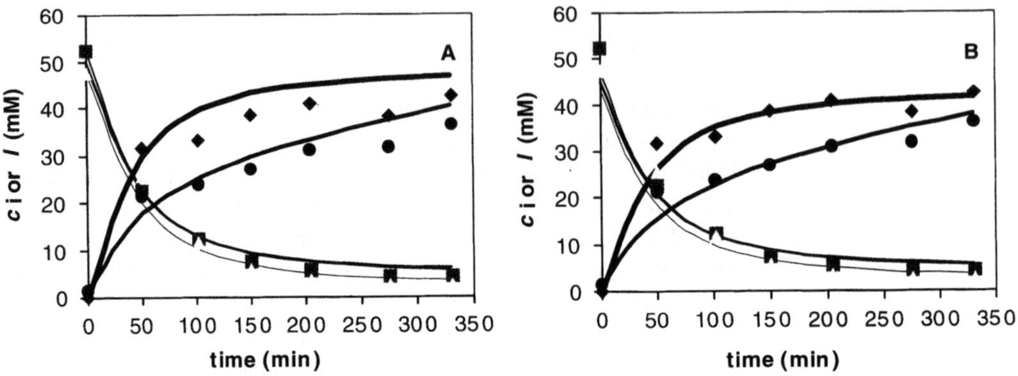

Figure IV.19 Course of full conversion reaction for the synthesis of Amox starting from 47 mM APA and 52 mM HPGM with AEH at pH 6.2 and 25°C using HPGM for titration. A. Original data. B. Predicted data corrected for a dilution error. Symbols: ◆ Amox; ● HPG; ▲ HPGM; ■ APA. Lines through the data points are model predictions.

Although the aforementioned extractive reaction leads to a suspension of APA with low levels of PAA and Pen G, these levels may be not low enough for the synthesis of Amox using AEH. However, a multifunctional bioreactor as discussed in a later section, may lead to the desired low levels of PAA and Pen G in an extractive reaction.

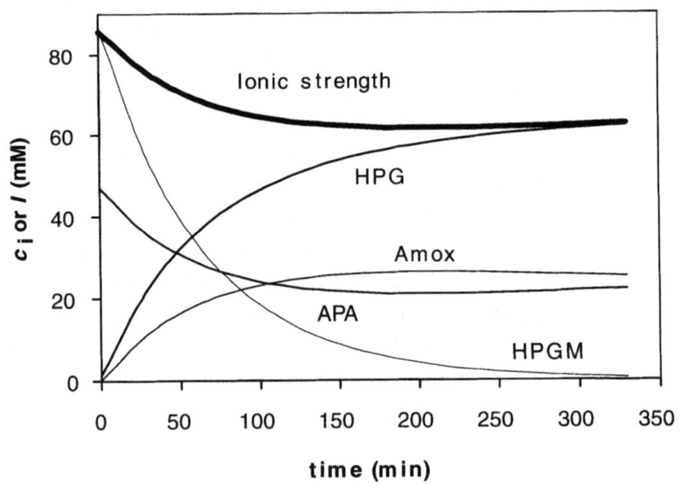

Figure IV.20 Prediction of course of full conversion reaction for the synthesis of Amox starting from 47 mM APA and 86 mM HPGM with AEH at pH 6.2 and 25°C using KOH for titration.

5.5 One-pot shortcut route in anhydrous organic solvent

If the synthesis reaction could be performed in the absence of water, the S/H ratio would be infinite, as there is no water available for substrate and product hydrolysis. A possible one-pot shortcut route from Pen G to Amox in anhydrous organic solvent may be based on the following reactions (Diender et al., 2001[c]):

$$\begin{array}{rcl} \text{Pen G} + NH_3 & \rightleftharpoons & \text{APA} + \text{PA-amide} \\ \text{APA} + \text{HPGA} & \rightleftharpoons & \text{Amox} + NH_3 \qquad + \\ \hline \text{Pen G} + \text{HPGA} & \rightleftharpoons & \text{Amox} + \text{PA-amide} \end{array}$$

A solution of Pen G acid is amidated with a catalytic amount of ammonia to form APA and phenylacetamide (PA-amide). Formed APA is coupled to hydroxyphenyl glycine amide (HPGA) in the synthesis reaction of Amox. (Alternatively, phenylglycine amide can be used in the synthesis of Ampicillin.) Because APA is used in the second reaction, the first reaction equilibrium is shifted to the product side and reaches a higher extent of conversion. In addition, if the solubility concentration of the products is exceeded, crystallization of those products might occur. This could also improve the yield considerably. In the second reaction, ammonia is formed which can be used again for the amidation reaction, making it possible to achieve a stoïchiometric conversion. As both reactions are performed in anhydrous organic solvent, no activated side chain is hydrolyzed. Recycling of the activated side chain is no longer required and an extra reduction in the number of process steps can be achieved.

There are two types of obstacles that might prohibit this one-pot shortcut route: thermodynamic reasons (the reaction equilibria are unfavorable) and kinetic reasons (the available enzymes are intrinsically not active and stable, or the solubility of the reactants is too low to obtain reasonable rates).

A thermodynamic cycle was designed to calculate the yield. This cycle included reaction equilibria in water, solubilities of the reactants in water and organic solvent, and partition coefficients of the reactants. The model predicted that a 99.2% yield of Pen G to Amox in anhydrous BuAc might be achieved, starting from 0.1 mol•kg-1 Pen G, 0.1 mol•kg^{-1} HPGA and 0.001 mol•kg^{-1} NH_3. For the synthesis reaction of Amox directly starting from 0.1 mol•kg^{-1} APA and 0.1 mol•kg^{-1} HPGA, a 100% yield was predicted. This synthesis reaction was tested experimentally, but no synthesis of Amox was observed. Probably this is caused by kinetic limitations of the enzyme under these conditions. Still, penicillin acylase cross-linked enzyme crystals showed activity in the hydrolysis of PA-amide in anhydrous BuAc. The one-pot shortcut route from Pen G in anhydrous BuAc is thermodynamically very attractive, but calculations show that enzymatic reaction rates in anhydrous butyl acetate are probably too slow due to low solubilities of the substrates (Diender et al., 2001c). Therefore, the one-pot shortcut route is not feasible in anhydrous butyl acetate, but the situation might be different in other organic solvents.

5.6 Notation

AEH	α-amino acid ester hydrolase	
APA	6-aminopenicillanic acid	
BuAc	butyl acetate	
HPG	D-(*p*-)hydroxyphenylglycine	
HPGM	D-(*p*-)hydroxyphenylglycine methyl ester	
PAA	phenylacetic acid	
Pen G	penicillin G	
c_i	dissolved concentration of species i	(mol•l^{-1})
F	fraction	(-)
I	ionic strength	(mol•l^{-1})
K_a	dissociation constant	(mol•l^{-1})
K_{eq}	enzymatic equilibrium constant	(mol•l^{-1})
K_p	partition coefficient	(-)
n_i	total amount of moles of component i	(mol)
S	solubility	(mol•l^{-1})
V	volume	(l)
Sub- or superscript		
app	apparent	
aq	aqueous	
bi	biphasic	
org	organic	
ref	reference	

§6 Multifunctional Bioreactors

Bioconversions at an industrial scale, although highly selective, are seldom complete in a single step or pass and often require recycling of the unconverted substrates. Also, while the product in the bioreactor is just waiting for full substrate conversion, it may degrade. These are the main motives to integrate biotransformation and separation technology. When products are withdrawn from the reactor, they cannot degrade and substrate conversion will be complete.

Several authors (Van der Wielen and Luyben, 1992; Freeman et al., 1993) gave overviews about the state of the art in *In-Situ Product Recovery* (ISPR). It can easily be demonstrated on the basis of elementary economic calculations that the benefits of ISPR-technology are (1) enhanced product yields by suppressing degradation via undesired reactions and by-product formation, (2) less unit operations leading to less but higher yielding steps and (3) reduced consumption of auxiliary materials.

ISPR-advantages seldom originate from the reduction of equipment volume or release of product inhibition. They are, however, potentially successful in the production of instable antibiotics, flavors and fragrances, and for sensitive pharmaceutical proteins.

The field of multifunctional bioreactors was mostly of academic interest in the past 30 years, but now seems to attract serious industrial interest due to its potential to enhance the performance of established processes. It may also proof to be of use for the production of SSAs or their precursors such as 6-APA. The enzymes involved are hydrolases that catalyze hydrolysis or reactions in aqueous environments. The hydrolysis reaction equilibria can -in principle- be shifted completely to the product side by dilution (increasing entropy of product formation). Synthesis reactions using the reverse action of hydrolases can be enhanced by using excess of the cheaper reactants or by removal of one of the products, in particular water. In both cases, this does lead to voluminous processes with recycle streams and an overall negative impact on economic feasibility. Therefore, possibilities to selectively remove reaction products from each other in (reactor) space or completely from the reactors are very attractive.

Successful in the sense of being implemented at industrial scale are the so-called crystallization reactors working according to the suspension-to-suspension concept. Other integrated enzymatic reactor concepts that rely on the complete and selective separation of one species from the reactor are so far less successful. The obvious reason is that reactants and products are essentially too identical, which is also the case for SSA-production. Only when e.g. specific charge effects can be exploited, one may find "simple" one-pot concepts that works. However, technologically more advanced fractionating concepts may work, even for relatively close distribution coefficients. We will demonstrate this fractionation reactor concept as is shown in Figure IV.21 for a hydrolysis reaction using the general model as is shown in the following section. Both reaction and partitioning over the two phases are assumed to be at equilibrium in this simplified approach. A much more detailed theoretical analysis of its behavior is given by Den Hollander et al. (2001a), as well as the demonstration for the hydrolysis of Penicillin G (Den Hollander et al., 2001bc).

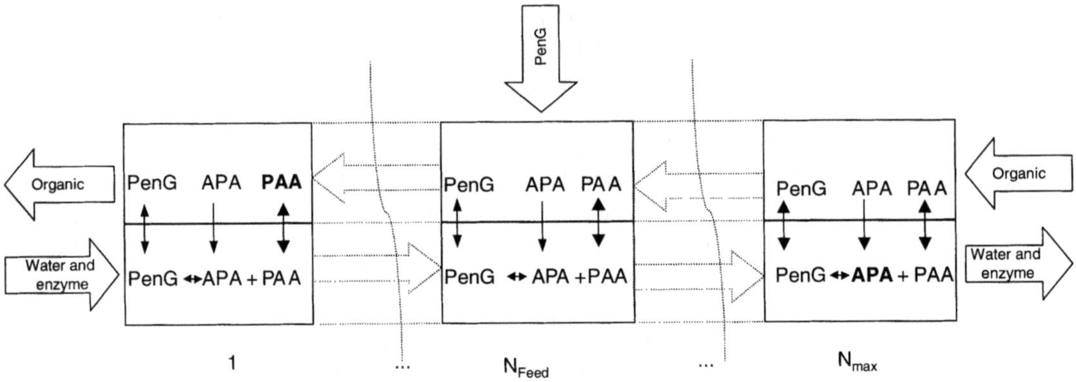

Figure IV.21 Fractionating bioreactor for the enzymatic hydrolysis of Pen G into 6-APA and phenylacetic acid (PAA).

6.1 A Model for Fractionating Reactors

A general procedure to describe fractionating contactors is by assuming a cascade of interconnected stages, numbered from 1 (top) to N (bottom). A typical stage is shown schematically in Figure IV.22. The stages can -in principle- have a feed stream F and withdrawal streams (*U*, *W*). A stage may be an actual tray (distillation, extraction) or be a theoretical tray, representing a certain length of bed. A feed stream at one stage may be an external feed, but may also include internal streams, withdrawn from other stages. This allows recycle flows. The effluent flows are assumed to be at thermodynamic equilibrium as is shown in Figure IV.22, although non-equilibrium approaches can be worked out (Taylor and Krishna, 1993).

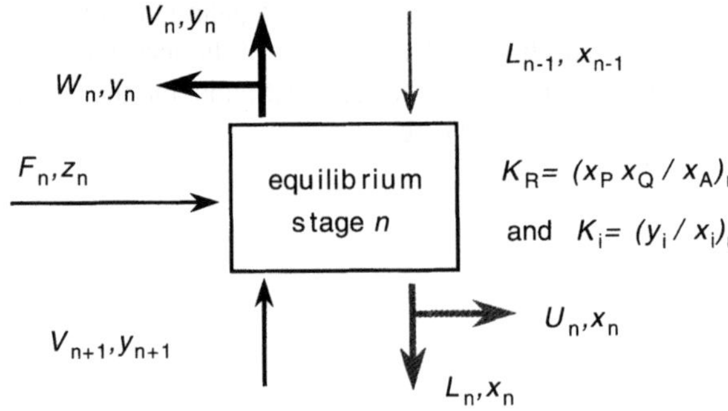

Fig. IV.22 Schematic representation of a reactive equilibrium stage with a feed (*F*) and side withdrawal streams (*U*, *W*).

Each stage can be described by a set of $2c+3$ MESH[1] equations, where c is the number of components. Sometimes, the set of equations can be reduced further, for instance by substituting the equilibrium relations in the species mass balances. For n stages, we have $(2c+3)n$ equations. For chromatographic systems, where typically $n = 100\text{-}500$ theoretical trays and $c = 3$ components (binary mixture in solvent), this leads to a system of 900-4500 equations. These have to be solved simultaneously using a multi-variate Newton method, with special matrix handling.
Models for fractionating reactors have the same structure as those of non-reactive systems. Reaction terms, however, are often highly non-linear and couple the various equations more intimately. We use the same equilibrium stage model as discussed above, but supplement the reaction details as well (Figure IV.22).
The mass balance for species i at stage n without W and U streams, now reads:

$$V_{n+1}y_{i,n+1} + L_{n-1}x_{i,n+1} + F_n z_{i,n} = V_n y_{i,n} + L_n x_{i,n} + v_i R_n \tag{23}$$

where v_i is the stoichiometric coefficient of reacting species i and R is the reaction rate. v_i is negative for reactants and positive for products. For equilibrium reactions (infinite rate), the reaction rate can be eliminated by adding mass balance equations pair wise for a substrate and a product. For the common equilibrium reaction of the type A = P + Q, this reduces the number of mass balances by one and extends the number of equilibrium relations by one. For constant distribution coefficients in a dilute system of species A, P and Q, we assume a mass action law type phase equilibrium. The resulting set of equations reads as follows:

$$\begin{aligned} &V_{n+1}\left(y_{A,n+1} + y_{P,n+1}\right) + L_{n-1}\left(x_{A,n-1} + x_{P,n-1}\right) + F_n\left(z_{A,n} + z_{P,n}\right)z_i \\ &= V_n\left(y_{A,n} + y_{P,n}\right) + L_n\left(x_{A,n} + x_{P,n}\right) \\[1em] &V_{n+1}\left(y_{A,n+1} + y_{Q,n+1}\right) + L_{n-1}\left(x_{A,n-1} + x_{Q,n-1}\right) + F_n\left(z_{A,n} + z_{Q,n}\right)z_i \\ &= V_n\left(y_{A,n} + y_{Q,n}\right) + L_n\left(x_{A,n} + x_{P,n}\right) \\[1em] &K_r x_{A,n} - x_{P,n}\, x_{Q,n} = 0 \end{aligned} \tag{24}$$

where the reaction takes place primarily in the L-phase. These equations can be solved with the same procedure as outlined before. The initial estimate of the profile is now much more crucial. In some cases, it is required to use analytical solutions as initial estimates for calculations and special numerical techniques to solve the problem.

[1] $2c+3$ MESH equations: species and overall Mass balances ($c+1$), Equilibrium relations (c), Sum-of-Mole fraction relations (1) and, when applicable, an enthalpy balance (1). For further details, see Taylor and Krishna (1993) or Seader and Henly (1998).

6.2 Simulation results for fractionating hydrolysis reactions

We have investigated the hydrolysis of A in P and Q. Examples are the hydrolysis of Pen G to give APA but also the enantioselective hydrolysis of L-acetyl amino acids in a DL-mixture which yields an enantiomerically pure amino acid, as well as the unhydrolyzed D-acetyl amino acid. In concentrated solutions, these hydrolysis reactions are incomplete due to the reaction equilibrium. It is evident that for an accurate analysis of weak electrolyte systems, the association-dissociation reactions and the related phase behavior of the reacting species must be accounted for precisely in the model (see a.o. Jansen et al., 1996, and Van der Wielen et al., 1996). We have simplified this example to neutral species A, P and Q. The distribution coefficients are $K_Q = 0.5$ and $K_P = K_A = 2$. The equilibrium constant for the reaction $K_r = x_{P,n}\, x_{Q,n} / x_{A,n} = 0.01$ where x is a measure for concentration (mass or mole fractions) compatible with the partition coefficients. The mole fraction of A in the feed, z_A, was 0.1, which corresponds to an extremely high aqueous feed concentration of approximately 5 M. We have simulated the hydrolysis conversion in the fractionating reactor with 50-100 equilibrium stages. A further increase in the number of stages did not improve the conversion or the selectivity to a significant extent. Depending on the initial estimate, the calculation requires typically less than 5 iterations.

A typical concentration profile for a 50 stage fractionating reactor, with a feed at stage $N_F = 35$ is given in Figure IV.23. V runs from top (stage 1) to the bottom (stage 50) and L in the opposite direction. The feed was concentrated $z_A = 0.1$, and the flow rate ratios $m_1 = 1.667$ and $m_2 = 1.833$ ($m = V/L$). The conversion under these conditions was 90.3% (in batch 10.5%), with purities for Q of 84.8%, and for P of 94.8%. Further diluting the feed stream increases the conversion further. Because the partition coefficients of A and P are equal, also their separation factors are equal and they move in the same directions (towards the bottom with V). This is visible in Figure IV.23 as the parallel concentrations profiles of A and P.

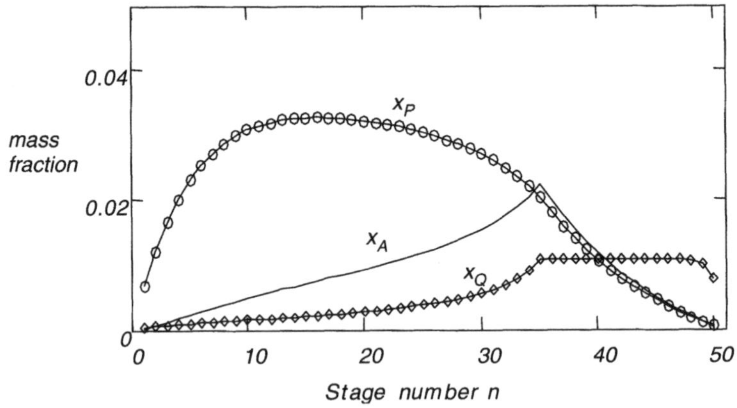

Fig. IV.23 Calculated composition profiles in a fractionating reactor. The conditions are shown in the text.

Varying the flow rate ratios in a systematic manner gives an insight in optimal conditions. The calculation results are summarized for the conversion in Figure IV.24 and for the purities of P and Q in Figure IV.25a,b. Close to the diagonal, conversion increases. This is partially a dilution effect: $m_2 - m_1 = F/V$ decreases to small numbers while approaching the diagonal. Figure IV.25a and b show the variation of purity of the products Q (IV.25a) and P (IV.25b) respectively, while varying (m_1,m_2) approximately parallel to the diagonal. The closer the flow rate ratios are to the distribution coefficient of a species, the better the criterion for the 'other' species is satisfied. This results in a more complete removal of the 'other' species and to a higher purity.

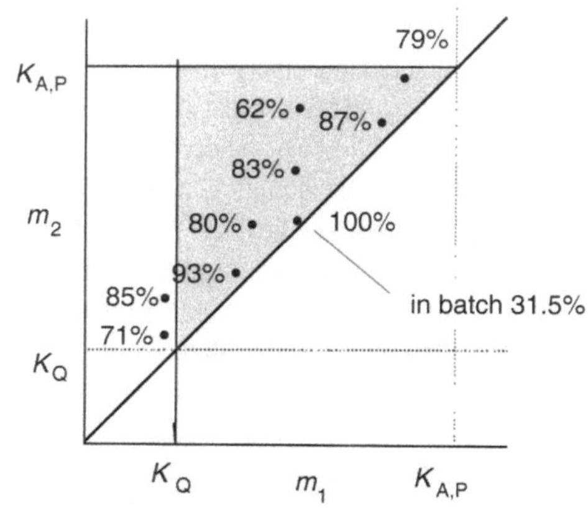

Fig. IV.24 Conversion (degree of hydrolysis) in a fractionating reactor. Dots represent (m_1,m_2), the nearby number is the calculated conversion for data in the text. The batch conversion, corresponding to the 100% conversion point, is limited to 31.5%.

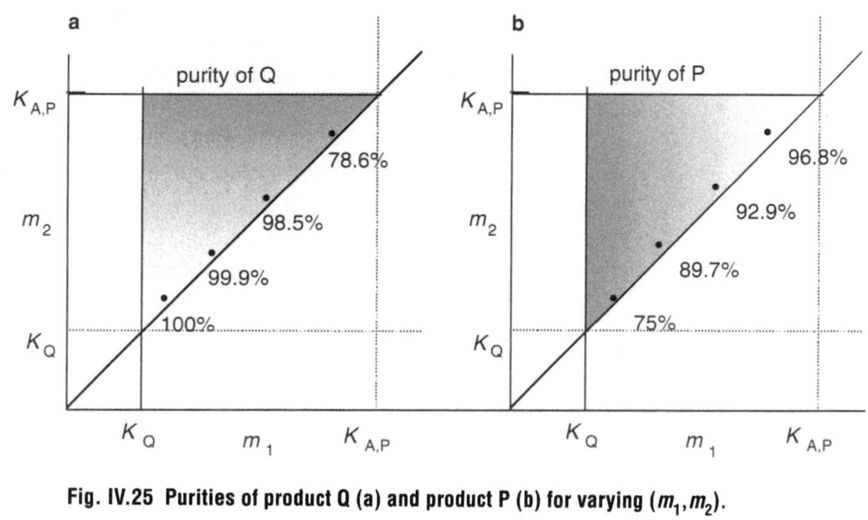

Fig. IV.25 Purities of product Q (a) and product P (b) for varying (m_1,m_2).

This example is most likely a worst case analysis. Systems with (1) a more dilute feed, (2) with $K_Q < K_A < K_P$ and (3) where the partition coefficients are more disinct, can lead to complete conversion in a few stages. Also manipulating the local partition coefficients in different stages by varying pH, salt concentrations or solvent composition, offers a large potential for further optimization. A last area that is practically unexplored is to use internal recycle streams (refluxes), that can lead to accumulation of specific products in specific sections.

6.3 Case study: hydrolysis of Pen G

Den Hollander et al. (2001b,c) investigated the enzymatic hydrolysis of penicillin G to phenylacetic acid and 6-aminopenicillanic acid in biphasic aqueous organic systems without pH-control. In a preliminary study, the two phases were countercurrently contacted in a discrete manner, so that equilibrium was reached in each stage. Sets of three and five shake flasks served to mimic equilibrium stages in the countercurrent set-up. It was shown, that countercurrent contact leads to significant improvement of the equilibrium conversion when compared to the batch or co-current situation. When penicillin G was fed in at an intermediate stage, either exit contained mainly one of the two products. This simplifies product recovery.

A mathematical model was used to calculate the concentrations of all components and the pH at every equilibrium stage. The pH and concentrations of the components at every equilibrium stage were predicted with reasonable accuracy. This model is based on dissociation and reaction equilibria of the compounds, stoichiometric balances and an electro-neutrality equation. Precipitation of 6-aminopenicillanic acid, which was observed at a combination of low pH and high 6-APA concentration in the aqueous phase, is not taken into account in the model.

Experimental conversions in this simple system without control of pH etc. could be as high as 98%, depending on the flow rate ratios. The conversion was typically 10-30% larger in the 3-stage and over 50% larger in the (simulated) countercurrent system relative to batch-wise conversion. A further increase in the number of stages seems attractive, but it can be demonstrated that adding stages to systems containing over 25-50 equilibrium stages does not notably improve the conversion. On the basis of these results, a countercurrent fractionating L-L reactor system with an increased number of stages can be investigated by model. An idealized reactor, equipped with an axial pH-control system and operated at pH 6, would lead to the following composition profiles with near-complete conversion and purification as is shown in Figure IV.26. A further optimization, for instance with respect to minimal flow of solvents, feed location and effect of crystallization, is not done here, but subject of future work.

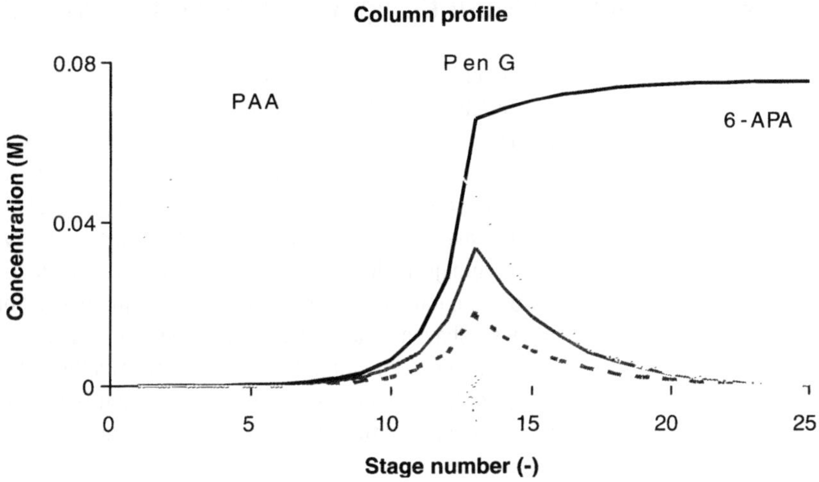

Figure IV.26 Near complete conversion and purification in an idealized 25 stage fractionating reactor. The PenG feed is introduced at the arrow (unpubl. results Den Hollander et al., 2001).

6.4 Fractionating synthesis reactor

At this moment, fractionating reactors are mostly studied and applied outside the fine chemical field. Examples are the production of MTBE and TAME via reactive distillation. Some biocatalytic studies have been performed. Kosugi et al. (1990) investigated the hydrolysis of esters by lipase in a countercurrent reactor. A number of synthesis reactions has been studied such as the esterification of ethanol and acetic acid to form ethyl acetate and water (Mazzotti et al., 1996) in a SMB-reactor with chemocatalysts (acidic ion exchange resins). Another, fairly similar application was presented by Kawase et al. (1996) to manufacture phenylethyl acetate. Carta and coworkers (Mensah and Carta, 1999) used fixed bed as well as SMB chromatography with lipases immobilized on resin to produce isoamyl propionate from propionic acid and isoamyl alcohol. Isoamyl alcohol is used to wash water from the catalytic adsorbent. The reactants are fed between a regeneration section and reaction section. Hexane is used as the continuous organic phase. The continuous desorption of water from the adsorbent material reduces enzyme deactivation and improves the productivity of the process. But these are isolated examples and this field is still wide open for further investigation. It seems worthwhile to study the feasibility of kinetically controlled enzymatic antibiotics synthesis by this technology.

§7 Single and Multi Component Crystallization of SSA

Conventional and new synthesis routes for SSA will always incorporate crystallization as essential step in the purification and formulation of the final product. Ideally, crystallization-for-formulation is mostly conducted in a single (water + solute) component system. In practice, crystallization processes contain multi-component systems where the presence of solutes, other than the targeted product, may be very influential upon the kinetics of crystallization (McPherson, 1999). Foreign molecules such as degraded products and by-products, additives and other contaminants may interfere with the nucleation as well as the growth process (Chayen et al., 1993). The crystal growth may be disrupted due to the incorporation of the impurities into the crystal lattice or to their adsorption at the surface of the crystal (Davey, 1976; Kubota and Mullin, 1995). The structure, size and morphology of the final product may consequently be modified by the presence of impurities.

In the following sections, we will discuss single component crystallization from aqueous solution as a base case, and give a preliminary discussion on possible effects of impurities upon the crystallization kinetics of semi-synthetic penicillins, and upon the product quality. The impact of impurities will therein be monitored by investigating the crystallization kinetics of the process, i.e. the induction time, de-supersaturation rate and growth rate. We will use a model to describe the induction time, de-supersaturation rate and growth rate. This delivers a framework in which the effects of impurities upon the SSA crystallization can be accurately analyzed and predicted in a wide range of operating conditions.

7.1 Model

A crystallization process can be defined by the variation with time of the solute concentration in the liquid (the so-called de-supersaturation curve) as well as of the Crystal Size Distribution (CSD). Both measurable curves can also be simulated using an appropriate crystallization model. A popular, generally applicable model is based on a population balance for crystals in a specific size class (Randolph and Larsson, 1972; Tavare 1995):

$$\frac{\partial Vn(t,x)}{\partial t} = -\frac{\partial Vn(t,x)G(t,x)}{\partial x} + Vb(t,x) - Vd(t,x) - \sum_k \Phi_k n_k(t,x) \qquad (25)$$

with V being the crystallizer volume in m^3, n the number of particles per volume per size class in # m^{-4}, t the time in s, x the length coordinate in m, G the growth rate in m s^{-1}, b the birth function in a certain crystal size class in # m^{-4} s^{-1}, d the death function in a certain crystal size class in # m^{-4} s^{-1}, Φ_k the flow rate of the k stream containing crystals entering or leaving the compartment in m^3 s^{-1}.

Usually, the following constraints apply for batch-wise crystallization:
- a constant volume,
- batch operation (no input and output),
- no agglomeration or no breakage ,
- no death and only birth in the lowest particle class, and
- a size independent growth.

Eq. 25 reduces to the following hyperbolic partial differential equation (PDE):

$$\frac{\partial n(t,x)}{\partial t} = -\frac{\partial [n(t,x)G(t)]}{\partial x} \tag{26}$$

To solve this equation, boundary and initial conditions are required. The common boundary condition at $x = 0$ is given by the common equation :

$$n(t,0) = \frac{J(t)}{G(t)} \tag{27}$$

with J being the nucleation rate in # m^{-3} s^{-1}. The initial condition at $t = 0$ is given by:

$$n(0,x) = 0 \tag{28}$$

The de-supersaturation curve can be obtained via a mass balance:

$$M_w V \frac{dC(t)}{dt} + \frac{dM(t)}{dt} = 0 \tag{29}$$

with M_W being the molar mass of the crystallizing component. The mass of crystals M formed is given by integrating over all crystal size classes:

$$M(t) = \int_0^\infty k_v \rho_v n(t,x) x^3 dx \tag{30}$$

The nucleation rate is described by the following equation (Tavare, 1995):

$$J(t) = J_0 \exp(-\frac{B}{\ln(S(t))^2}) \tag{31}$$

The growth rate is given by power law kinetics (Tavare, 1995):

$$G(t) = k_g (S(t)-1)^n \tag{32}$$

Both nucleation rate and growth rate depend on the supersaturation ratio of the crystallizing component:

$$S(t) = \frac{C(t)}{C_S} \tag{33}$$

We use a numerical solution method to solve the PDE, which is described in more detail elsewhere (Ottens et al., 2001). The kinetic parameters are estimated by plotting the results of the crystallization experiments in linearized forms of Eq. 31 and Eq. 32.

7.2 Batch-wise experiments

A supersaturated situation is required for crystallization. An important measure for the driving force is the difference between the actual concentration and the solubility. The latter has to be known very accurately, since it provides an essential control parameter in the crystallization process. The solubility of SSA and their precursors can be influenced by dosing a concentrated salt solution, anti-solvents or by changing pH. Because the solubility is a strong function of pH, this may be the most popular method. Changing the pH of the aqueous solution to the pI of the solute gives rise to a substantial drop in the solubility and introduces the supersaturation and subsequent crystallization of the semi-synthetic antibiotic.

Kinetic experiments can be performed in an experimental system as shown in Figure IV.27 (Lebreton et al., 1999; Ottens et al., 2001). The de-supersaturation curve is determined by analyzing the liquid phase concentration versus time. The crystal growth rate is determined via an adapted Single-Crystal Growth Analysis method. Frequent crystal samples were taken from the batch-crystallizer and introduced into a glass cell. The crystal growth was monitored using a phase contrast microscope (Olympus, Japan) by acquiring images at various time intervals (from 20 to 200 s). A few crystals were selected from the large number of crystals observed per image, and their length measured via image analysis. The growth rate was thus determined. The operation was repeated for a series of samples corresponding to distinct supersaturations, decreasing in time during the batch experiment. The advantage of this method is that the growth rate is related to the actual liquid composition, which seems to reduce the experimental error significantly.

The *induction time* is defined as the period of time that elapses between the achievement of supersaturation and the appearance of crystals having a "detectable" size. The induction time depends on the initial supersaturation as well as on the detection method. For instance, if a concentration measure is used, the induction time depends on conversion. However, when light scattering is used, the induction time depends on the crystal surface area produced. We have used laser reflection, which depends on surface area x^2. As is shown elsewhere (Ottens et al., 2001), the induction time is related to the initial supersaturation S_0 as follows:

$$t_{ind} \propto \left[(S_0 - 1)^n\right]^{\frac{-1}{i+1}} \exp\left(\frac{B}{(i+1)(\ln(S_0))^2}\right) \quad \text{with} \quad i = 2 \qquad (34)$$

Since the exponential term will dominate, a plot of $\ln(t_{ind})$ versus $(\ln(S_0))^{-2}$ for the different crystallization experiments will yield $B/(i+1)$ as the slope.
The parameters k_g and n in the growth rate equation Eq. (32) are obtained from batch growth rate experiments by plotting $\ln(G)$ vs. $\ln(S(t)^{-1})$ as the abscissa and the slope respectively. Since Ampicillin is a rather large molecule the crystallization rate is relatively low and the desupersaturation period turned out to be long enough

to determine the growth kinetics from the second part of the batch experiment. The power number in the growth rate equation was for a wide range of experiments close to 2 (1.7). This was also observed for the growth kinetics of similar molecules such as aspartame and indicates a spiral growth mechanism (Frank, 1949). The other parameters for Ampicillin were $B = 0.376$ and $k_g = 45.8$ nm/s.

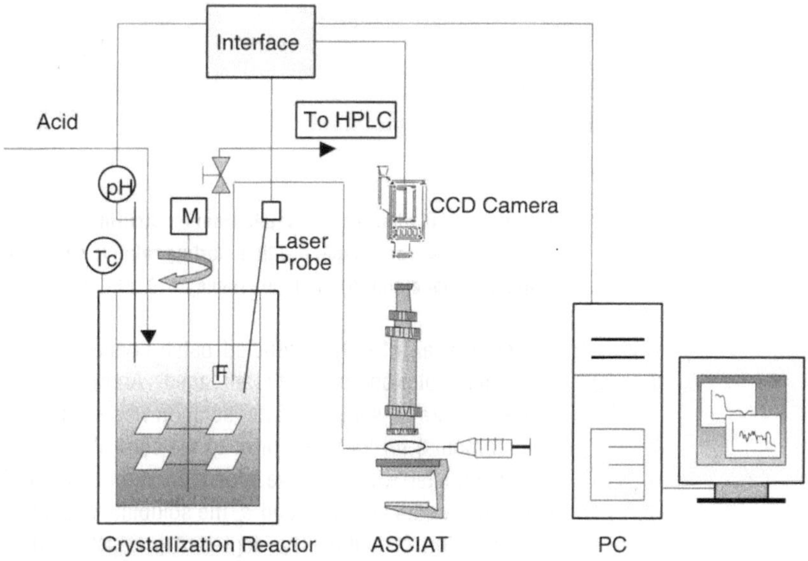

Fig. IV.27 Experimental setup for SSA crystallization experiments, with the Adapted Single Crystal Growth Analysis technique.

7.3 Crystallization: Morphology and Kinetics

SEM pictures of typical Ampicillin crystals are shown in Figure IV.28. The Ampicillin crystals are boat-like crystals, and tapered with mixed blunt and sharp ends. Furthermore they show some limited twinning and limited breakage. The mean aspect ratio ranges between 5 and 8. The left panel shows the crystals of 'pure' Ampicillin with relatively smooth surfaces. The right panel shows crystals of Ampicillin, obtained in a batch crystallization with Ampicillin precursors, at otherwise identical conditions as the 'pure' Ampicillin. The crystal surfaces are covered with much more irregularities. This is a clear indication that the presence of contaminants interferes with the crystallization process.

a)

b)

Fig. IV.28 SEM pictures of crystals from Ampicillin batch-crystallizations. The experiments were conducted with a starting relative supersaturation of 1.8. (a) Ampicillin alone and (b) in the presence of 10 mM of 6-APA and 30 mM of PG.

Lebreton et al. (1999) conducted batch-crystallizations, with a freshly prepared Ampicillin solution and with an "aged" Ampicillin solution where degradation products were present. The latter solution was aged at pH 5.0 (pH adjusted with HCl and ammonia) for 24 hours prior to the start of the experiment. Both experiments were conducted with a relative supersaturation σ of 1.8 (defined as $\sigma = (c-c_s)/c_s$ with c the solute concentration and c_s the solubility at pH 5.0, T = 25°C with equivalent concentration of salts). The degraded products had a clear influence upon the kinetics of Ampicillin crystallization as shown in Figure IV.29. For similar initial supersaturation, the desupersaturation phenomenon was retarded. The laser reflection indicated an increased induction time, from 25 to 35-40 min. The growth rate appears unchanged by the presence of the degraded products relative to the desupersaturation.

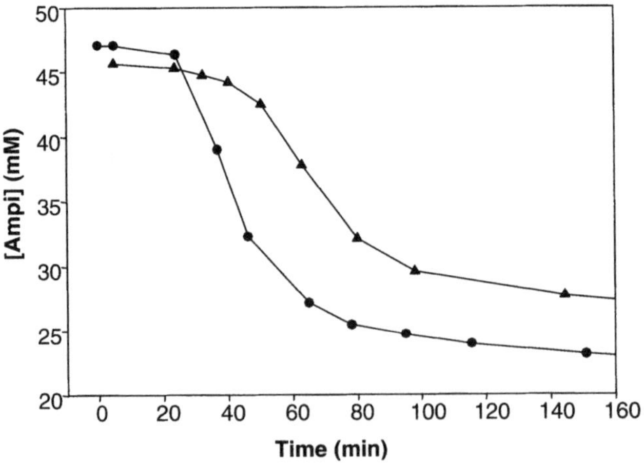

Fig. IV.29 De-supersaturation curves for Ampicillin (●) and "aged" Ampicillin (▲).

Also here, the surface of the needle-shape crystals displayed more asperities for "aged" Ampicillin as compared to "fresh" Ampicillin solutions. The X-ray diffraction pattern indicated that both final products of each crystallization were optically pure, and the re-dissolution of these materials did not indicate the presence of extra components in the crystal lattice. There seems to be no evidence for direct incorporation of contaminants in the crystal lattice, so possible effects are likely to occur by adsorption at the crystal surface or in the nucleation phase of the process. This assumption is supported by the increased induction time for the crystallization of the 'aged' Ampicillin, as is shown in Figure IV.29.

7.4 Simulation of batch-wise crystallization process

The model is used to simulate de-supersaturation profiles as well as the CSD in time of 'pure' Ampicillin crystallization. A characteristic example of the simulated growth and nucleation rates is given in Figure IV.30, and compared to experimentally determined growth rates using the Adapted Single-Crystal Growth method. At S_0, the crystal growth rate and especially the nucleation rate are fast. After the initial period, the nucleation rate drops dramatically because of its strong dependence on supersaturation. The crystal growth rate decreases because of the decreased supersaturation and finally becomes zero. Also the induction period is well described by the model and the obtained kinetic data (Ottens et al., 2001).

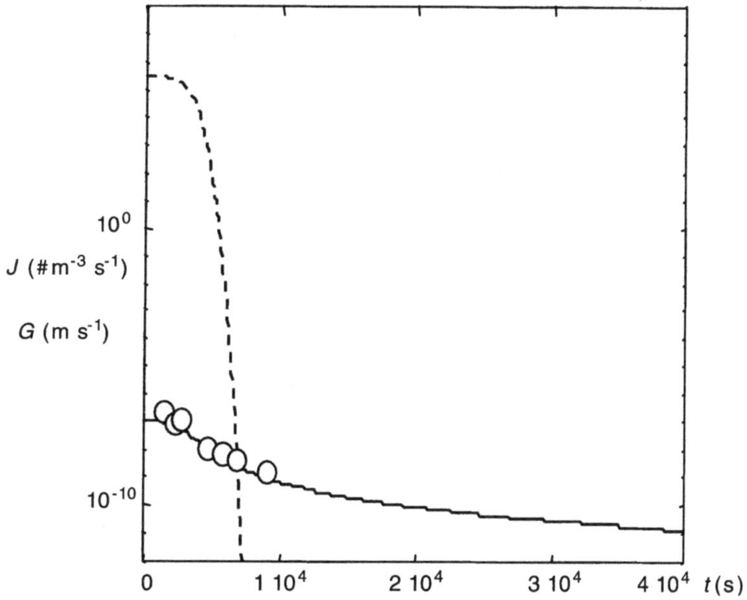

Fig. IV.30 Predicted nucleation (dotted line) and growth rates (solid line; markers are experimental data). Initial supersaturation ratio of 2.45; ρ_v = 1500 kg m^{-3} (after Ottens et al., 2001).

Figure IV.31 shows measured and calculated final crystal size distributions for a typical experiment (Ottens et al., 2001). The full CSD as well as the average crystal size is predicted reasonably well with a somewhat overprediction of the average crystal size. This may be due to several factors, such as breakage of the needle like crystals.

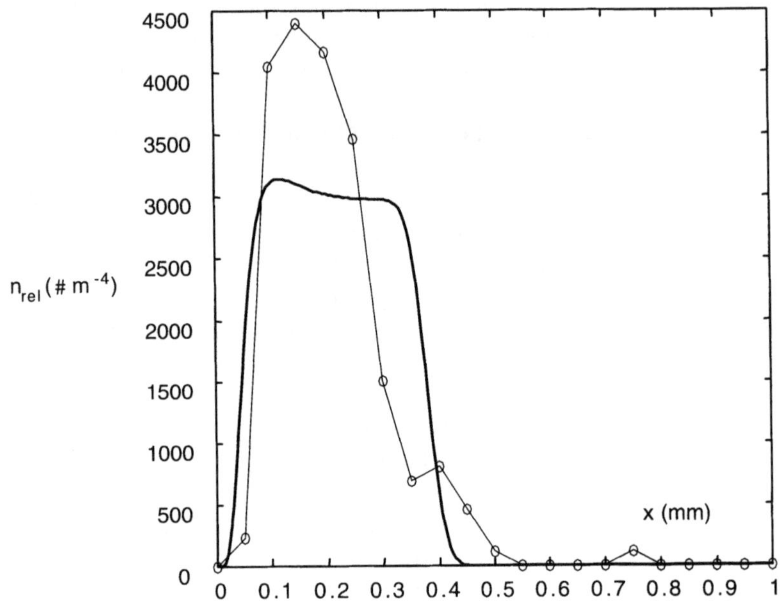

Figure IV.31 Comparison of predicted final CSD (solid curve) and experimental results (connected circles); 172 crystals, size classes of 50 (μm). Pure Ampicillin crystallization at pH = 5 and T = 298 K.

The model can be used to test alternative relations for growth *G* and nucleation *J* and to apply these *G* and *J* relations to the entire experimental database. A parametric optimization should be performed, but that falls beyond the scope of the present paper. We anticipate that the model can be extended for multicomponent effects primarily in the nucleation rates. Controlled experiments, comparable to those shown in figure IV.29 for the 'aged' Ampicillin solution, have been performed with accurate control of contaminant type and concentrations. A preliminary analysis seems to suggest that increasing contaminant concentrations lead to an increasing inhibition of the nucleation mechanism, possibly by adsorption. This seems logical, given the strong structural resemblance between target molecules and contaminants.

7.5 Notation

	description	unit
roman		
V	volume	m^3
n	number of particles per volume per size class	$\# \, m^{-4}$
t	time	s
x	length	m
G	growth rate	$m \, s^{-1}$
b	birth function	$\# \, m^{-4} \, s^{-1}$
d	death function	$\# \, m^{-4} \, s^{-1}$
J	nucleation rate	$\# \, m^{-3} \, s^{-1}$
C	concentration	$mol \, m^{-3}$
M	mass	kg
k_v	shape factor	-
S, σ	supersaturation ratio	-
k	Boltzmann constant = 1.83 10^{-23}	$J \, K^{-1}$
T	temperature	K
k_g	growth rate coefficient	$m \, s^{-1}$
greek		
Φ	flow rate	$m^3 \, s^{-1}$
ρ_v	crystal density	$kg \, m^{-3}$
ν	molecular volume	m^3
subscript		
0	initial	
S	saturated	

§8 Conclusions

The processes for the enzymatic production of semisynthetic penicillins may be simplified considerably by integration of the hydrolysis of penicillin G and the kinetically controlled coupling of 6-APA to an activated side chain. About 7 unit operations may be bypassed. This would also reduce the salt formation significantly, because the number of pH shifts may be decreased.

To exploit the full potential of the aforementioned integration, the further development of the concept of fractionating reactors seems particularly valuable. A condition for a successful development is that enzymes that have a high activity, selectivity and stability at the conditions dictated by this technology, will become available.

More direct processes, based on the thermodynamically controlled coupling of 6-APA with unactivated side chain, do not seem to be feasible. The feasibility of a very direct shortcut route, the one-pot synthesis of Amoxicillin from Pen G acid in organic solvent, is not clear yet.

For the evaluation of such process options, one relies heavily on phase equilibrium data of antibiotics and their precursors. Such data are relatively scarce. A generalized framework has been presented that allows the estimation of phase behavior of net uncharged molecules, and the extension of such a framework to charged molecules would be valuable.

Crystallization of antibiotics will play a role in any conventional purification or in any new process, and will be particularly important when suspension-to-suspension reactions are involved. It has been shown that the rate of crystallization can be predicted reasonably well using models that include conventional expressions for nucleation and growth. Such predictions can be used to control the crystallization rate in a process, or even to prevent crystallization if it is undesired in a process step.

§9 Acknowledgements

The authors wish to acknowledge M.B. Diender, S. Rudolph, B. Lebreton, J.L. den Hollander, M. Gude, F.E.D. van Halsema, M. van Berlo, T. van der Does, O.S.L.Bruinsma, R. Grimbergen, M. Rijkers and J.J. Heijnen for their valuable input.

§10 References

Andersson E, Mattiasson B, Hahn-Hägerdal B. 1994. Enzymatic conversion in aqueous two-phase systems: deacylation of benzylpenicillin to 6-aminopenicillanic acid with penicillin acylase. Enzyme Microb Technol 6:301-306.

Barenschee T, Scheper T, Schügerl. 1992. An integrated process for the production and biotransformation of penicillin. J Biotech 26:143-154.

Blinkovsky AM, Markaryan AN. 1993. Synthesis of β-lactam antibiotics containing α-aminophenylacetyl group in the acyl moiety catalyzed by D(-) α-phenylglycyl-β-lactamide amidohydrolase. Enzyme Microb Technol 15:965-973.

Bruggink A, Roos EC, De Vroom E. 1998. Pencillin acylase in the industrial production of β-lactam antibiotics. Org Proc Res Dev 2:128-133.

Chayen, N.E., Radcliffe, J.W. and Blow, D.M. 1993. Control of nucleation in the crystallization of lysozyme, Protein Science, 2: 113-118.

Clausen K, Dekkers RM. 1996. Process for preparation of β-lactams at constantly high concentration of reactants. Int Patent Appl WO 96/02663.

Collander, R. 1951. Acta Chem. Scan. 5, 774-780, 1951

Davey, R.J. 1976. The effect of impurity adsorption on the kinetics of crystal growth from solution, Journal of Crystal Growth, 34:109-119.

Den Hollander J.L., A.J.J. Straathof and L.A.M. van der Wielen. 2001a. Design of fractionating enzymatic reactors for hydrolysis reactions. Submitted.

Den Hollander J.L., A. Aversente, M.B. Diender, A.J.J. Straathof and L.A.M. van der Wielen. 2001b. Enzymatic penicillin G hydrolysis in discretely contacted countercurrent water - butyl acetate biphasic systems. Submitted.

Den Hollander J.L., M. Zomerdijk, A.J.J. Straathof and L.A.M. van der Wielen. 2001c. A continuous countercurrent enzyme reactor for the hydrolysis of penicillin G. Submitted.

Diender MB. 2001. New process concepts for the enzymatic synthesis of Amoxicillin from penicillin G. PhD thesis, Delft University of Technology.

Diender MB, Straathof AJJ, Van der Wielen LAM, Ras C, Heijnen JJ. 1998a. Feasibility of the thermodynamically controlled synthesis of Amoxicillin. J Mol Catal B: Enz 5:249-253.

Diender MB, Straathof AJJ, Heijnen JJ. 1998b. Predicting enzyme catalyzed reaction equilibria in cosolvent-water mixtures as a function of pH and solvent composition. Biocatal Biotransform 16:275-289.

Diender MB, Straathof AJJ, Van der Does T, Zomerdijk M, Heijnen JJ. 2000. Course of pH during the formation of Amoxicillin by a suspension-to-suspension reaction. Enzyme Microb Technol 27:576-582.

Diender MB, Straathof AJJ, Van der Does T, Ras C, Heijnen JJ. 2001a.Equilibrium modelling of extractive enzymatic hydrolysis of penicillin G with concomitant 6-aminopenicillanic acid crystallization. Submitted.

Diender MB, Straathof AJJ, Van der Does T, Heijnen JJ. 2001b. Modeling the influence of ionic strength on α-amino acid ester hydrolase in the salt-free production of Amoxicillin. Submitted.

Diender MB, Straathof AJJ, Van der Does T, Heijnen JJ. 2001c. Feasibility of a one-pot shortcut route from penicillin G to amoxicilin in anhydrous organic solvent. Submitted.

Dye, S.R., J.P. DeCarli II and G. Carta. 1990. Equilibrium sorption of amino acids by a cation exchange resin. Ind. Eng. Chem. Res. 29:849-857

Eggers DK, Blanch HW, Prausnitz JM. 1989. Extractive equilibria: solvent effects onequilibria of enzymatic reactions in two-phase systems. Enzyme Microb Technol 11:84-89.

Eiteman, M.A., and J.L. Gainer. 1991. Predicting partition coefficients in polyethylene glycol - potassium phosphate aqueous two-phase systems. J. Chrom. 586:341-346.

Eiteman, M.A. and J.L. Gainber. 1992. A correlation for predicting partition coefficients in aqueous two-phase systems. Sep. Sci. Technol. 27:313-324.

El-Tayar, N., R.S. Tsai, P. Vallet, C. Altomare and B. Testa. 1991. Measurement of partition coefficients by various centrifugal partition chromatographic columns. J. Chrom. 556: 181-194.

Erbeldinger M, Ni X, Halling PJ. 1998. Enzymatic synthesis with mainly undissolved substrates at very high concentrations. Enzyme Microb Technol 23:141-148.

Erbeldinger M, Ni X, Halling PJ. 1999. Kinetics of enzymatic solid-to-solid peptide synthesis: intersubstrate compound, substrate ratio and mixing effects. Biotechnol Bioeng 63:316-321.

Fernandez-Lafuente R, Rosell CM, Piatkowska B, Guisán JM. 1996. Synthesis of antibiotics (cephaloglycin) catalyzed by penicillin G acylase: Evaluation and optimization of different synthetic approaches. Enzyme Microb Technol 19:9-14.

Fernandez-Lafuente R, Rosell CM, Guisán JM. 1991a. Equilibrium controlled synthesis of cephalothin in cosolvent-water systems by stabilized penicillin G acylase. Appl Biochem Biotechnol 27:277-290.

Fernandez-Lafuente R, Rosell CM, Guisán JM. 1991b. Enzyme reaction engineering: synthesis of antibiotics catalysed by stabilized penicillin acylase in the presence of organic cosolvents. Enzyme Microb Technol 13:898-909.

Flaschel E, Crelier S, Schulz K, Huneke FU, Renken A. 1992. Process development for the optical resolution of phenylalanine by means of chymotrypsin in a liquid-liquid-solid three-phase reaction system. In: Tramper J, Vermuë MH, Beeftink HH, von Stockar U, editors. Biocatalysis in non-conventional media. Elsevier Science Publishers, Amsterdam p. 163-170.

Freeman A, Woodley JM, Lilly MD. 1993. In situ product removal as a tool for bioprocessing. Bio/Technology 11:1007-1012.

Frank, F.C. 1949. Disc. Faraday Soc. 48: 67.

Gmehling , J. and U. Onken. 1977. Vapour-Liquid Equilibrium Data Collection. Aqueous-Organic Systems. Vol. 1., Chem. Data Ser., Dechema, Frankfurt.

Grant, D.J.W. and T. and Higuchi. 1990. Solubility behavior of organic compounds. Techniques of Chemistry, vol. XXI. Wiley-Interscience Publ.

Gude, M.T., H. Meuwissen, L.A.M. van der Wielen and K.Ch.A.M. Luyben. 1996a.Partition coefficients and solubilities of α-amino acids in aqueous 1-butanol solutions. Ind. Chem. Eng. Res. 35:4700-4712.

Gude, M.T., L.A.M. van der Wielen and K.Ch.A.M. Luyben. 1996b. Phase behavior of α-amino acids in multicomponent aqueous alkanol solutions. Fluid Phase Equilibria 116:110-117.

Halling PJ, Eichhorn U, Kuhl P, Jakubke HD. 1995. Thermodynamics of solid-to-solid conversion and application to enzymic peptide synthesis. Enzyme Microb Technol 17:601-606.

Hernadez-Justiz O, Fernandez-Lafuente R, Terrini M, Guisán JM. 1998. Use of aqueous two-phase systems for in situ extraction of water-soluble antibiotics during their synthesis by enzymes immobilized on porous supports. Biotechnol Bioeng 59:73-79.

Hersbach GJM, Van der Beek CP, Van Dijck PWM. 1984. The penicillins: properties, biosynthesis and fermentation. In: Vandamme EJ, editor. Biotechnology of industrial antibiotics. New York: Marcel Dekker, Inc. p45-140.

Homandberg GA, Mattis JA, Laskowski JM. 1978. Synthesis of peptide bonds by proteinases. Addition of organic cosolvents shift peptide bond equilibria toward synthesis. Biochemistry 17:5220-5227.

Jansen ML. 1996. Integration of ion exchange chromatography with an enzymatic reaction. PhD thesis, Delft University of Technology.

Janssen AEM, Van der Padt A, Van 't Riet K. 1995. Prediction of the reaction equilibrium of biocatalytic reactions in aqueous-organic two-phase systems. Biocatal Biotransform 12:223-240.

Johansson HO, G Karlström, F Tjerneld and CA Haynes. 1999. Driving forces for phase separation and partitioning in aqueous two-phase systems. J Chrom B 711: 3.

Kasche V. 1986. Mechanism and yields in enzyme catalysed equilibrium and kinetically controlled synthesis of β-lactam antibiotics, peptides and other condensation products. Enzyme Microb Technol 8:4-16.

Kasche V, Galunsky B. 1994. Enzyme catalyzed biotransformations in aqueous two -phase systems with precipitated substrate and/or product. Biotechnol Bioeng 45:261-267.

Kato K, Kawahara K, Takahashi T, Igarasi S. 1980. Enzymatic synthesis of Amoxicillin by the cell-bound α-amino acid ester hydrolase of *Xanthomonas citri.* Agric Biol Chem 44:821-825.

Kawase M, Suzuki TB, Inoue K, Yoshimoto K, Hashimoto K. 1996. Increased esterification conversion by application of the simulated moving-bed reactor Chem Eng Sci, 51: 2971-2976.

Kim MG, Lee SB. 1996. Penicillin acylase-catalyzed synthesis of beta-lactam antibiotics in water-methanol mixtures: effect of cosolvent content and chemical nature. J Mol Catal B: Enz 1:201-211.

Kosugi Y, Tanaka H, Tomizuka N. 1990. Continuous hydrolysis of oil by immobilized lipase in a countercurrent reactor. Biotechnol. Bioeng. 36: 617-622

Kubota, N. and Mullin, J.W. 1995. A kinetic model for crystal growth from aqueous solution in the presence of impurity, Journal of Crystal Growth, 152: 203-208.

Lebreton, B., Zomerdijk, M., Ottens, M., Rijkers, M., and L.A.M. van der Wielen. 1999. Effects of impurities upon crystallization kinetics of beta-lactam antibiotics, paper at the AIChE Annual Meeting, Dallas, USA.

Lee DC, Kim HS. 1998. Optimization of a heterogeneous reaction system for the production of optically active D-amino acids using thermostable D-hydantoinase. Biotechnol Bioeng 60:729-738.

Martinek K, Semenov AN, Berezin IV. 1981[a]. Enzymatic synthesis in biphasic aqueous-organic systems I. Chemical equilibrium shift. Biochim Biophys Acta 658:76-89.

Martinek K, Semenov AN. 1981[b]. Enzymatic synthesis in biphasic aqueous-organic systems II. Shift of ionic equilibria. Biochim Biophys Acta 658:90-101.

Mazzotti, M., A. Kruglov, B. Neri, D. Gelosa and M. Morbidelli. 1996. A continuous chromatographic reactor: SMBr. Chem. Engng. Sci. 51:1827-1836.

McPherson, A.. 1999. Impurities, defects and crystal quality, in: Crystallization of biological macromolecules, Cold Spring Harbor Laboratory Press.

Mensah, P. & Carta, G. 1999. Adsorptive control of water in esterification with immobilized enzymes. Continuous operation in a periodic counter-current reactor. Biotechnol Bioeng 66:137-146.

Mwangi SM. 1994. Reactive precipitation of 6-aminopenicillanic acid: application of a crystallization methodology. Ph.D. thesis. University of Manchester, UK.

Nakanishi K, Matsuno R. 1986. Kinetics of enzymatic synthesis of peptides in aqueous/organic biphasic systems. Thermolysin-catalyzed synthesis of N-(benzyloxycarbonyl)-L-phenylalanine methyl ester. Eur J Biochem 161:533-540.

Nakanishi K, Kimura Y, Matsuno R. 1986. Kinetics and equilibrium of enzymatic synthesis of peptides in aqueous/organic biphasic systems. Thermolysin-catalyzed synthesis of N-(benzyloxycarbonyl)-L-aspartyl-L-phenylalanine methyl ester. Eur J Biochem 161:541-549.

Nara T, Misawa M, Okachi R. 1977. Verfahren zur Herstellung von α-Aminobenzylpenicillin. Patent Kyowa Hakko Kogyo Co., Ltd Tokyo.

Ottens, M., Lebreton, B., Zomerdijk, M., Rijkers M.P.W.M., Bruinsma D., and Van der Wielen, L.A.M. 2001. Crystallization kinetics of Ampicillin. Submitted.

Rekker, R.F. 1977. The Hydrophobic Fragmental Constant- Its derivation and application - A means of characterizing membrane systems. Elsevier, Amsterdam.

Reschke M, Schügerl K. 1984. Reactive extraction of penicillin. I: Stability of penicillin G in the presence of carriers and relationships for distribution coefficients and degrees of extraction. Chem Eng J 28:B1-B9.

Rindfleisch D, Syska B, Lazarova Z, Schügerl K. 1997. Integrated membrane extraction, enzymic conversion and electrodialysis for the synthesis of Ampicillin from penicillin G. Proc Biochem 32:605-616.

Rudolph, E.S.J, K.Ch.A.M. Luyben and L.A.M. van der Wielen. Phase behavior of some semi-synthetic antibiotics and their precursors in aqueous alkanol solutions. Fluid Phase Eq. 158-160, 903-912, 1999.

Rudolph, E.S.J, M. Zomerdijk, M.Ottens and L.A.M. van der Wielen. 2001a. Solubilities and partition coefficients of semi-synthetic antibiotics in water + 1-butanol systems. Ind. Eng. Chem. Res.

Rudolph, E.S.J., Hamelink, J.M., Zomerdijk, M., Rijkers, M.P.W.M., Ottens, M.,Vera, J.H. and van der Wielen, L.A.M. 2001b. Influence of electrolytes on the phase behaviour of water + β-lactam antibiotics and their precursors. To be submitted.

Ryu YW, Ryu DDY. 1987. Semisynthetic β-lactam antibiotics synthesising enzyme from Acetobacter turbidans: catalytic properties. Enzyme Microb Technol 10:239-245.

Sasagawa et al. 1990, in D.Teller. CRC Handbook of High-Performance Liquid Chromatography of Amino Acids, peptides and Proteins.

Schroën CGPH, Nierstrasz VA, Kroon PJ, Bosma R, Janssen AEM, Beeftink HH, Tramper J. 1999. Thermodynamically controlled synthesis of ß-lactam antibiotics. Equilibrium concentrations and side-chain properties. Enz Microb Technol 24:498-506.

Seader, J.D., and E.J. Henley. 1998. Separation process principles , Wiley and Sons.

Straathof AJJ, Litjens MJJ, Heijnen JJ. 2001. Enzymatic transformations in suspensions. In: Enzymes in nonaqueous solvents, EN Vulfsson, PJ Halling, HL Holland (eds.), Humana Press, Totowa, NJ, in press.

Svedas VK, Margolin AL, Berezin IV. 1980. Enzymatic synthesis of β-lactam antibiotics: a thermodynamic background. Enzyme Microb Technol 2:138-144.

Tavare, N.S., 1995, Industrial crystallization. Process simulation analysis and Design; The Plenum Chemical Engineering Series; Plenum Press, New York.

Taylor, R. and R. Krishna. 1993. Multi-component Mass Transfer, Wiley and Sons, New York.

Tewari YB, Goldberg RN. 1988. Thermodynamics of the conversion of penicillin G to phenylacetic acid and 6-aminopenicillanic acid. Biophys Chem 29:245-252.

Tewari YB, Schantz MM, Pandey PC, Rekharsky MV, Goldberg RN. 1995. Thermodynamics of hydrolysis of N-acetyl-L-phenylalanine ethyl ester in water and in organic solvents. J Phys Chem 99:1594-1600.

Tsuji, A., O. Kubo, E. Miyamoto and T. Yamana. 1977. Physicochemical properties of ß-lactam antibiotics: Oil-water distibution, J. Pharm. Sci. 66:1675

Van Berlo, M., M.T. Gude, L.A.M. van der Wielen and K.Ch.A.M. Luyben. 1997. Solubilities and partitioning of glycine in water+ethanol+butanol solutions. Ind Eng Chem Res, 36:2474-2482.

Van Buel, M.J., M. Gude, L.A.M. van der Wielen and K.Ch.A.M. Luyben. 1997. Gradient elution in CPC. J. Chrom. 773, 13-22.

Van der Wielen, L.A.M., M.J.A. Lankveld and K.Ch.A.M. Luyben. 1995. Anion exchange equilibria of Penicillin G, Phenylacetic acid and 6-Aminopenicillanic Acid versus Cl- in IRA400 Ion Exchange Resin. J. Chem. Eng. Data 41:239-243.

Van der Wielen, L.A.M., and S.J.G. Rudolph. 1998. On the generalization of thermodynamic properties for selection of bioseparation processes. J Chem Biochem Technol 74: 275-283.

Van der Wielen L.A.M. and K.Ch.A.M. Luyben. 1992. Integrated product formation and recovery in fermentation. Review. Current Opinions in Biotechnology 3:130-138.

Van der Wielen, L.A.M., P.J. Diepen, J. Houwers and K.Ch.A.M. Luyben. 1996. A countercurrent adsorptive reactor for acidifying bioconversions. Chem. Engng. Sci. 51:2315-2325.

Van de Sandt E.J.A.X., De Vroom E. 2000. Innovations in cephalosporin and penicillin production: painting the antibiotics industry green. Chimica oggi/ Chemistry today 18:72-75.

Van Halsema, F.E.D., L.A.M. van der Wielen and K.Ch.A.M. Luyben. 1997. The modelling of carbon dioxide aided extraction of carboxylic acids from aqueous solutions. Ind. Eng. Chem. Res. 37:748-758.

Van Ness, H.C. and M.M. Abbott. 1982. Classical thermodynamics of Non-electrolyte solutions. McGraw-Hill Book Co., New York.

Wandrey C, Flaschel E. 1979. Process development and economic aspects in enzyme engineering. Acylase L-methionine system. In: Ghose TK, Fiechter A, Blakebrough N, editors. Advances in Biochemical Engineering. Vol. 12. Springer-Verlag, Berlin p. 147-218.

Woodley JM, Lilly MD. 1990. Extractive biocatalysis: the use of two-liquid phase biocatalytic reactors to assist product recovery. Chem Eng Sci 45:2391-2396.

Chapter V Biocatalytic production of semi-synthetic cephalosporins: process technology and integration *207*

§1 Background ***207***

§2 Decision strategy ***207***

§3 The biocatalytic reaction ***209***
3.1 Adipyl-7-ADCA hydrolysis with immobilized enzyme *212*
3.2 Cephalexin synthesis with immobilized enzyme *215*

§4 The biocatalyst itself: form ***217***
4.1 Cell or enzyme ? *219*
4.2 Free or immobilized ? *219*
4.3 Free or immobilized biocatalyst: Adipyl-7-ADCA hydrolysis *220*
4.4 Free or immobilized: Cephalexin synthesis *221*
4.5 Diffusion limitation *225*

§5 The biocatalyst and its surroundings: reaction medium ***226***
5.1 Liquid / liquid ? *226*
5.2 Solid / solid ? *227*
5.3 Thermodynamic model *228*

§6 The biocatalytic reactor` ***229***
6.1 Standard or novel ? *229*
6.2 Non-isothermal bioreactor *229*
6.3 Batch or continuous ? *230*
6.4 Batch or continuous: Adipyl-7-ADCA hydrolysis *230*

§7 Process integration ***233***
7.1 Cephalexin synthesis and in-situ product removal *235*
7.2 Integration of adipyl-7-ADCA hydrolysis and Cephalexin synthesis *239*
7.3 Integration of reactions and product removal *240*
7.4 Adipyl-7-ADCA hydrolysis and downstream processing *241*
7.5 Process concept *243*

§8 Future prospects ***244***

§9 Acknowledgements ***245***

§10 References ***246***

Chapter V Biocatalytic production of semi-synthetic cephalosporins: process technology and integration

J. TRAMPER, H.H. BEEFTINK, A.E.M. JANSSEN, L.P. OOIJKAAS, J.L. VAN ROON, M. STRUBEL AND C.G.P.H. SCHROËN. (WAGENINGEN UNIVERSITY, FOOD AND BIOPROCESS ENGINEERING GROUP)

§1 Background

Developments in biotechnology have been extremely rapid the last three decades. This is particularly true for recombinant-DNA technology and genetic engineering, which have strongly stimulated the study of biocatalysis. Consequently, many new (applications of) biocatalysts exist and biocatalysts are becoming more and more shelf reagents in synthetic chemistry. Still, there is some uneasiness in their use in an industrial environment, which may largely be attributed to unfamiliarity with these catalysts from Nature.

Recently, a generic strategy has been introduced (Tramper, 1996) to facilitate an answer to the main question '*When and How to use Biocatalysts.*' Unfortunately, these twin questions are mutually dependent and the answer is thus complex by nature: many decisions have to be made simultaneously to allow a more or less rational judgment of the feasibility of a biocatalytic process. Obviously, such a judgment should involve a comparison to alternative, non-biocatalytic processes.

As an aid in assessing this feasibility, the strategy proposed identifies a number of key issues that should be resolved to facilitate a rational and optimal deployment of biological catalysts (if at all). Many examples were given (Tramper, 1996) to illustrate this decision process; more recent ones are provided in the various editions of Holland Biotechnology, in which the strategy is shortly discussed.

In the contribution of the Wageningen University to the cluster project with DSM, this strategy was applied to the synthesis of semi-synthetic cephalosporins. In this context, a number of key research subjects is presented. Characteristically, these subjects aim at very different scales, ranging from molecular to macroscopic. The chapter concludes by reviewing the future prospects for a biocatalytic set-up in cephalosporin production.

§2 Decision strategy

As indicated above, the assessment of the feasibility of a biocatalytic production scheme is neither straightforward nor simple. Ideally, such an assessment would involve a strictly quantitative performance evaluation in which all configurations and variables with respect to design and operation would be considered. As a result, the unique answer to the key question 'When and How' would be obtained. Clearly, this answer would be relative, i.e. it would be a comparison to an alternative process set-up (e.g. biocatalysis vs. chemistry). A similarly thorough analysis for the alternative process should thus be available also. Such a direct quantitative comparison, however, is outside the scope of the present chapter. Moreover, in view of the plethora of processing options and conditions, the abovementioned exhaustive process analysis is impractical at least. In addition, several process related aspects, such as social and psychological factors, may not be readily amenable to quantitative analysis. As a result, practical process selection and design is only partly quantitative. For actual selection, design and operation of a process, purely rational and quantitative criteria are commonly supplemented with (or, indeed, superseded by) more qualitative arguments concerning for instance the history and status of production facilities and market position.

The strategy identifies a number of key issues that will provide a rational basis for process selection. Ultimately, it aims at answering the question: When is biocatalysis an attractive alternative to the far more familiar chemical conversions? The strategy is outlined in Table V.1 and groups the subjects of key decisions according to scale: it ranges from a molecular and mesoscopic scale (the reaction itself and the biocatalyst facilitating the reaction) to macroscopic scales (the reactor and the overall process).

Table V.1 Key issues in the decision 'When and how to use biocatalysts'

level		*consideration*
Reaction		Thermodynamics and kinetics
Biocatalyst	type:	Single enzyme vs. whole cell
	form:	Free vs. immobilized biocatalyst
	environment:	Homogeneous vs. heterogeneous
Reactor	type:	Standard vs. novel
	operation:	(Fed-)batch vs. continuous
Process	overall design:	Level of integration

Prime aspects to be considered at the molecular scale obviously are the thermodynamic and kinetic characteristics of the actual reaction. In addition, the availability, purity and costs of reactants and of other medium components, the need for co-factors, the number of reaction steps involved, selectivity and yield, reaction conditions, ease of product purification, final product quality, and environmental aspects ('greenness') all should be evaluated carefully. Experience learns that biocatalysis will be favored only if most of these issues are clearly positive for biocatalysis (as compared to chemical processing).

A more mesoscopic level of decision-making concerns the form of the biocatalyst, i.e. whether whole cells, cell organelles, enzyme complexes, crude enzyme preparations, or isolated enzymes should be used. Catalyst availability and price are characteristic points to consider at this level. At the same level the question of using a free or an immobilized biocatalyst should be addressed. Obviously the costs associated with the immobilization itself should be earned back through the possibility of developing a more efficient process.

Also at the mesoscopic level, decisions on the homogeneity of the biocatalyst surrounding should be made. Should it be a conventional homogeneous aqueous environment, allowing the label 'green', or are non-conventional heterogeneous media to be preferred? Such heterogeneous media could involve additional liquid or solid phases and may have convincing advantages with respect to compound solubility, substrate and product inhibition, equilibrium reactant composition and product purification.

At a macroscopic scale, a reactor configuration should be chosen: should it be standard, or novel or dedicated? Here, hardware availability and available know-how are important considerations, in addition to technological considerations such as desired degree of mixing and mass transfer properties. At the same level is the question of mode of operation of the bioreactor, which can be run in a batch, fed-batch or continuous manner. In this context, kinetics, stability and form of the biocatalyst, desired substrate conversion and product concentration, and need of process control, all play a role.

At the highest macroscopic level, the need and degree of process integration with respect to reactions (if more than one is involved), process steps, and overall processing should be taken into account.

A number of key research items in deciding about the feasibility of biocatalytic production of semi-synthetic cephalosporins are worked out in more detail below. Each of these items represents a specific part of the scale mentioned above.

§3 The biocatalytic reaction

In particular when the number of reaction steps in an existing process can be drastically reduced by introduction of a biocatalyst, chances are in favor of biocatalysis. Even then, however, conventional chemistry is a strong competitor due to the availability of extensive knowledge and historic experience. Therefore, other aspects such as fewer and cheaper reagents and other essential compounds, higher yields, faster reaction rates and milder reaction conditions should speak for biocatalysis as well. Enantio- and stereo-selectivity often are strong points in favor of biocatalysis, but enormous progress has recently been made in this respect in chemical catalysis as well. If 'natural' and 'green' can help in selling a product, this can further help biocatalysis to become the production procedure of choice (Fig. V.1).

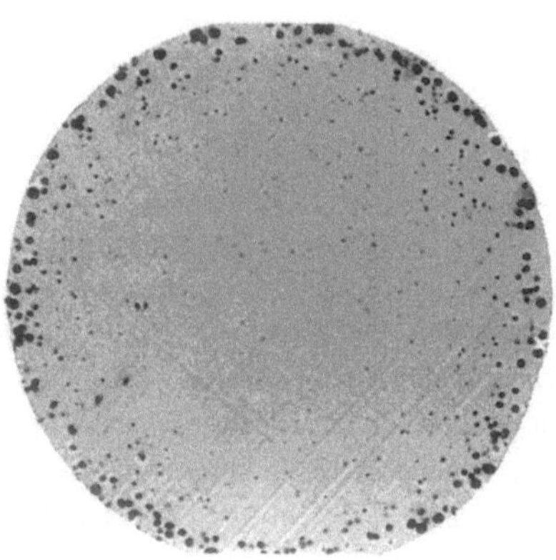

Fig. V.1 Biocatalysis is considered 'green' chemistry. An example is the immobilized micro-algae in this picture. They are studied in view of production of functional foods, energy and fine chemicals from carbon dioxide and light. Also selective removal of heavy metals from polluted waters is a potential application.

The (un)certainty that a process will run efficiently via a particular route is another important consideration. Furthermore, a certain fitting to existing processes and equipment is also desirable. Obviously this short list of considerations is far from exhaustive and not given in any order of priority. Additionally, each application will have its own specific questions to be answered in the decision-making process. Irrespective of the particular application, however, the thermodynamic and kinetic characteristics of the reactions involved always play a major role in the ultimate design decision. Consequently, the focus for the present level of decision-making (molecular scale c.q. reaction) will be on these aspects (§ 3.1 and 3.2). First, however, an outline will be given of the current state of affairs of the biocatalytic production of semi-synthetic β-lactam antibiotics.

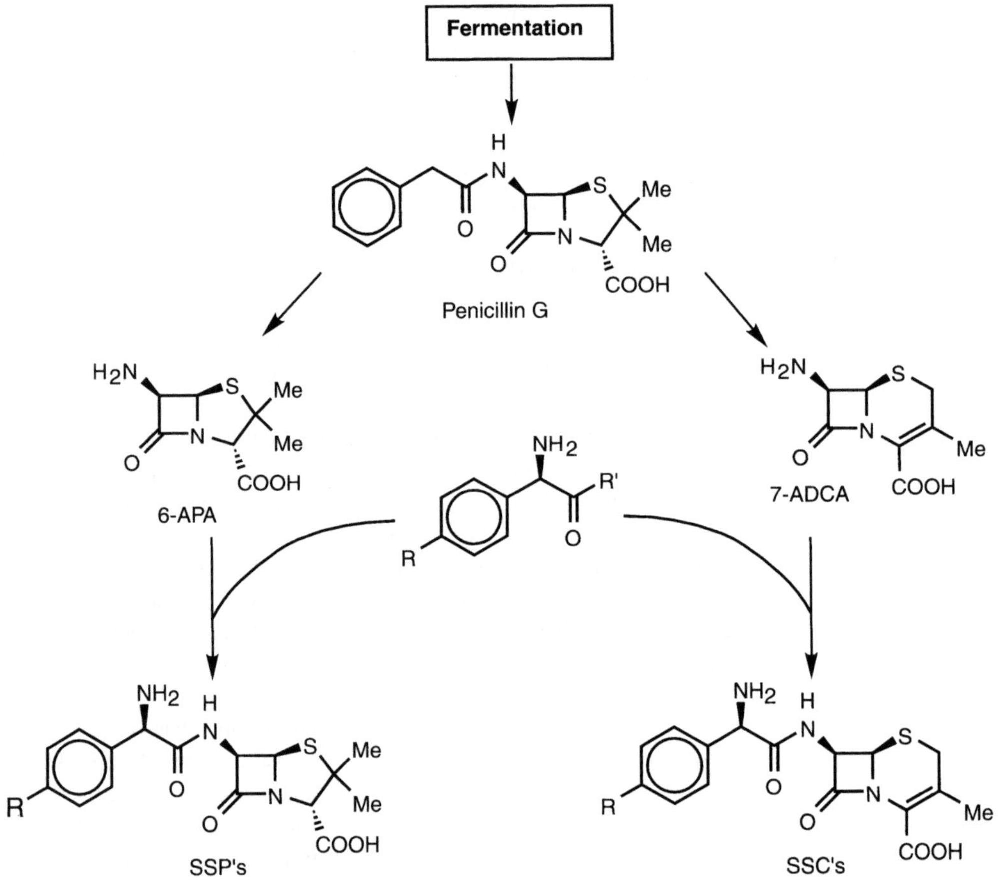

Fig. V.2 General production chart of penicillin-derived antibiotics (semi-synthetic penicillins, SSP's, e.g. Ampicillin (R = H) and Amoxicillin (R = OH)) and cephalosporin-derived antibiotics (semi-synthetic cephalosporins, SSC's, e.g. Cephalexin (R = H) and Cefadroxil (R = OH)). Intermediates: 6-aminopenicillanic acid (6-APA) and 7-amino-desacetoxy-cephalosporanic acid (7-ADCA). Reproduced from Bruggink *et al.*, 1998.

Traditionally, The Netherlands are strong in 'penicillins,' and the recent merger of DSM and Gist-brocades has further strengthened this position. Production of Penicillin G by fermentation, synthesis of 'nuclei' from this fermentation product (e.g. 6-aminopenicillanic acid, 6-APA, and 7-aminodesacetoxycephalosporanic acid, 7-ADCA), manufacturing of 'side-chains' (e.g. phenylglycine and 4-hydroxyphenylglycine), and the coupling of these nuclei with these side-chains to yield semi-synthetic penicillins (SSP's, e.g. Ampicillin and Amoxicillin) and cephalosporins (SSC's, e.g. Cephalexin and Cefadroxil) are all in one hand now (Fig. V.2).(Bruggink *et al.*, 1998; Van de Sandt & De Vroom, 2000).
The industrial production of these β-lactam antibiotics and their intermediates is undergoing a remarkable transformation. Enzyme-catalyzed processes are replacing traditional chemical conversions based on stoichiometry. The starting materials for these transformations were and still are fermentatively obtained chiral structures such as cephalosporin C, penicillin G and penicillin V. The ever-increasing insight into biochemical pathways and the possibility to genetically modify these, with optimalized fermentation processes as a result, will make applications of the chiral pool even more profitable in the years to come.

In the last two decades, many companies have replaced chemical side-chain hydrolysis (left second step in Fig. V.2), requiring hazardous chemicals and solvents such as phosphorous pentachloride and dichloromethane, by penicillin-acylase-catalyzed hydrolysis in an aqueous environment. Although microbial conversion of Penicillin G into 6-aminopenicillanic acid (6-APA) has been known for almost five decades, industrial application of the enzyme involved, penicillin acylase, has been introduced successfully only since 1980. This primarily is the result of the fact that efficient enzyme production and recovery has become available, mainly through genetic engineering.
The main difference between SSP's and SSC's is that the former class of antibiotics has a skeleton consisting of a five-membered ring fused to a four-membered ring whereas in the SSC's the five-membered ring is expanded to include six atoms. The conversion of the penicillin to the cephalosporin nucleus is a complicated process from a chemical perspective (right second step in Fig. V.2). With the aid of metabolic-pathway engineering a large step forward has been made in simplifying the production of 7-ADCA. The conversion of the five-membered into the six-membered ring can now be accomplished in the first step, that is the fermentation, yielding adipyl-7-ADCA, which can be hydrolyzed by the enzyme glutaryl-acylase to form 7-ADCA and adipic acid. New plants for this process are under construction at DSM's Delft site and expected to come on stream by the end of the year 2000. Compared to the old process for 7-ADCA, the major advantages of this process are higher purity of the end product, much greater energy efficiency and the almost complete absence of organic solvents and reagents.
Currently, the majority of N-deacylations in β-lactam production processes is thus carried out enzymatically using enzymes of the group of penicillin acylases. Present developments indicate that the same enzymes can also be exploited successfully in a synthetic fashion. Consequently, processes for the large-scale enzyme-catalyzed

production of valuable antibiotics such as Ampicillin, Amoxicillin, Cephalexin, Cefadroxil and Cefaclor, based on the condensation of the appropriate D-amino-acid derivative with a β-lactam nucleus are implemented or in a well-advanced stage of development (Fig. V.2).

3.1 Adipyl-7-ADCA hydrolysis with immobilized enzyme

Hydrolysis of adipyl-7-ADCA (a-7-ADCA) is an equilibrium reaction of which the yield is determined by initial concentrations and the equilibrium constant. The reaction rate, in its turn, is determined by the substrate and product concentrations and reaction rate constants. As indicated above, both the thermodynamic and kinetic aspects are highly relevant in reactor design and were thus studied extensively (Schroën *et al.*, 2000). Some of the results are summarized below.

Equilibrium position

Adipyl-7-ADCA-hydrolysis is an equilibrium reaction, of which the yield can be calculated using the following approach if the system behaves ideally. From experiments the apparent equilibrium constant, K_{app}, can be calculated using the total concentrations at equilibrium of adipic acid (AA), 7-ADCA and a-7-ADCA at a certain pH and temperature (e.g. Svedas et al., 1980):

$$K_{app} = \frac{[\mathrm{AA}]^{tot} \cdot [\mathrm{7\text{-}ADCA}]^{tot}}{[\mathrm{a\text{-}7\text{-}ADCA}]^{tot}} \qquad [1]$$

From the apparent equilibrium constant, the thermodynamic equilibrium constant can be calculated if the dissociation constants of the substrate and products are known (Svedas et al., 1980; Spieß *et al.*, 1999).

$$K_{app} = K_{thermo} \cdot K_{ion} \qquad [2]$$

With K_{ion}:

$$K_{ion} = \frac{\left(1 + \frac{10^{-pK_1,\,\mathrm{AA}}}{10^{-pH}} + \frac{10^{-pK_2,\,\mathrm{AA}}}{10^{-pH}}\right) \cdot \left(1 + \frac{10^{-pK_1,\,\mathrm{7\text{-}ADCA}}}{10^{-pH}} + \frac{10^{-pH}}{10^{-pK_2,\,\mathrm{7\text{-}ADCA}}}\right)}{\left(1 + \frac{10^{-pK_1,\,\mathrm{a\text{-}7\text{-}ADCA}}}{10^{-pH}} + \frac{10^{-pK_2,\,\mathrm{a\text{-}7\text{-}ADCA}}}{10^{-pH}}\right)} \qquad [3]$$

The *pK*'s of adipic acid (4.4 and 5.4; Weast, 1971) and 7-ADCA (2.95 and 4.87;Yamana and Tsuji, 1976; Yamana *et al.*, 1974) are known. The *pK*-values of adipyl-7-ADCA are expected to be equal to the respective groups of 7-ADCA and adipic acid (2.95 and 5.3; see also equilibrium position results). For the temperature dependence of the thermodynamic equilibrium constant (K_{thermo}), equation 4 is valid (e.g. Svedas *et al.* 1980):

$$\Delta G_c^0 = -R \cdot T \cdot \ln(K_{thermo}) \qquad [4]$$

ΔG_c^0 is the (pH-independent) standard Gibbs energy change. If this value is known, also the equilibrium concentrations are known as a function of temperature and pH. If no products are present at the beginning of the reaction the yield of a reaction carried out with an initial concentration of adipyl-7-ADCA$_0$ is given by:

$$Yield = \frac{0.5 \cdot (-K_{\text{app}} + \sqrt{K_{\text{app}}^2 + 4 \cdot [\text{a-7-ADCA}]_0 \cdot K_{\text{app}}})}{[\text{a-7-ADCA}]_0} \qquad [5]$$

From equation 5, it can be deduced that the yield is a function of pH, temperature and initial adipyl-7-ADCA concentration. The yield of the reaction can be predicted if the standard Gibbs energy change (ΔG_c^0) in combination with all dissociation constants, is known. The ΔG_c^0 is calculated from 14 independent apparent equilibrium constants (Eq. 4). The dissociation constants and reaction pH and -temperature can be used to calculate the reaction yield (Eq. 4, 3, 2, 1, 5).

Based on a large number of equilibrium measurements, a standard Gibbs energy change of 19.8 kJ/mol was calculated. The effect of *pH*, temperature and initial substrate concentration could be estimated and is illustrated in Fig. V.3. From Fig. V.3, it can be concluded that high *pH*'s lead to high yields. Also high temperatures lead to improvement of yield, although the difference is less pronounced as for *pH*. Although the increase in yield is not much, it has to be noted that it is a significant increase and that a difference of 1% in yield is very important for the economics of the ultimate process. The yield is also a function of the initial concentration as stated in equation 5, and, as expected, lower yields are found for higher initial substrate concentrations.

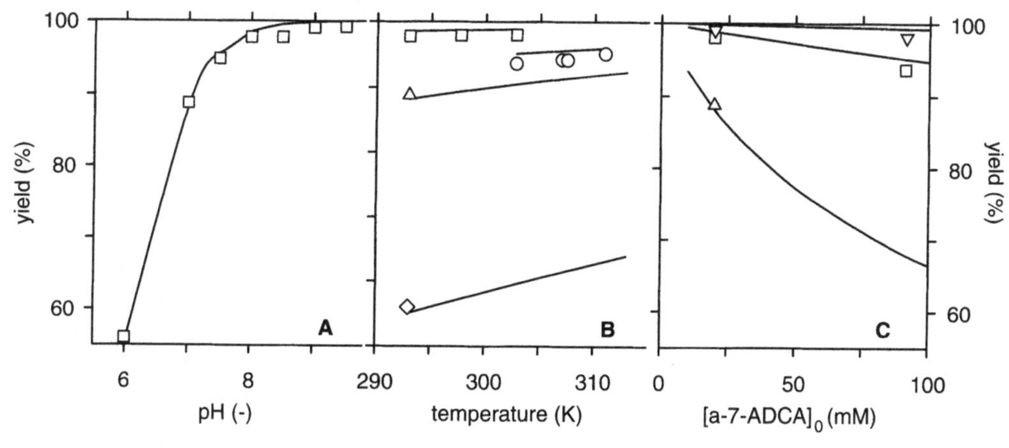

Fig. V.3 Hydrolysis of adipyl-7-ADCA. A. Yield on conversion as a function of *pH* at 293 K and 20 mM adipyl-7-ADCA. B. Yield on conversion as a function of temperature at *pH* 6 (◇) and *pH* 7 (△) both at 20 mM adipyl-7-ADCA, and at *pH* 8 for 20 (□) or 100 mM (○) adipyl-7-ADCA. C. Yield on conversion as a function of initial adipyl-7-ADCA concentration at *pH* 7 and 293 K (△), *pH* 8 and 303 K (□), and *pH* 9 and 293 K (▽). The lines in all Figures are based on a ΔG_c^0 of 19.8 kJ/mol.

Enzyme kinetics

The mechanism of the glutaryl-acylase enzyme for adipyl-7-ADCA hydrolysis has not been studied in detail but it seems reasonable to assume that the kinetic mechanism is similar to that of other acylases like penicillin G acylase (Duggleby *et al.*, 1995). In that case, the simplified reaction scheme shown in Fig. V.4 is valid.

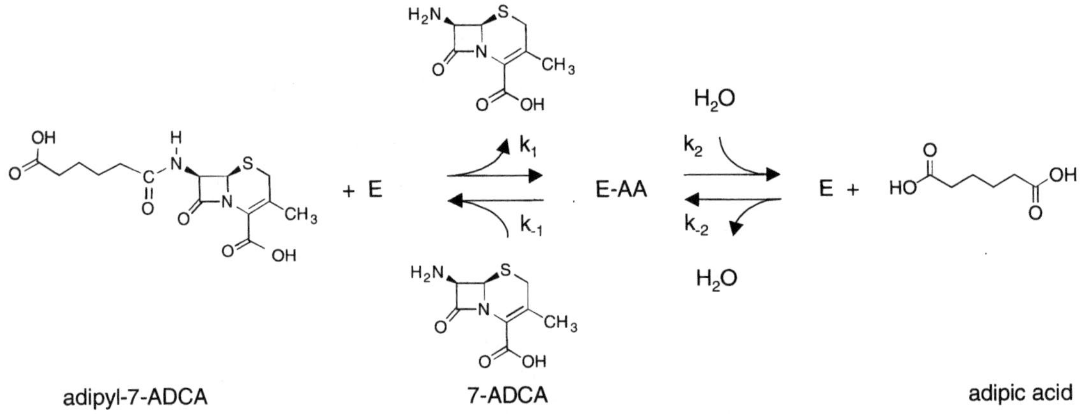

Fig. V.4 Scheme for adipyl-7-ADCA hydrolysis by glutaryl acylase. E: enzyme; AA: adipic acid.

The adipyl-7-ADCA binds to the enzyme (via reaction rate constant k_1) and an enzyme-acyl complex is formed, which is subsequently hydrolyzed (k_2). Both reaction steps are reversible; adipic acid binds to the enzyme (k_{-2}) to form the enzyme-acyl complex that can react with 7-ADCA (k_{-1}) to give adipyl-7-ADCA. As mentioned in the previous paragraph, this is a simplified approach. The actual reaction scheme will, most likely, consist of more steps (e.g. reversible formation of an enzyme-substrate complex). However, the constants connected to these individual steps could not be calculated separately. Therefore, it was decided to consider these steps as one; the constants connected to these lumped steps can be calculated with appreciable accuracy.

Because the water concentration is constant, k_2 can be replaced by k'_2 corresponding to k_2 multiplied by the water concentration. At equilibrium the synthesis rate equals the hydrolysis rate and for this reason the quotient of the reaction rate constants equals the apparent equilibrium constant (equation 1) or:

$$K_{\mathrm{app}} = \frac{k_1 \cdot k_2'}{k_{-1} \cdot k_{-2}} \quad [6]$$

Using the King-Altman-method (Cornish-Bowden, 1995), the following reaction rate (r) equations were derived for the scheme given in Fig. V.4.

$$r_{\mathrm{a\text{-}7\text{-}ADCA}} = \frac{-k_1 \cdot k_2' \cdot [\mathrm{a\text{-}7\text{-}ADCA}] + k_{-2} \cdot k_{-1} \cdot [\mathrm{7\text{-}ADCA}] \cdot [\mathrm{AA}]}{k_1 \cdot [\mathrm{a\text{-}7\text{-}ADCA}] + k_2' + k_{-1} \cdot [\mathrm{7\text{-}ADCA}] + k_{-2} \cdot [\mathrm{AA}]} \cdot E_0 \quad [7]$$

$$r_{\text{a-7-ADCA}} = -r_{\text{7-ADCA}} = -r_{\text{AA}} \qquad [8]$$

In equations 7 and 8, the concentrations are always total concentrations. E_0 denotes the initial enzyme concentration (g enzyme/ g total). The dissociation constants of all reacting species are at least 1.5 *pH* unit below the *pH* range that is investigated in this paper (7-9.5). Therefore, it can be assumed that all the reacting species are merely in one charged form, which may or may not be the correct form to bind to the enzyme. The carboxylic-acid groups will be largely negatively charged while the amino group will be mainly uncharged (see *pK*-values in previous paragraph). As shown in the results, the hydrolysis rate is independent of the *pH* and it can be expected that glutaryl acylase accepts adipyl-7-ADCA as a substrate independent of it's charge. For the reverse reaction (adipyl-7-ADCA-synthesis), the amino-group of 7-ADCA should be uncharged, as is the case, but the carboxylic-acid group of adipic acid should be uncharged as well, and this is not the case. Therefore the reaction rate constants: k_1, k'_2 and k_{-1} are independent of *pH*, while k_{-2} is calculated using the apparent equilibrium constant and is based on the total adipic-acid concentration. Herewith, the conditions for equilibrium are also obeyed by kinetics.
The temperature dependency of reaction rate constants: k_1, k'_2 and k_{-1} is described using the Arrhenius equation (e.g. Van 't Riet and Tramper, 1991).

$$k_{n,T} = k_{n,T_{\text{ref}}} \cdot e^{-\left(\Delta H_n^*/R\right)\cdot\left(T^{-1}-T_{\text{ref}}^{-1}\right)} \qquad [9]$$

In this equation $k_{n,T}$ is the reaction rate constant at temperature T, $k_{n,T\text{ref}}$ is the reaction rate constant at reference temperature (T_{ref} = 293 K) and ΔH_n^* is the activation enthalpy change (further called the reaction enthalpy) related to k_1, k'_2 or k_{-1}. The temperature dependency of k_{-2} is again calculated using the apparent equilibrium constant at the required temperature and *pH*.
Based on the model described earlier, the model parameters were calculated from experimental data. With the calculated set of constants, we were able to describe the course of the reaction for a *pH* interval of 6.5-9, a temperature interval of 293-313 K, and initial adipyl-7-ADCA concentrations between 20 and 100 mM. The initial reaction rate is slightly lower at low *pH* but in general it is constant as a function of *pH*. The initial reaction rate is strongly dependent on temperature and initial substrate concentration, as expected. The model is used to optimize the adipyl-7-ADCA hydrolysis process and evaluate reactor and process options (see §7).

3.2 Cephalexin synthesis with immobilized enzyme

Enzyme kinetics
Duggleby *et al.* (1998) have stated a mechanism for penicillin G acylase. From this, we deduced a simplified reaction scheme for Cephalexin synthesis with immobilized penicillin G acylase at *pH* 8 and at various temperatures (Schroën *et al.*, 2000, 2001[f]).

During the kinetic synthesis of Cephalexin, three enzyme-catalyzed reactions occur simultaneously (Fig. V.5; Schroën *et al.*, 2000, 2001[f]). One of these is Cephalexin synthesis from activated phenylglycine and 7-ADCA (cf. k_1 and k_3 in Fig. V.5). The activated phenylglycine can be an amide (Bruggink *et al.*, 1998; Schroën *et al.*, 2001[f]) or an ester (Blinkovsky and Markaryan, 1993; Kasche, 1985; Kasche 1986, Nam *et al.*, 1985). The other two reactions are the hydrolysis of activated phenylglycine (k_1 and k_2) and of Cephalexin (k_{-3} and k_2). Further the enzyme may be inhibited by phenylacetic acid (PAA: k_4 and k_{-4}), which may be present in 7-ADCA.

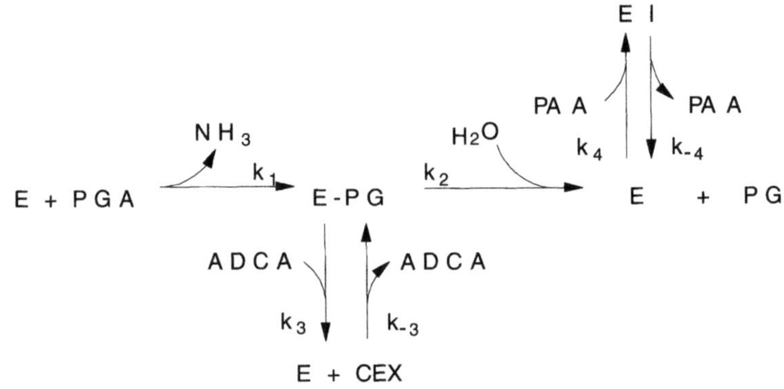

Fig. V.5 Reaction scheme for Cephalexin synthesis; rate equations for this scheme are given in Eq. 10-14. Further explanation is given in the text.

Using the King-Altman approach (Cornish Bowden, 1995), the reaction rate equations were derived for the scheme given in Fig. V.5. Because the water concentration is constant, k_2 is replaced by k'_2 corresponding to k_2 multiplied by the water concentration. All concentrations are in millimoles per kilogram total.

$$r_{\mathrm{CEX}} = \frac{k_1 \cdot k_3 \cdot [\text{7-ADCA}] \cdot [\mathrm{PGA}] - k_2' \cdot k_{-3} \cdot [\mathrm{CEX}]}{\Sigma} \cdot E_0 \qquad [10]$$

$$r_{\mathrm{PG}} = \frac{k_1 \cdot k_2 \cdot [\mathrm{PGA}] + k_2 \cdot k_{-3} \cdot [\mathrm{CEX}]}{\Sigma} \cdot E_0 \qquad [11]$$

$$r_{\mathrm{PGA}} = \frac{-k_1 \cdot k_2' \cdot [\mathrm{PGA}] - k_1 \cdot k_3 \cdot [\mathrm{PGA}] \cdot [\text{7-ADCA}]}{\Sigma} \cdot E_0 \qquad [12]$$

$$r_{\text{7-ADCA}} = -r_{\mathrm{CEX}} \qquad [13]$$

$$\Sigma = k_1 \cdot [\mathrm{PGA}] + k_2 + k_3 \cdot [\mathrm{ADCA}] + k_{-3} \cdot [\mathrm{CEX}] + k_4 / k_{-4} \cdot [\mathrm{PAA}] \cdot (k_2 + k_3 \cdot [\mathrm{ADCA}]) \qquad [14]$$

Here, E_0 denotes the initial enzyme concentration (g enzyme per g total). In case no product is present at time zero (subscript 0), the mass balances for the nucleus and for the side-chain are:

$$[\mathrm{ADCA}]_0 = [\mathrm{ADCA}] + [\mathrm{CEX}] \qquad [15]$$

$$[\mathrm{PGA}]_0 = [\mathrm{PGA}] + [\mathrm{PG}] + [\mathrm{CEX}] \qquad [16]$$

For the temperature-dependency of all reaction rate constants, the Arrhenius equation is used as described above for adipyl-7-ADCA hydrolysis. With the calculated sets of constants, the time course of reactions could be described accurately for a wide range of substrate concentrations (0-500 mM) and temperatures (273-303 K). Various reaction conditions were compared using the yield on 7-ADCA ([Cephalexin] / [7-ADCA]0) and the synthesis/hydrolysis ratio ([Cephalexin] / [phenylglycine]) as criteria. In general, high substrate concentration and low temperatures lead to high yields and high synthesis/hydrolysis ratios (see also Kaasgaard and Veitland, 1992; Boesten *et al.*, 1996; Boesten *et al.*, 1997 and other related DSM patents). This model was used to optimize the synthesis reaction and to evaluate reactor and process options (see §7).

§4 The biocatalyst itself: form

In addition to molecular-scale considerations (*cf.* Table V.1) about reaction type and characteristics (§3 above), a more mesoscopic scale is highly relevant in ultimate process selection. The main decision at this scale is: How to use a biocatalyst? It involves questions about the form of the biocatalyst, *i.e.* whether whole cells (which may be growing or non-growing), 'engineered' cells (again: growing or non-growing), cell organelles, an enzyme complex, a crude enzyme preparation, or an isolated enzyme should be used. For each of these catalysts, it should be decided whether it should be used freely suspended or immobilized (Fig. V.6).

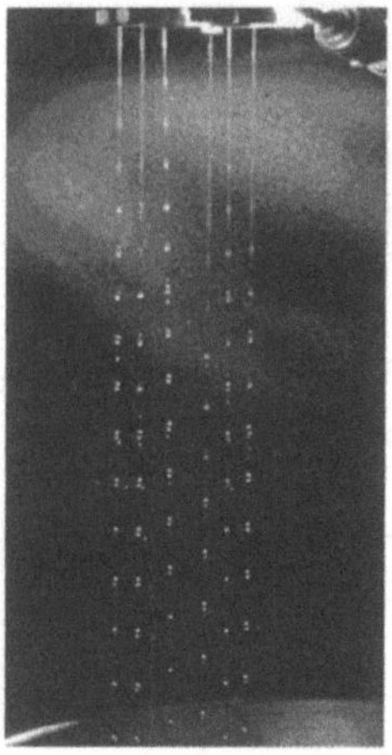

Fig. V.6 Immobilization is the binding of a biocatalyst to a solid support or the entrapment in a gel or membrane; the aim is to facilitate a more efficient use. The picture shows the large-scale production of immobilized biocatalyst by means of a multi-nozzle resonance device.

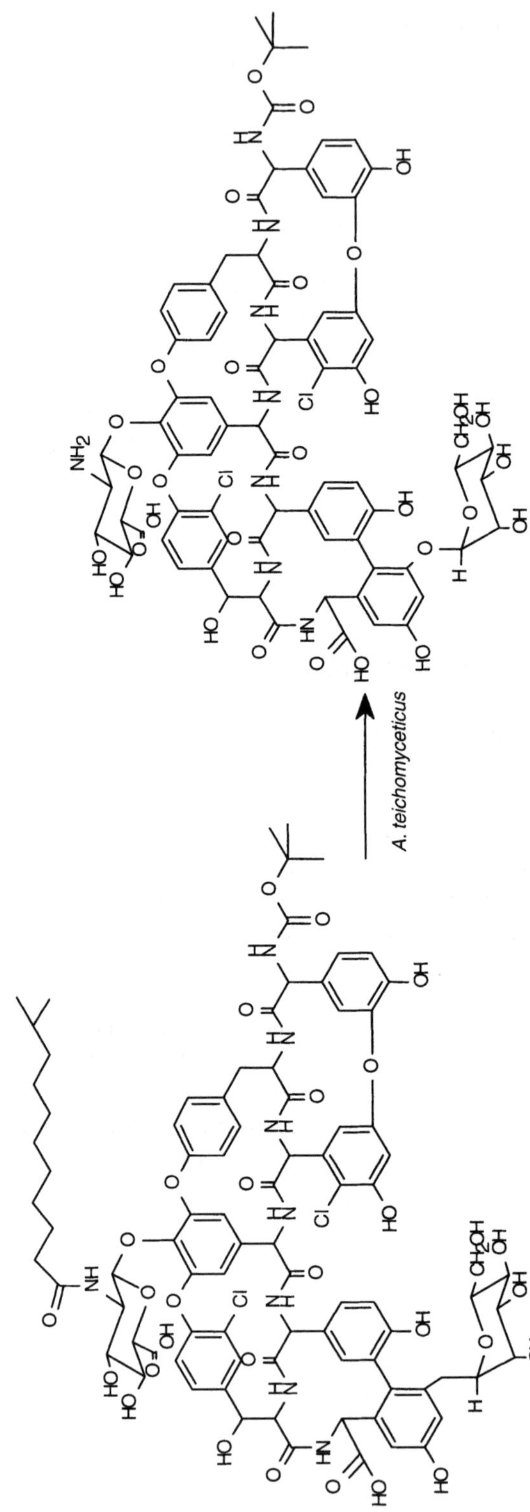

A40926

deacyl-A40926

Fig. V.7 Biotransformation of A40926 to deacyl-A40926 by *Actinoplanes teichomyceticus*.

4.1 Cell or enzyme ?

The source of biocatalysts can be of either microbial, plant, or animal origin. In particular in the latter two cases, enzyme purification usually is unavoidable, as tissue is generally not suitable in biocatalysis. Plant and animal cells are not cultivated for enzyme production to any considerable extent, nor are they genetically modified with the purpose of developing a whole-cell biocatalyst. This is in sharp contrast to microbial cells, which are genetically engineered both for the purpose of increased enzyme production and for use as biocatalyst. Intermezzo 1 shortly describes a selective deacylation reaction requiring the use of whole cells.
Factors to consider in deciding whether an enzyme or a whole cell should be used are availability, number of steps involved, yield, cofactor need, by-product formation, product purity, price, etc. If, for instance, a biocatalyst is not available, it should be made in-house or developed/manufactured externally on contract basis, in other words, at least one extra step is needed.

Intermezzo 1: A40926 (Jovetic* et al.*, 1998)
Compound A40926 (Fig. V.7) is a lipoglycopeptide antibiotic with a strong inhibitory activity against Gram-positive bacteria and Neisseria gonorrhoea. *It is produced in a fermentation process using* Actinomadura *ATCC 39727. New and often more specific and more active analogs of this attractive antibiotic can be obtained by synthetic re-acylation of its de-acylated derivative. To this end A40926 is selectively de-acylated by* Actinoplanes teichomyceticus *to yield the glycopeptide nucleus. These proprietary compounds are currently under development at Biosearch Italia. In cooperation with this company we have demonstrated the feasibility of immobilized* A. teichomyceticus *for deacylation of A40926 (Fig. V.7).*

4.2 Free or immobilized ?

Biocatalyst costs are often relatively important. Therefore, one of the most important reasons to consider the immobilization of a biocatalyst is the possibility of facilitated reuse or continuous utilization. Furthermore, the stability of biocatalysts, in particular of enzymes and recombinant cells, can be improved by immobilization. Immobilization also creates the possibility of using the biocatalyst in a packed-bed or fluidised-bed reactor. Due to easy retention of the immobilized biocatalyst in the reactor, high volumetric activities can be realized and, in case of immobilized growing cells, operation can be done under washout conditions with respect to free cells. Obviously, the costs associated with immobilization should be earned back through the possibility of developing a more efficient process. The synthesis of Cephalexin and its precursors is described below as an example. It includes two enzymatic steps in which immobilization of the enzymes is essential (Schroën *et al.*, 2001).

Cephalexin is a cephalosporin antibiotic with a substantial market as shown in chapter I. It is produced by DSM from 7-ADCA and activated phenylglycine in an enzymatic reaction. This is not only a very good example of the competition between chemistry and biocatalysis, but also an example where both whole cells as well as immobilized enzymes are used. In paragraph 3 the production of adipyl-7-ADCA by a genetically engineered micro-organism was introduced, followed by an enzymatic hydrolysis to form 7-ADCA, one of the two precursors of Cephalexin. The other precursor, activated phenylglycine, on the other hand is an interesting example where biocatalysis has not replaced chemistry. This in contrast to 4-hydroxyphenylglycine (precursor for Amoxicillin and Cefadroxil), which is fully made by biocatalysis. The production of phenylglycine occurs with an established chemical process in optimized, dedicated equipment, while market growth is limited. For the hydroxy-derivative there is still a substantial market growth and in particular in such cases biocatalysis can be an attractive competitor to traditional chemistry. Besides, the chemical characteristics of phenylglycine make biocatalysis also less advantageous as compared to the hydroxy-compound.
The last step in the production of Cephalexin, the coupling of 7-ADCA with activated phenylglycine, again is carried out with an immobilized enzyme. As long as genetic engineering has not resulted in a micro-organism that can do the whole job, combination of fermentation, chemistry and immobilized-enzyme technology, is an economic route to Cephalexin.

4.3 Free or immobilized biocatalyst: Adipyl-7-ADCA hydrolysis

For a single reaction (*e.g.* hydrolysis), the effect of immobilization is most clearly reflected in the apparent enzyme activity. In immobilized biocatalysts, substrate and product diffusion will be (partly) hindered and therewith the observed reaction rate will become lower (*e.g.* Van ' t Riet and Tramper, 1991). This effect can even be more pronounced if a pH-gradient is generated in the immobilization material as described in literature (Spieß et al., 1998; Guisan *et al.*, 1994).

For adipyl-7-ADCA-hydrolysis with immobilized enzyme, we investigated these (*pH*) effects (Schroën *et al.*, 2001[b]). For hydrolysis reactions, it is known that a *pH*-gradient of 2-3 can exist in the immobilized biocatalyst (Spieß *et al.*, 1998), the *pH* inside the biocatalyst being lower then in the surrounding liquid. The activity of glutaryl acylase is lower at low pH; therefore, the reaction rate of immobilized glutaryl acylase might be enhanced by use of an appropriate buffer to eliminate the *pH*-gradient. In Fig. V.8, as an example, the reaction rate is plotted as a function of the concentration of phenylglycine amide (the substrate for Cephalexin synthesis), which also acts as a buffer. The reaction rate increases up to 3.5 times (right y-scale) compared to the situation without buffer (horizontal line). For other buffers, a linear relationship between the reaction rate and the buffering capacity was found. This indicates that the buffer levels of the *pH*-gradient inside the immobilized biocatalyst and therewith the biocatalyst can be used more effectively.

For hydrolysis of adipyl-7-ADCA, the *pH* gradient within the immobilized biocatalyst leads to over-dosage of enzyme (compared to free enzyme), although addition of an efficient buffer can keep the reaction rates at an acceptable rate.

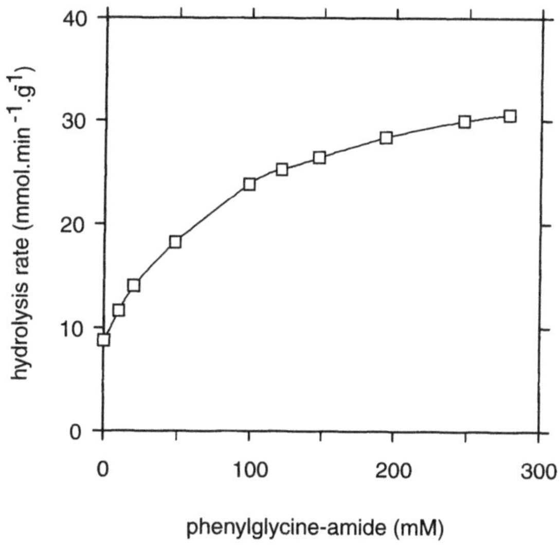

Fig. V.8 Initial adipyl-7-ADCA hydrolysis rate as a function of the added phenylglycine amide concentration. Conditions: 100 mM initial adipyl-7-ADCA concentration, 1% glutaryl-acylase, *pH* 8.0 and 293 K.

4.4 Free or immobilized: Cephalexin synthesis

The situation becomes more complicated if more reactions occur simultaneously together with diffusion limitations. The rates of the various reactions will be influenced differently by diffusion limitation and therewith the formation of one of the products may be enhanced.

During kinetic Cephalexin synthesis two products are formed, Cephalexin and the by-product phenylglycine. Using the model described in the enzyme kinetics part together with the calculated constants for the respective biocatalysts the course of the reaction can be described accurately (Schroën *et al.*, 2001[a]). For a comparison between the biocatalysts two criteria were used, the Cephalexin yield on ADCA and the ratio between Cephalexin and phenylglycine concentration (called Synthesis/ Hydrolysis-ratio or S/H-ratio). They are both indicators for the processes that occur inside the immobilized biocatalyst, and more specifically, for the effect of internal diffusion limitation (from now on called diffusion limitation) on the course of the reaction.

In Fig. V.9, the course of reactions carried out with free enzyme and the immobilized forms Assemblase® 3750 and Assemblase® 7500 at pH 8.0, 293 K and 100 mM substrate concentration is shown. Assemblase® 7500 carries approximately double the enzyme loading of Assemblase® 3750. When comparing the reactions shown in

Fig. V.9, it becomes clear that the course of the reaction is very different. Assemblase® 3750 produces more phenylglycine and less Cephalexin as compared to the free enzyme and Assemblase® 7500 produces even more phenylglycine and less Cephalexin compared to Assemblase® 3750. If the kinetic properties of the enzyme did not change upon immobilization, as we assume, this effect must be a result of diffusion limitation.

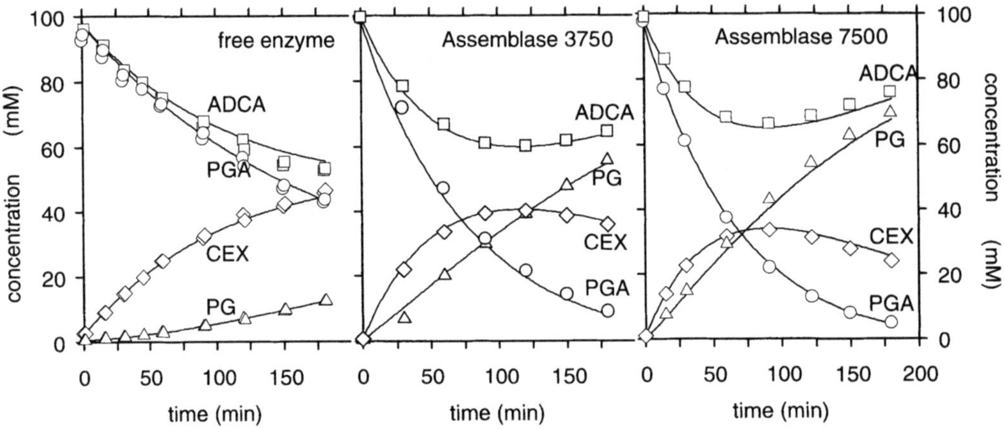

Fig. V.9 Cephalexin synthesis with free enzyme, Assemblase® 3750, and Assemblase® 7500 at *pH* 8.0, 293 K and 100 mM of both substrates. Lines represent model predictions with Eq. 1-7 and with model parameters from Schroën *et al.*, 2001[a].

Diffusion of ADCA inside the biocatalyst may be hindered, and therewith, synthesis can be reduced and hydrolysis of the enzyme-phenylglycine complex may be enhanced. Cephalexin diffusion from within the biocatalyst to the bulk solution also may be hindered, resulting in increased Cephalexin hydrolysis and subsequent formation of phenylglycine. Further, diffusion of phenylglycine amide may be hindered resulting in slower reactions and possibly a shift in synthesis/hydrolysis ratio. Which of these effects is most pronounced is investigated, by varying the temperature.

If the enthalpy of a reaction with an immobilized enzyme is (considerably) lower than that of the same reaction with free enzyme, the reaction with immobilized enzyme is considered to be diffusion limited (see *e.g.* Tramper, 1979). For Assemblase® 7500 and free enzyme, the reaction enthalpies were calculated as described in Schroën *et al.* (2001[f]). For Assemblase® 3750, experimental values for the reaction enthalpies were available (Schroën *et al.* (2001[f]). The results are shown in Table V.2 together with statistical 95% confidence intervals. The average deviation between model and measured concentrations was less than 1.8 mM for all enzymes. The reaction enthalpies of free enzyme are in the order of magnitude that can be expected for enzyme-catalyzed reactions.

Table V.2 Calculated reaction enthalpies for free enzyme, Assemblase® 3750 and Assemblase® 7500 at *pH* 8.0 and their statistical reliability.

Reaction	Estimated value (kJ/mol)	95% confidence interval lower	upper
Free enzyme			
ΔH_1	53.4*	38.6*	68.2*
ΔH_2	46.2*	40.6*	52.6*
ΔH_3	66.9	46.7	87.1
ΔH_{-3}	115.3	86.9	143.8
ΔH_{4-4}	52.4	33.8	70.9
Assemblase® 3750			
ΔH_1	54.8	42.2	67.3
ΔH_2	45.3*	41.3*	49.3*
ΔH_3	13.1	-18.8	44.9
ΔH_{-3}	48.2	15.0	81.4
ΔH_{4-4}	6.6	-3.5	16.6
Assemblase® 7500			
ΔH_1	31.3*	29.6*	33.2*
ΔH_2	40.5*	40.0*	40.9*
ΔH_3	22.7	17.5	28.0
ΔH_{-3}	57.7	44.9	70.5
ΔH_{4-4}	25.2	8.7	41.8

* *Calculated from phenylglycine-amide hydrolysis experiments only.*

When comparing free enzyme with Assemblase® 3750, it becomes clear that the 95% confidence intervals of both ΔH_3 and ΔH_{-3} do not overlap. This indicates that both coupling of ADCA to the enzyme-complex and binding of Cephalexin to the enzyme are diffusion limited or put differently, both Cephalexin synthesis and hydrolysis are influenced by diffusion effects. When comparing Assemblase® 7500 with free enzyme, it becomes clear that besides Cephalexin synthesis and -hydrolysis also amide binding to the enzyme is diffusion limited (ΔH_1), because again, the 95% confidence intervals do not overlap. In Assemblase® 7500, three steps are influenced by diffusion effects and at low substrate concentrations these effects result in an undesirable situation for the synthesis reaction.

The kinetic constants are obviously a function of temperature but their temperature-dependency differs. For free enzyme, the hydrolysis of Cephalexin is suppressed at low temperatures (see Table V.2: highest value for the reaction enthalpy). For the immobilized enzymes, the synthesis reaction is less temperature sensitive than the hydrolysis reactions (reaction enthalpy of synthesis is considerably smaller than the reaction enthalpy for hydrolysis). In both cases the results are higher S/H-ratio's and better yields at low temperature, which is beneficial for the process (Kaasgaard and Veitland, 1992; Boesten *et al.*, 1996; Boesten *et al.*, 1997 and other related DSM patents).

For all three biocatalysts the yield and S/H ratio were calculated at different temperatures, the result is shown in Fig. V.10. The difference in yield between the three biocatalysts becomes smaller at low temperature although it is still clear. In spite of the slower reactions, at 278 K the reaction rates are approximately 3-4 times lower, the diffusion rates apparently still could not keep up with the reaction rates resulting in diffusion limitation and lower S/H-ratio's and yields for the immobilized enzymes. The effects described in Fig. V.10 become smaller at higher substrate concentrations but the difference still remains (Schroën *et al.*, 2001[a]).

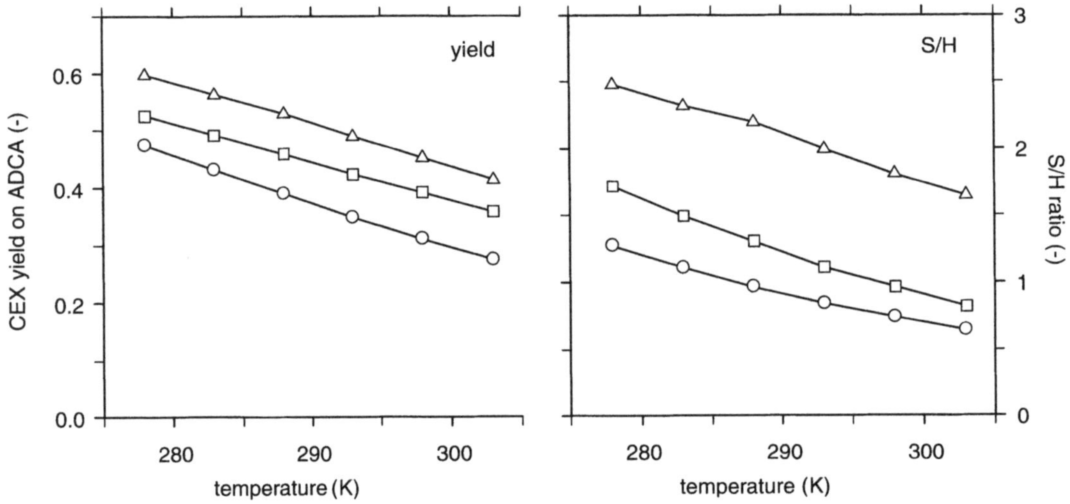

Fig. V.10 Calculated Cephalexin yield and S/H-ratio for reactions carried out at *pH* 8.0, starting from 100 mM of both substrates and at various temperatures for free enzyme, Assemblase® 3750 and 7500 (from top to bottom).

As far as the reaction yield and S/H-ratio are concerned, the free enzyme is the biocatalyst of choice for the Cephalexin synthesis process. On the other hand, it is more difficult to remove the free enzyme from a reaction mixture. Because both the yield and the S/H-ratio for all biocatalysts are relatively close together at high substrate concentration, the eventual choice of biocatalyst may still be the immobilized one because of the ease of removal and recycling.

4.5 Diffusion limitation

Various strategies to overcome diffusion limitation are discussed by Mohy Eldin *et al.* (2001). These strategies can be divided into the following groups:

(1) Immobilization of the enzymes onto soluble-insoluble matrices.

Here the enzyme is immobilized onto a matrix that is reversible in a soluble or insoluble state. During the enzymatic reaction, the immobilized enzymes are in a soluble form. After completion of the reaction, the enzyme can be recovered in the insoluble form by changing the pH, ionic strength, temperature and/or salt concentration of the reaction medium.

(2) Immobilization of the enzymes in thermally reversible hydrogels.

Hydrogels exhibiting a lower critical solution temperature, deswell when warmed up to this temperature. Above this temperature the gel collapses. Reversibly, the gel expands and reswells when it is cooled below the critical temperature. The thermal cycling acts like a 'hydraulic pump' which enhances mass transfer of the substrate into and the product out of the gel, thereby minimizing the diffusion limitation and product inhibition problems.

(3) Immobilization of the enzymes in pressure-sensitive gels.

Pressure-sensitive gels swell or collapse sharply upon the rise or fall of environmental pressure. Under pressure cycling operation, the gel is analogous to a cylinder and a piston, such as in a 'micro-pump'. The piston pushes up and draws back, corresponding to the swelling and collapsing of the gel. In addition to the diffusive flow, a convective flow occurs in this way, which enhances mass transfer within the gel and reduces diffusion limitation and product inhibition.

(4) Processing the enzymatic reaction under non-isothermal conditions.

The reaction can be carried out under non-isothermal conditions in a membrane bioreactor. The temperature difference across the membrane causes thermal diffusion for both water and salts, which is known as 'thermo dialysis effect'. This effect brings the substrate in and the product out through the membrane, thus minimizing both the diffusion-limitation and the product-inhibition problems. The net result is enhancement of the reaction rate.

Among these techniques to overcome the diffusion limitation problem, not one of them is suitable for all enzymes. It is essential that a careful choice of strategy is made for each enzyme.

§5 The biocatalyst and its surroundings: reaction medium

An important question concerns the reaction medium. Will it be the conventional aqueous environment with the label 'green', or non-conventional media such as organic solvents or even no liquid phase at all, such as in solid-to-solid biocatalysis and solid-substrate fermentations (Fig. V.11). The latter can have convincing advantages with respect to solubility, substrate/product inhibition, reaction equilibrium, and product purification.

Fig. V.11 Solid-state fermentation is the cultivation of micro-organisms, mainly fungi, on moist solid substrates in the absence of free-flowing water. It is a possible alternative for submerged fermentation for the production of fine chemicals, e.g. polyphenols, polypeptides, polysaccharides, etc.

5.1 Liquid / liquid ?

Water as reaction medium for biocatalysis has long been advocated as one of the major advantages in the application of biocatalysts. At the same time, this advantage severely limits the number and scope of applications. In particular, when poorly water-soluble substrates and/or products are involved, or when they are inhibitors of the biocatalyst, the eventual product concentrations are generally very low, making product recovery laborious and expensive. In those cases, an organic-solvent phase or another non-conventional medium can be used as substrate reservoir and/or product extractant. This will enhance the overall end concentration of the product, which facilitates downstream processing. It is also possible to shift equilibriums in the direction of synthesis if a solvent can be found which selectively extracts the desired product into the organic phase. Complete replacement of the free water by a suitable organic solvent, thereby lowering the water activity, can reverse hydrolytic reactions (*e.g.* transesterification instead of fat hydrolysis by lipases). Lower water activities can also greatly enhance stability of enzymes. Finding a biocompatible solvent is obviously the first requirement in cases where the use of water is more of a limitation than an advantage.

An aspect which also has to be dealt with, is the more complex system created in cases in which only some of the free water is replaced by a water-insoluble organic solvent, yielding a so-called two-liquid-phase bioreactor. An example of reactors especially developed for such systems is given in Fig. V.12.

Fig. V.12 The liquid-impelled loop reactor is a bioreactor that is specifically designed for biocatalysis in non-conventional media. In addition to the aqueous reaction phase, a water-immiscible organic solvent phase serves as reactant reservoir or product sink.

5.2 Solid / solid ?

For hydrophilic compounds the advantages of the introduction of an organic solvent in case of hydrophobic substrates and/or products, can be gained in systems with solid substrate and/or solid product as a second phase. On a commercial scale, such bioconversions are already applied; examples are the production of aspartame, acryl amide and D-malate.

Intermezzo 2: D-malate (Michielsen, 1999)

We are currently developing a model bioreactor for the conversion of solid Ca-maleate to solid Ca-D-malate by permeabilized whole cells of Pseudomonas pseudoalcaligenes *in a continuous process. D-malate is an optically active α-hydroxy acid. In contrast to L-malate, it sporadically occurs in nature. However, it is becoming a valuable source of chirality for various applications in organic chemistry, e.g. synthesis of antibiotics, vitamins and pheromones. D-malate is produced chemically or enzymatically. Both routes result in D-malate of high optical purity, but the molar yield depends very much on the method used. With chemical methods yields are low, since more than one reaction step is required. Enzymatic production can result in high yields. However, cofactor regeneration is needed, making the process very complex. The process with permeabilized whole cells of* Pseudomonas pseudoalcaligenes *seems to be the most promising. This process does not require cofactor regeneration and the substrate maleate, a cheap bulk chemical, is converted into D-malate in a simple one-step bioconversion. We have developed a mechanistic model for this bioconversion, featuring the kinetics of the maleate dissolution, the actual bioconversion and the D-malate crystallization. Using this model we calculated the economic feasibility of this process and concluded that D-malate production by permeabilized* Pseudomonas pseudoalcaligenes *cells in a solid-to-solid system can be very profitable, especially in a continuous bioreactor.*

5.3 Thermodynamic model

A key feature of biocatalysis in substrate suspensions is that the reaction thermodynamics exhibit a switch-like behavior giving either very low or very high yields. If the product concentration at equilibrium is lower than the solubility limit, it accumulates in the liquid phase up to the equilibrium level and the overall yield is low. Alternatively, if the product concentration at equilibrium is higher than the solubility limit, the product will precipitate before equilibrium is reached. This process will continue until the concentration of one of the reactants falls below saturation. Only then can equilibrium in the liquid phase be established.
To be able to predict which type of 'switch behavior', low yield or high yield, a particular reaction would exhibit, a thermodynamic model has been developed (Ulijn *et al.*, 2001). In this model a mass action ratio is calculated, which only requires the melting points of reactants and products and in certain cases the pKa of ionisable groups as input parameters. If the mass action ratio is larger in value than the equilibrium constant (K_{eq}), the product will not precipitate. In case the mass action ratio is smaller than K_{eq}, solid-to-solid reaction is feasible and high yields can be obtained.
Ulijn *et al.* (2001) have predicted the feasibility of a number of peptide and antibiotic synthesis reactions and they have compared this with data from literature. In case of Amoxicillin (from hydroxy-PG and 6-APA) and Cephalexin (from PG and 7-ADCA) product precipitation is not feasible according to the model predictions. This is in agreement with the experimental data (Diender *et al.* 1998, Schroën *et al.* 1999). The synthesis of Cefamandole (from PG and TET-ACA) is feasible according to model predictions, however this was not confirmed experimentally (Nierstrasz *et al.* 1999).
In cases where the thermodynamics are unfavorable for precipitation driven synthesis, other methods are necessary. Ulijn *et al.* (2001) have shown that metal complexation might be a valuable tool to improve the yield. In case of the thermodynamically controlled Amoxicillin synthesis the yield in a substrate suspension is 2 mmol/l. In the presence of 0.1 M $ZnSO_4$ the same suspension gives a yield of 31 mmol/l. Optimization of *pH* and zinc concentration could probably improve this yield even further.

§6 The biocatalytic reactor

On the level of the bioreactor we have to make two general decisions. Can we suffice with the standard bioreactor or is it advantageous to use/develop a more novel and dedicated type? Also, it should be decided whether the process is going to be operated batch wise, in a fed-batch mode or continuously.

6.1 Standard or novel ?

The stirred-tank reactor has been in existence for a long time and still is the 'workhorse' in the fine-chemical industry. Operators have 'grown up with it' and there is a great reluctance to change to a different configuration. However, the characteristics of biocatalysts sometimes call for other bioreactor types. Michaelis-Menten kinetics in combination with product inhibition, for instance, call for a plug-flow reactor such as the packed-bed reactor. For this, immobilization of the biocatalyst is required. This is also true for fluidised-bed reactors, which can be advantageous if, for example, the medium feed contains particulate materials. The reactor shown in Fig. V.12 is a novel bioreactor especially designed for biocatalysis in non-conventional media. When isolated, co-factor requiring enzymes are involved, the membrane can be a good choice. Tonnage is another decision factor in the choice of the multipurpose standard vessel *versus* the dedicated novel bioreactor; the larger the scale the more appropriate a dedicated plant becomes.
In short, kinetics, form of the biocatalyst, availability, experience, tonnage, desired mixing and mass-transfer properties all play a dominant role.

6.2 Non-isothermal bioreactor

The non-isothermal bioreactor has been developed to overcome diffusion limitation. In this reactor, the enzyme is immobilized onto a membrane and a temperature gradient is applied over this membrane (Fig. V.13). Under these conditions, the activity of the catalytic membrane is higher as compared to the activity under isothermal conditions. Under non-isothermal conditions a volume flow of liquid is observed together with differential solute fluxes (Mita, 2000).
The non-isothermal bioreactor has been applied for a number of enzymatic reactions e.g. the β-galactosidase hydrolysis of lactose and the penicillin G acylase catalyzed hydrolysis of Cephalexin (Mita, 2000, Mohy Eldin *et al.*, 2000). In these reactions, an increase in reaction rate was found with increasing temperature gradient, indicating that diffusion of the substrates was enhanced by the applied gradient.

The non-isothermal bioreactor has also been tested for the synthesis of Cephalexin (Schroën *et al.*). This exploring study shows that application of a temperature gradient across the membrane resulted in less by-product formation and therewith in a favorable situation compared to isothermal synthesis. The immobilized enzyme could be used repeatedly. Thermodialysis is an interesting new technique that has potential for application on a larger scale. Prerequisite is that the membrane surface area per volume of reactor can be improved.

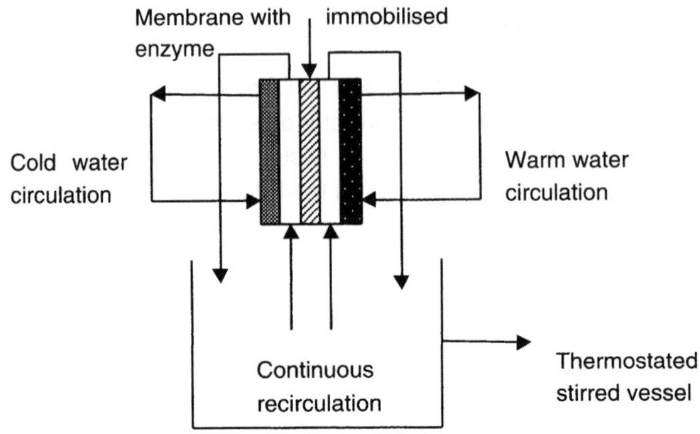

Fig. V.13 Schematic representation of the thermodialysis reactor.

6.3 Batch or continuous ?

Many factors play a role at this decision level. Type of kinetics, operational stability of the biocatalyst, form of the biocatalyst, type of bioreactor, bioreactor operating costs, necessity of process control, compatibility with the downstream processing, tonnage or scale of operation, existing facilities and experience, feedstock, type of product, and other factors are all involved in an intricate manner. Economy, however, determines the eventual decision. In the chemical industry, in particular, bulk products with a huge market are manufactured in fully continuous, specially designed processes. In the fermentation industry, on the contrary, the products with the largest market volumes are made in batch and especially fed-batch fermentors, although biological wastewater treatment plants are continuous.

Cheetham (2000) discusses the 15 most successful commercial biotransformations (excluding fermentative conversions and steroid modifications). Five to six of these are performed in continuous bioreactors with immobilized biocatalysts. Mostly, these processes yield relatively low-value, large-volume products. Even small improvements in production, therefore, have significant economic consequences, which justifies the necessary research and development in process optimization. Several of the batch processes also featured immobilized biocatalysts, among others for the production of 6-APA.

As stated above, many factors play a role in an intricate manner at this decision level and in fact no unifying criteria exist on which basis the mode of operation can be chosen.

6.4 Batch or continuous: Adipyl-7-ADCA hydrolysis

The adipyl-7-ADCA hydrolysis kinetics were modeled and the course of the reaction could be described accurately for substrate concentrations between 20 and 100 mM, *pH*'s between 6.5 and 9 and temperatures between 293 and 313 K (Schroën *et al.*, 2000). With the kinetic model described before, batch reactors and continuous stirred tank reactors (CSTR) were evaluated.

For a batch reaction, the maximum reaction yield (from now on called yield) is mainly determined by the equilibrium position (see kinetics), which was found to be a function of *pH*, temperature and initial adipyl-7-ADCA concentration. High *pH*'s, high temperatures and low initial substrate concentrations all rendered high yields. For a CSTR, the concentrations of substrate and products and the temperature and *pH* determine whether a certain yield can be achieved or not (Schroën *et al.*, 2001[9]).

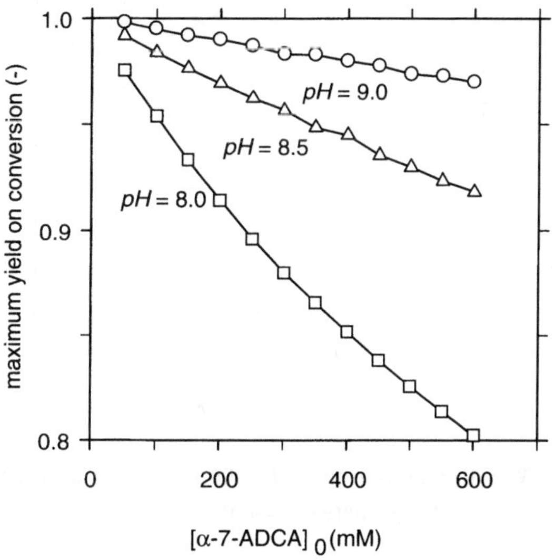

Fig. V.14 Hydrolysis of adipyl-7-ADCA: theoretical maximum yield in a batch process as depending on *pH* (*T* = 308 K).

Because both adipyl-7-ADCA and 7-ADCA are prone to chemical degradation, this effect has to be taken into account in an actual process. Therefore, chemical stability of adipyl-7-ADCA and 7-ADCA was incorporated in the model and this was used to evaluate reaction and process conditions.

Batch

At a temperature of 308 K and a *pH* of 8, 8.5 and 9, the maximum yield was calculated for substrate concentrations between 50 and 600 mM using the extended model including chemical degradation (Fig. V.14). The result is practically the same as for the model without chemical degradation. At these *pH* values and this temperature, less then 0.1% of the initial amount of adipyl-7-ADCA is affected by chemical degradation. High yields are found at high *pH* (beneficial apparent equilibrium constant: K_{app}: see also kinetics); also, the yield increases with decreasing initial adipyl-7-ADCA concentration. Consequently, high yields are not possible for a batch process, since such a process is likely to involve high initial substrate concentrations.

At *pH* 8.5, the maximum yield was calculated for 308, 318, 328 and 338 K, again including chemical degradation (Fig. V.15). A higher temperature has a positive effect on the yield because the apparent equilibrium constant increases (see Eq. 4).

Although the effects of temperature and initial substrate concentration are there, they are relatively small. At 338 K, a yield of >95% can even be achieved at an initial substrate concentration of 600mM.

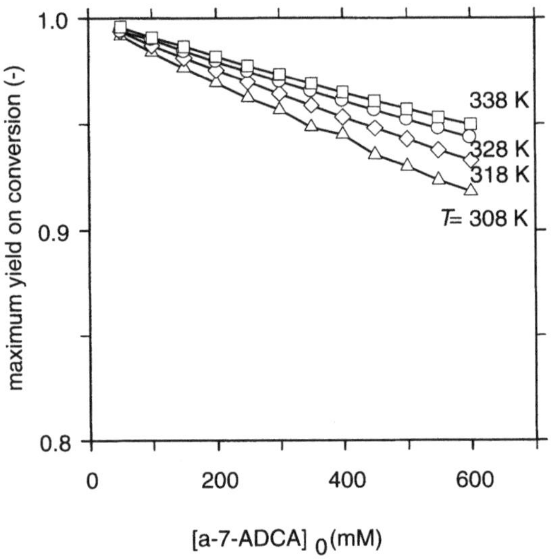

Fig. V.15 Hydrolysis of adipyl-7-ADCA: theoretical maximum yield in a batch process as depending on temperature (*pH* = 8.5).

Chemical degradation is a function of temperature and substrate concentration. The temperature enhances both the reaction rate of the enzyme and the chemical degradation rate and clearly the effect on chemical degradation is more pronounced. At a higher substrate concentration, the reaction takes longer to reach completion, and therewith more chemical degradation may take place. At the highest temperature and substrate concentration, only 0.6% of the 7-ADCA was lost to chemical degradation. Therefore, it can be concluded that chemical degradation is not a problem for the batch process.

Continuous

For continuous processing, similar calculations were carried out as for the batch process. Continuous conversion of adipyl-7-ADCA to 7-ADCA is assumed to have 95% yield and a daily production of 5 kmol 7-ADCA. For initial adipyl-7-ADCA concentrations between 50 and 600 mM, the percentage of 7-ADCA lost to chemical degradation was calculated for a *pH* of 8.5 and temperatures between 303 and 328 K (Fig. V.16).

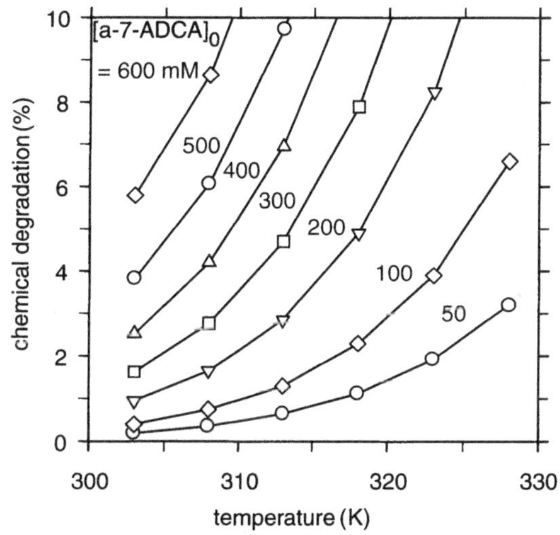

Fig. V.16 Theoretical extent of chemical 7-ADCA degradation in a CSTR as a function of temperature and initial adipyl-7-ADCA concentration (*pH* 8.5).

With respect to chemical degradation, continuous processing is quite different from batch processing. Chemical degradation does play an important role in a CSTR. At high temperatures and high substrate concentrations, chemical degradation is very pronounced because of the longer residence times that are required in continuous processes. However, there are possibilities for continuous operation; at 303 K and low substrate concentrations, chemical degradation is less than 1%. Chemical degradation can even be further reduced using a cascade of CSTR's in which the over-all residence time is considerably lower. Another option is to use a cascade of CSTR's of which the last one is operated at a slightly elevated *pH* in order to enhance the reaction rate. Whether these are genuine process options, will be discussed later.

Cephalexin synthesis

For the Cephalexin process similar calculations to those for the adipyl-7-ADCA hydrolysis process can be carried out (Schroën *et al.*, 2001[f]),based on the kinetic model presented earlier. In summary, high *pH*'s and temperatures lead to high losses due to chemical degradation (and also to undesirable reaction conditions from a viewpoint of enzyme kinetics). It was shown that both batch and continuous processes are feasible.

§7 Process integration

Finally, throughout the whole decision-making, the need and degree of process integration with respect to reactions (if more than one is involved), process steps and overall integration should be taken into account. In general, integration is an important, multifaceted issue in the development of a biocatalytic process. As always, in biotechnology, integration of biological and engineering sciences immediately from the start is essential, but will not be discussed further here. If more than one reaction step is involved, integration of these steps into a 'one-pot' synthesis should be considered, because in that case intermediate products do not have to be isolated.

D-(-)-*p*-hydroxyphenylglycine

An interesting example of integration of reaction steps is the production of D-(-)-*p*-hydroxyphenylglycine, one of the two building blocks in the synthesis of Amoxicillin and Cefadroxil. Racemic hydantoins are converted by microbial cells to D-*N*-carbamoyl amino acids. Under the pH conditions of the conversion, the unwanted L-enantiomer undergoes rapid racemization, thus allowing a theoretical yield of 100%. The resulting D-*N*-carbamoyl amino acid can then be converted into product by treatment with nitrous acid. Kanegafuchi has operated this process since 1983, rapidly followed by Recordati, who further integrated the reaction steps. Recordati uses *Agrobacterium radiobacter*, which contains in addition to D-hydantoinase, N-carbamoyl-D-amino acid amidohydrolase as well. Thus, direct formation of product is possible without the nitrous-acid treatment required after the D-hydantoinase reaction. The amidohydrolase reaction is especially high yielding because gaseous carbon dioxide is formed as a by-product, pulling the reaction to completion. The analogous process to produce D-phenylglycine is mechanistically and commercially inferior to the chemical process for reasons given in paragraph 4.2.

Intermezzo 3: S-2,2-dimethylcyclopropanecarboxamide (Birch* et al. *1994)

An industrial process for the production of S-2,2-dimethylcyclopropane-carboxamide (an intermediate for the production of Cilastatin, which is used in conjunction with certain antibiotics to prevent their degradation in the kidney), based on an integrated chemical and biocatalytic approach, has been developed by Lonza (Birch et al., *1994). The two biotransformation steps are executed subsequently by a strain containing a highly active nitrile hydratase and by an* E. coli *strain containing a cloned stereoselective amidase. It is claimed that the biocatalytic route has a higher enantioselectivity than the chemical alternatives (less recycle) and that it is a one-pot synthesis, at least the two biotransformation steps. The process runs efficiently on an industrial scale (15 m3) and allows cost-effective production of a pure product in a high yield.*

Considering the possibility of physical integration of process steps is also of importance, e.g. part of the downstream processing in the actual biocatalysis. An example of this is two-liquid-phase biocatalysis (see *medium design* above), whereby the organic-solvent phase is used as extractant for the product. Similarly, solid-to-solid biocatalysis is a pronounced example of integration of a purification step (crystallization or precipitation) in the actual biocatalysis.

Overall process integration goes further; each individual process step is optimized, not on its own, but with respect to the whole process, taking into account all interactions. It is a current topic, both in chemical and biochemical engineering, but illustrative industrial examples are scarce and outside the scope of this chapter.

7.1 Cephalexin synthesis and in-situ product removal

The simultaneous reactions during kinetic Cephalexin synthesis (Verweij and De Vroom, 1993; Schroën *et al.*, 2001[f]) are schematically given in Fig. V.5. It illustrates that if either one of the two hydrolysis reactions can be suppressed, higher Cephalexin concentrations can be reached, less by-product will be formed (Schroën *et al.*, 2001[e]) and downstream processing may be facilitated. Because hydrolysis of activated phenylglycine is catalyzed by water, this reaction can only be suppressed if for instance a water miscible solvent was used. However, addition of solvents is not an option here, therefore, Cephalexin removal was our goal (Schroën *et al.*, 2001[d]). If Cephalexin can be removed from solution it will be protected from enzymatic hydrolysis (*e.g.* Hernandez-Justiz *et al.*, 1998) and from chemical degradation (Yamana and Tsuji, 1974; Yamana *et al.*, 1976). Complexation (Clausen, 1995; Bruggink, 1996; Kemperman *et al.*, 1999), adsorption using e.g. amberlite XAD (Chaubal *et al.*, 1995) and aqueous two-phase systems (Hernandez-Justiz, *et al.*, 1998) are known from literature as product removal techniques in Cephalexin production.
In one of our publications (Schroën *et al.*, 2001[e]), we describe *in situ* adsorption. The main disadvantage of this technique is that it is not specific enough; the substrates also bind to the adsorbent, therewith resulting in lower substrate concentrations and consequently lower product concentrations (Schroën *et al.*, 2001[e]). This is also the case in the aqueous two-phase system presented by Hernandez-Justiz and co-workers (1998); substrate is lost to the extracting polymer-phase resulting in a lower product concentration.
In contrast to adsorption and aqueous two-phase extraction, complexation is a very specific removal technique; only the Cephalexin is bound. Together with our cluster project colleagues from Nijmegen University we selected the following complexing agents for further testing during Cephalexin synthesis: β-naphthol, 1,5-dihydroxy-naphthalene, 2,3-dihydroxy-naphthalene and 2,7-dihydroxy-naphthalene. The effect of the complexing agents on the Cephalexin reaction was evaluated using the complexation rate, the residual Cephalexin concentration in solution and the residual enzyme activity as indicators. The results are summarized in Table V.3.

Table V.3 Complexation of 30 mM Cephalexin at pH 7.5 and 293 K and residual enzyme activity.

	complexation rate constant (1/min)	$[CEX]_{eq}$ (mM)	Residual activity (%)
1,5-dihydroxy-naphthalene	0.7	3.2	100
β-naphthol	0.4	6.0	28
2,3-dihydroxy-naphthalene	0.6	6.9	29
2,7-dihydroxy-naphthalene	1.8	8.0	6

Ideally, the complexing agent should have a residual Cephalexin concentration of 0, a high complexation rate and not inactivate the enzyme. For all complexing agents it is expected that the complexation rate is fast enough to keep up with the reaction rate during synthesis. But none of the complexing agents seems to be optimal on all three points. The most promising ones are 1,5-dihydroxy-naphthalene (least influence on enzyme activity) and β-naphthol (highest complexation rate) and these two were further tested in synthesis experiments.

For both complexing agents the Cephalexin concentration in solution (Fig. V.17;(◇) first increased and started to decrease after 15 minutes of incubation. After 30 minutes of incubation, the concentration in solution did not change anymore. The average of all Cephalexin concentrations in solution measured after 30 minutes of incubation, were used for the model predictions.

The results are quite different for 1,5-dihydroxy-naphthalene and β-naphthol, but in both cases, more Cephalexin was formed than in reactions without complexing agent. A thorough comparison is given later in this chapter. First the results in Fig. V.17 were analyzed using an existing model for enzymatic Cephalexin synthesis with Assemblase® 3750 (Schroën *et al.*, 2001[f]) and free enzyme (Schroën *et al.*, 2001[a]). For both enzymes, the model was extended with a fast reversible complexation step, to incorporate the effect of complexation in the model.

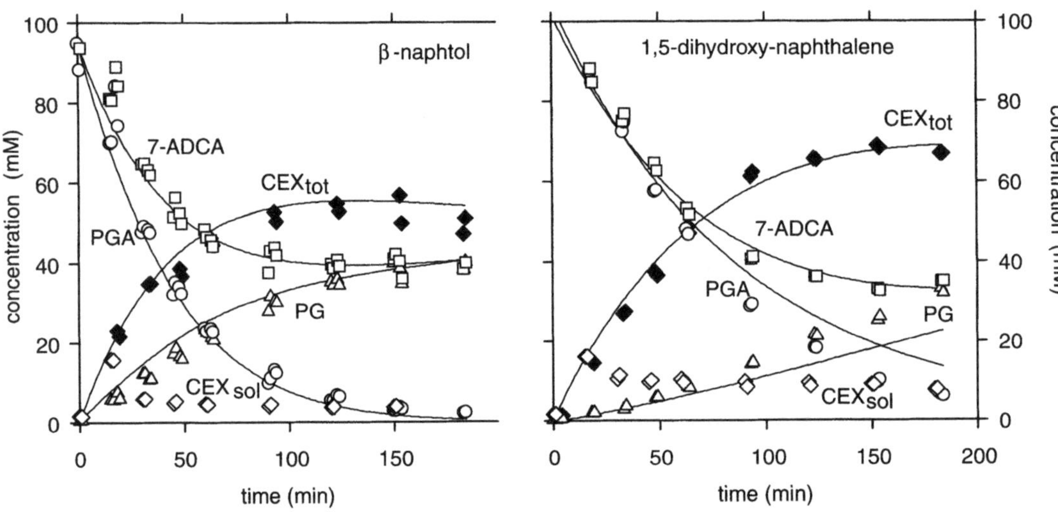

Fig. V.17 *In situ* complexation with 1,5-dihydroxy-naphthalene or β-naphthol during enzymatic Cephalexin synthesis at *pH* 7.5 and 293 K. For 1,5-dihydroxy-naphthalene, the lines in the graph are predictions obtained with the Cephalexin-synthesis model for Assemblase® 3750, for β-naphthol the lines are obtained with the model for free enzyme. The substrate concentrations are approximately 100 mM.

For 1,5-dihydroxy-naphthalene, the course of the reaction is described accurately by the model for Assemblase® 3750, which is also the enzyme used in these experiments. For β-naphthol, this is not the case (result not shown), but the course of the reaction is very similar to that predicted for free enzyme (lines in the graph). It was checked whether the same reaction, but now carried out with free enzyme, could be described by the model for free enzyme and this was found to be the case.

This implies that β-naphthol does not influence enzyme kinetics. This is a strong indication that diffusion limitation in the immobilized enzyme (as described in the part on choice of biocatalyst), may be partly lifted by inactivation of the enzyme by β-naphthol (see Table V.3). This results in higher Cephalexin concentrations (see Fig. V.17) because of the higher S/H-ratio of the 'free' enzyme. This finding also opens another option; combination of complexation with 1,5-dihydroxy-naphthalene and Cephalexin synthesis with free enzyme should lead to an even higher yield because this complexing agent leaves a lower residual Cephalexin concentration in solution. This was checked experimentally and the result is shown in Fig. V.18. Again, the course of the reaction could be described accurately with the model for the free enzyme.

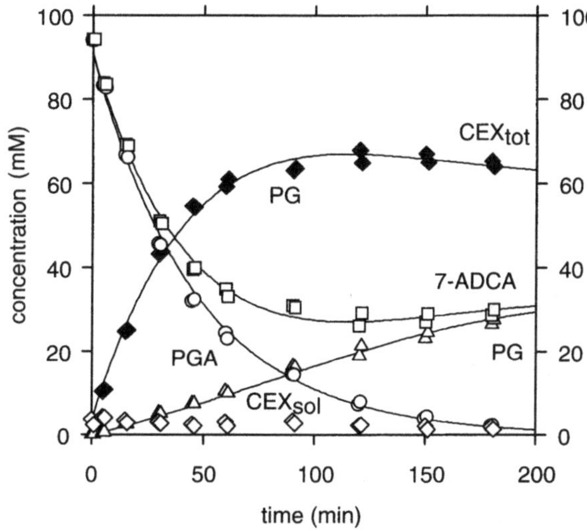

Fig. V.18 ***In situ*** **complexation with 1,5-dihydroxy-naphtalene during Cephalexin synthesis at 293 K, *pH* 7.5 and 0.5% free enzyme. Substrate concentrations are approximately 100 mM. Lines are constructed with the model for free enzyme.**

As shown, the models for Assemblase® and free enzyme can be used to describe the course of the Cephalexin synthesis reaction in the presence of a complexing agent. These models are now used to evaluate the effect of complexing agents that leave different Cephalexin concentrations in solution (equilibrium concentration). The calculations were done for substrate concentrations of 100 mM for both phenylglycine amide and 7-ADCA; the Cephalexin concentration in solution was varied between 0 mM (ideal complexing agent) and 50 mM (maximum Cephalexin concentration for synthesis with free enzyme in absence of complexing agent). Down-stream processing of the complex was not considered. The result is shown in Fig. V.19.

At high equilibrium Cephalexin concentrations (> 42 mM for Assemblase® and > 50 mM for free enzyme), the concentrations of Cephalexin and phenylglycine obviously level off to the values found for no complexation. If the equilibrium concentration is decreased, the Cephalexin concentration increases because part of the Cephalexin

is protected against enzymatic hydrolysis. The phenylglycine concentration shows a maximum. At intermediate equilibrium Cephalexin concentrations, the maximum Cephalexin concentration is reached after a longer time, in which also more hydrolysis of phenylglycine amide takes place compared to the situation without complexing agent. At low equilibrium Cephalexin concentrations, the phenylglycine concentration decreases again. The complexing agent is very effective and most of the Cephalexin is protected against enzymatic hydrolysis, therewith also resulting in lower phenylglycine concentrations.
For as far as the choice of complexing agent is concerned, it is clear that the complexing agent should have an equilibrium Cephalexin concentration that is as close to zero as possible, and preferably forms complexes fast. However, also complexing agents that are less effective still give considerably higher product yields as compared to the situation without complexing agent.

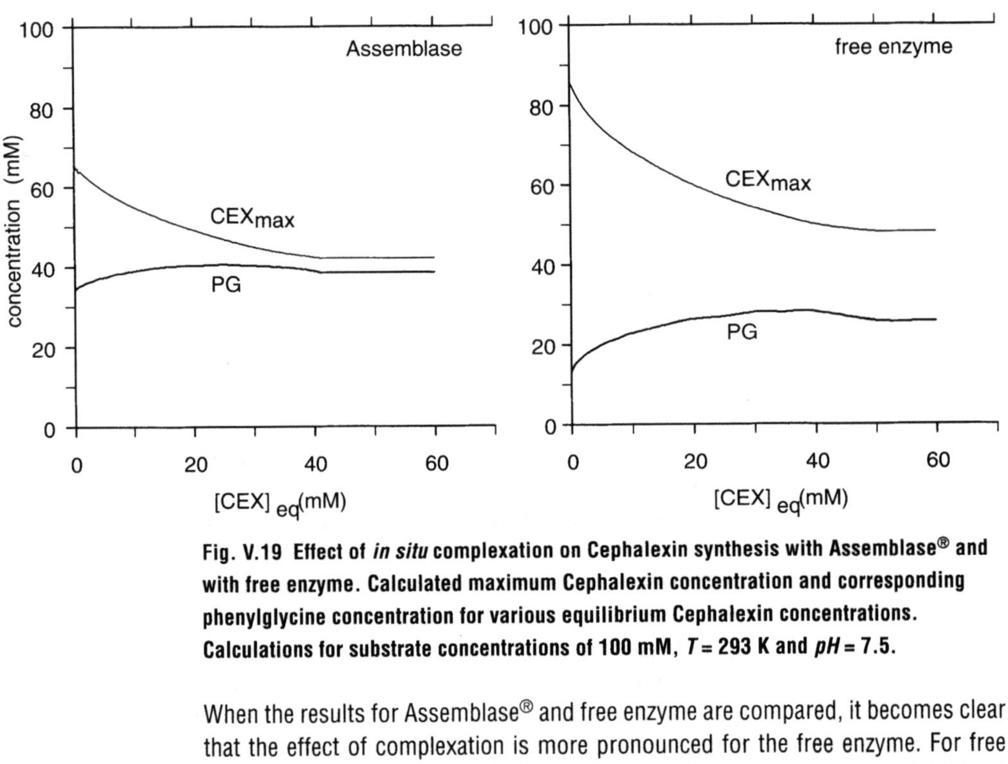

Fig. V.19 Effect of *in situ* complexation on Cephalexin synthesis with Assemblase® and with free enzyme. Calculated maximum Cephalexin concentration and corresponding phenylglycine concentration for various equilibrium Cephalexin concentrations. Calculations for substrate concentrations of 100 mM, T = 293 K and pH = 7.5.

When the results for Assemblase® and free enzyme are compared, it becomes clear that the effect of complexation is more pronounced for the free enzyme. For free enzyme, the maximum Cephalexin concentration increases from 48 mM (without complexing agent) to 85 mM (with the ideal complexing agent that leaves no Cephalexin in solution). Therewith the yield on 7-ADCA increases with 80%. For Assemblase®, the increase in yield can be as high as 60%. The effect on the ratio between maximum Cephalexin concentration and phenylglycine concentration is even more pronounced. For free enzyme, this ratio increases more than three-fold in the presence of the 'ideal complexing agent' while for Assemblase® the increase is less than two-fold.

For Cephalexin synthesis, *in situ* product removal by complexation is an interesting tool to increase the yield and to increase the ratio between Cephalexin and phenylglycine. However, also the enzyme properties are important; the effect of *in situ* product removal can be considerably larger if an enzyme with beneficial kinetic properties is used. In this light, it is very interesting to use free enzyme in combination with a complexing agent. If enzyme and product can be separated as easily as was the case for the immobilized enzyme and the enzyme activity is not influenced too much this is the combination to be preferred. Otherwise, an immobilized enzyme with a lower enzyme loading is the way to go.

7.2 Integration of adipyl-7-ADCA hydrolysis and Cephalexin synthesis

Based on the kinetics of both reactions (Schroën *et al.*, 2000, 2001[f]) it seemed possible to carry out both reactions simultaneously, in one pot. Therewith a reduction in the number of reaction steps can be achieved. Several reactions were carried out starting from 100 mM of adipyl-7-ADCA and phenylglycine amide concentrations ranging from 30 to 200 mM. Examples of experiments and model predictions (Schroën *et al.*, 2001[e]) based on the kinetic models for the separate reactions (see kinetics) extended with the increase in reaction rate of glutaryl acylase in the presence of phenylglycine amide (Schroën *et al.*, 2001[b]) are shown in Fig. V.20.

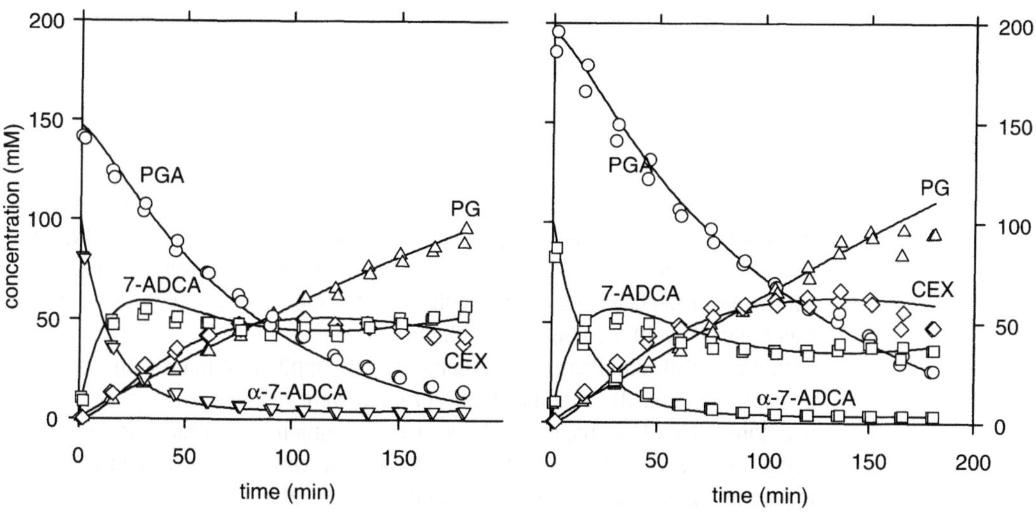

Fig. V.20 Integration of adipyl-7-ADCA hydrolysis and Cephalexin synthesis at initial phenylglycine-amide concentrations of 150 and 200 mM (Schroën *et al.*, 2001[e]).

It was found that adipyl-7-ADCA hydrolysis and Cephalexin synthesis could indeed be carried out simultaneously. The yield of the combined process was only slightly lower then the yield of the separate processes carried out under the same reaction conditions. This implies that the number of reaction steps in the Cephalexin process can be reduced (Schroën *et al.*, 2001[e]).

7.3 Integration of reactions and product removal

It has to be mentioned that the yields (and the S/H-ratio) of the adipyl-7-ADCA hydrolysis process and Cephalexin synthesis process can be considerably higher when carried out under their respective optimal conditions (see kinetics). To further increase the yield of the combined process, it was decided to also incorporate Cephalexin removal. This was done by adsorption (Schroën *et al.*, 2001[e]) and by complexation with β-naphthol (Schroën *et al.*, 2001[d], 2001[e]; Kemperman *et al.*, 1999). The latter will be discussed in more detail. Because complexation is more effective at *pH* 7.5 then at 8.0, and the course of the Cephalexin reaction at *pH* 7.5 is not different from that at *pH* 8.0 (Schroën *et al.*, 2001[f]), it was decided to study complexation at *pH* 7.5.

The yields and S/H-ratio's of the separate reactions at 293 K and *pH* 7.5 were compared to those for the combined process and the process with Cephalexin removal by β-naphthol complexation (Table V.4).

Table V.4 Comparison of Cephalexin concentration and phenylglycine concentration at maximum antibiotic concentration for various reactor concepts starting with 100 mM phenylglycine amide, 100 mM of adipyl-7-ADCA or 7-ADCA. If β-naphthol was added this was in a molar ratio of 1:1 with the initial 7-ADCA or adipyl-7-ADCA concentration.

	Cephalexin synthesis	Cephalexin synthesis + removal β-naphthol	simultaneous reactions + removal with β-naphthol
[CEX] mM	42	66	47
[PG] mM	38	21	40
S/H ratio	1.1	3.1	1.2

In situ product removal can give a considerable improvement in yield if the removal method specifically targets the product, as is the case for complexation. With β-naphthol (column 2), a 50% higher concentration of Cephalexin was found as compared to Cephalexin synthesis only and the amount of phenylglycine formed was reduced by 40% compared to column 2.

Combination of simultaneous reactions with β-naphthol complexation was also possible (column 3), although both enzymes were negatively influenced by β-naphthol, which is a major drawback of complexation with β-naphthol. Because adipyl-7-ADCA hydrolysis was too slow to keep up with Cephalexin synthesis, the maximum Cephalexin concentration and the S/H ratio decreased compared to Cephalexin synthesis in combination with β-naphthol complexation (column 2). However, even in this case, in which the enzyme concentrations were not in tune, a higher Cephalexin concentration and a slightly higher S/H ratio were found in comparison to the base case. This indicates that a highly integrated process in which two reactions are carried out simultaneously in combination with in situ product removal is possible, although further optimization is clearly required.

7.4 Adipyl-7-ADCA hydrolysis and downstream processing

As discussed before, the adipyl-7-ADCA hydrolysis reactor can either be operated batch-wise or continuously. For batch reactors, chemical degradation of 7-ADCA and adipyl-7-ADCA can be neglected, while for continuous reactors it has to be taken into account. It was shown, however, that the reaction conditions and reactor configuration could be chosen in such a way that also for continuous operation chemical degradation is not a problem.
For down stream processing, membrane separation and crystallization were considered (Schroën *et al.*, 2001[9]). Separation of adipic acid, adipyl-7-ADCA and 7-ADCA by nanofiltration, however, was not possible at the reaction *pH*; therefore, crystallization was attempted. The effect of efficiency of crystallization (the residual 7-ADCA concentration in solution is considered lost) on the calculated over-all yield of a batch process is shown in Fig. V.21.

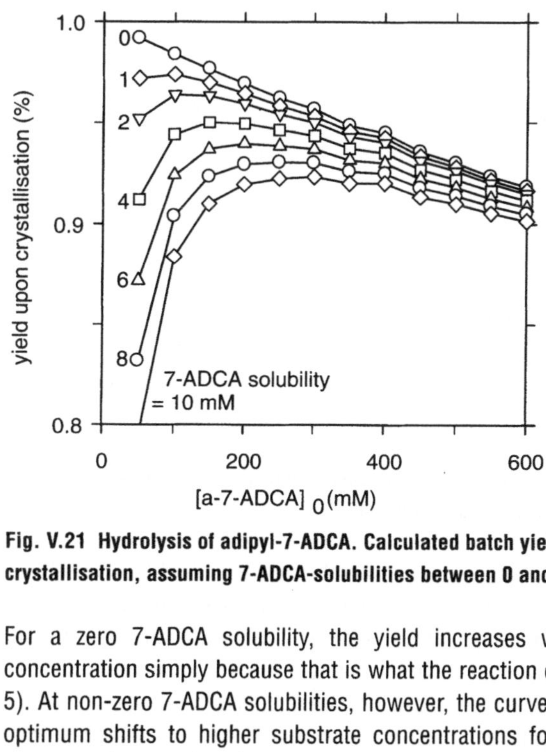

Fig. V.21 Hydrolysis of adipyl-7-ADCA. Calculated batch yield upon 7-ADCA crystallisation, assuming 7-ADCA-solubilities between 0 and 10 mM.

For a zero 7-ADCA solubility, the yield increases with decreasing substrate concentration simply because that is what the reaction dictates (see also Eq 1 and 5). At non-zero 7-ADCA solubilities, however, the curve exhibits an optimum. The optimum shifts to higher substrate concentrations for higher residual 7-ADCA concentrations or, in other words, less efficient crystallization. In that case, the positive effect of the low substrate concentration on reaction yield is outweighed by the larger loss due to inefficient crystallization.
The situation depicted in Fig. V.21 corresponds to a situation in which the reaction liquid is used directly in the crystallization step. However, if the reaction liquid can be concentrated prior to crystallization then a high reaction yield can be combined with a relatively low loss due to less efficient crystallization. For a CSTR, a

concentration step after the reaction seems to be a necessity because high substrate concentrations in the reactor are not a good option because of the chemical degradation as described in the operation part.

Membrane separation after crystallization

After isolation of the crystallized 7-ADCA, some of the 7-ADCA (2 mM as experimentally determined) and the residual amounts of adipyl-7-ADCA and of adipic acid remain in solution. Preferably, these components should be recycled. Adipic acid can be re-used in the fermentation. Adipyl-7-ADCA should ideally be returned to the hydrolysis reactor and 7-ADCA to the crystallization vessel. Therefore, it was investigated whether these components can be separated by membrane separation. Several membranes were tested and the best results were obtained with the Nanomax -50 membrane at *pH* < 5 (Fig. V.22).

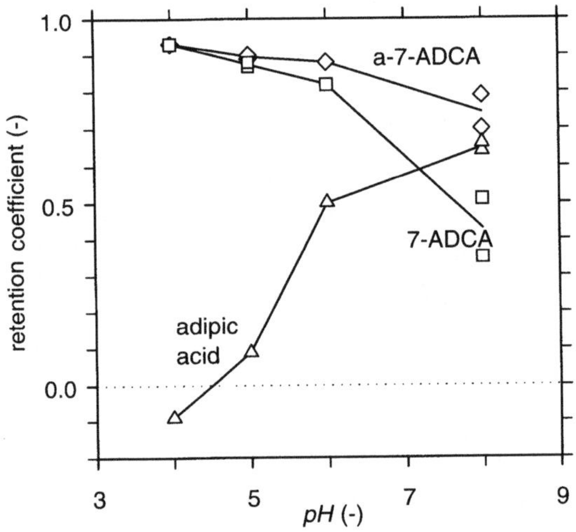

Fig. V.22 Retention coefficients of the Nanomax-50 membrane for adipyl-7-ADCA, 7-ADCA, and adipic acid.

The Nanomax-50 membrane showed retention coefficients that were a function of the *pH*. It was investigated whether membrane swelling or shrinkage could cause this effect. Therefore, the flux of the membrane was measured as a function of *pH*. It was found that the membrane becomes **more** permeable at low *pH*. If the separation would be based on size exclusion only, the retention coefficient would become smaller for all components. This is not the case (see Fig V.22). For the largest components, adipyl-7-ADCA and 7-ADCA, the retention coefficient even increases at low *pH*. Apparently, also the charge of the components and of the membrane plays a role.

The dissociation constants for adipic acid are 4.4 and 5.4 (Weast, 1971) and for 7-ADCA 2.9 and 4.9 (Tsuji and Yamane, 1976). For adipyl-7-ADCA, the dissociation constants were estimated to be equal to the dissociation constants of the respective constituents: 2.9 and 5.5. This implies that at low *pH* (<4) adipic acid is uncharged, adipyl-7-ADCA carries a negative charge and 7-ADCA is Zwitter-ionic. The

membrane is known to repulse charged species. This implies that at low *pH*, adipic acid can pass the membrane freely. At high *pH* all species are charged and therefore partly retained by the membrane.
We can conclude that the properties of the Nanomax-50 membrane are such that it can be used to separate adipic acid from 7-ADCA and adipyl-7-ADCA at low *pH*. Although some of the 7-ADCA and adipyl-7-ADCA will permeate as well, application of membranes seems to be promising.

7.5 Process concept

Based on the results described in the previous paragraphs, a process concept for 7-ADCA was derived (Fig. V.23). In the concept, the reaction is combined with crystallization and membrane separation. Both batch and continuous operation (CSTR) are possible for adipyl-7-ADCA hydrolysis. However, as shown in the part on reactor operation, continuous operation based on a single reactor becomes only possible if the substrate concentrations are low and the *pH* is not too high, otherwise chemical degradation has a big influence on the process yield.

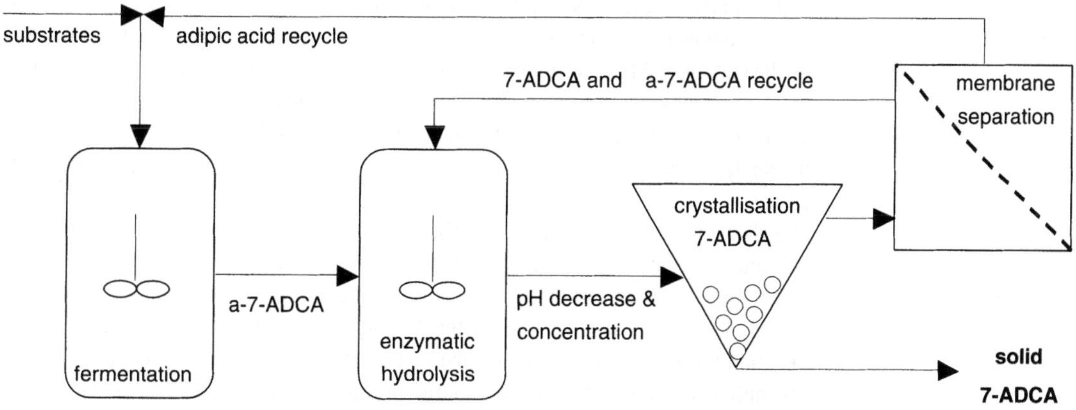

Fig. V.23 Schematic representation of the adipyl-7-ADCA hydrolysis process.

If more reactors are placed in series, this problem can largely be circumvented (Schroën, submitted for publication c). For example, if a series of 5 CSTR's were to be used, the residence time decreases with a factor of approximately five and therewith also chemical degradation reduces with the same factor. The residence time can even be reduced further if the *pH* in the last vessel of the series is slightly higher. In that case, that part of the reaction in which equilibrium is nearly reached is considerably faster because the equilibrium is shifted by the *pH* increase. At high *pH*, the chemical degradation constant is of course also higher but because of the reduction in over-all residence time, this effect is likely to be completely overcome. A concentration-step after the hydrolysis reaction is beneficial because losses of 7-ADCA in the subsequent crystallization step are reduced (Fig. V.23). The liquid obtained after crystallization can be separated efficiently by membrane filtration. The adipic acid can completely be recycled to the fermentation vessel together with small amounts of 7-ADCA and adipyl-7-ADCA. It is not an option to recycle the entire liquid to the hydrolysis vessel because adipic acid would accumulate. For the 7-ADCA and adipyl-7-ADCA containing liquid, recycling to the hydrolysis vessel is the best option because substrate and product losses are minimized (Fig. V.21).

For an actual process, reaction conditions like *pH* and temperature will partly be determined by the chemical stability of the reacting components. However, the stability of the enzyme is expected to play an even bigger role because of its influence on long-term productivity. In the current discussion, this effect is not considered.

Compared to the current 7-ADCA-production from penicillin G, the process shown in Fig. V.23 shows great potential. It consists of less reaction steps and waste streams can be minimized by the downstream-processing options shown. Eventually, economics will determine whether the suggested process will become a reality.

§8 Future prospects

In the preceding sections many questions, which have to be answered to come to an optimal application of a biocatalyst, have been addressed. However, it will ultimately be the economy that determines whether a biocatalytic application will be realized in practice or whether a conventional chemical synthesis wins the competition after all. Experience learns that only when most or all odds are greatly in favor of biocatalysis, or when there is no chemical alternative, it will be the process of choice.

The decision scheme depicted in Table V.1 is not a once-through strategy but a continuous process of simultaneous decisions at various levels or scales. On all levels, new findings can lead to developments elsewhere in the decision process. Depending on the level at which the finding takes place, this may lead to re-designing of the process. If biocatalysis (as opposed to chemistry) is adopted as the option of choice for cephalosporin production, the main level of decisions probably is the level of the biocatalyst.

The most far-reaching changes will have to take place if the current 'isolated-enzyme'' process can be replaced by direct fermentation. The group of Heijnen currently investigates whether direct fermentation of antibiotics is feasible. On the same level, a change in the form of the biocatalyst can lead to drastic changes in the process. In the discussion concerning this subject, it was shown that the free enzyme shows both higher yields and higher synthesis/hydrolysis ratio's then the immobilized enzyme. Further, it was also shown that the use of free glutaryl acylase could be beneficial. Herewith, a new field of research based on the difference between the free and immobilized enzyme was opened and is currently still under investigation. The use of free enzymes has considerable implications for the decisions at the levels following level 2.

Besides the use of the free enzyme, also design of the immobilized biocatalyst in such a way that diffusion limitation does not occur, (resulting in an immobilized

enzyme with the same 'kinetic' properties as the free enzyme), is an option that is currently investigated. The use of a designed biocatalyst is expected to lead to fewer changes in the further process as compared to the use of free enzyme. In fact, no drastic changes have to take place in the current process because the down-stream processing trajectory can remain the same (level 4).
If the free enzyme is to be used, this also results in a different bioreactor because separation of the biocatalyst from the product has to take place in a different way. For the immobilized enzyme, a sieve bottom reactor is used, but the sieve is too coarse to keep the free enzyme in the reactor. Therefore, a membrane bioreactor (depicted in continuous mode in Fig. V.24) was chosen. The membrane is impermeable to the enzyme; it is retained within a confined compartment of the process. Strictly speaking, this is a different way of 'immobilizing' the enzyme. The reactor can either be operated in batch or in continuous mode, and depending on the product concentrations the down stream-processing trajectory can remain the same as in the current process or require modifications.

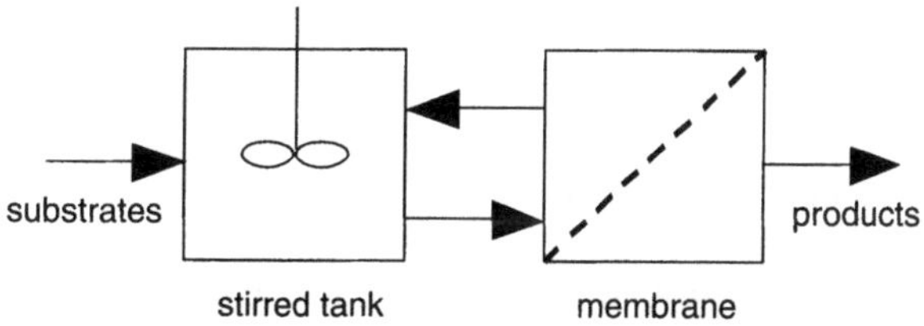

Fig. V.24 Concept for a continuous membrane bioreactor.

For the production of other antibiotics, like e.g. Cefadroxil, Amoxicillin and Ampicillin the same approach can be used as described in this chapter. But in these cases, several sort-cuts can be taken. Because of the similarities between these antibiotics, the decision-making process can be speeded up considerably, making good use of the learning-process undertaken to design and improve the Cephalexin process.

§9 Acknowledgements

Within the cluster project at Wageningen University, many people have participated in many different ways. During the first three years of the project, Vincent Nierstrasz, Rouke Bosma and Pieter Kroon participated in the project; many of their results are shown here. In addition, contributions were made by Harm Baten, Wouter Berendsen, Arjon Beulens, Valérie de Bruin, Yvette Deirkauf, Zwannet Dijkstra, Martijn Fox, Claudia Fretz, Marisca Hoogschagen, Michiel Joerink, Frank Laumen, Renate Okker, Peter Tjeerdsma, Sabine van de Wiel and Jan van Riel.
The project throve on a significant input from industrial participants: Sander Arendsen, Alle Bruggink, Erik de Vroom, Hans Kierkels, Gerard Kwant, Harold Moody, Marijn Rijkers, Eric Roos, Jan-Metske van der Laan and Emile van de Sandt.

§10 References

Birch, O.M., Böhlen, E., Gilligan, T., Hoeks, F.W.J.M.M., Malanik, V., Robins, K., Shaw, N.M., Zimmermann, T. (1994). Development of an industrial process for S-2,2-dimethylcyclopropanecarboxamide. Chiral Europe '94.

Blinkovsky, A.M., and A.N. Markaryan (1993). Synthesis of β-lactam antibiotics containing α-aminophenylacetyl group in the acyl moiety catalyzed by D-(-)-α–phenylglycyl-β-lactamide amidohydrolase. *Enzyme Microb.Technol.* **15**: 965-973.

Boesten, W.H J., W.J.J. van den Tweel, and R. M. Dekkers (1997). Process for the preparation of an antibiotic. International Patent Application to Chemferm V.O.F.

Boesten, W.H.J., T.J.G.M. van Dooren, and J.C.M. Smeets (1996). Process for the preparation of a β-lactam antibiotic. International Patent Application to Chemferm V.O.F.

Boogers, I.A.L.A., E. J. A. X. van de Sandt, and D. Schipper (1999). Novel process for the fermentative production of cephalosporin. International patent to DSM. WO 99/50271.

Bruggink, A. (1996). Biocatalysis and process integration in synthesis of semi-synthetic antibiotics. *Chimia* **50** (9):431-432, 1996.

Bruggink, A., E.C. Roos, and E. de Vroom (1998). Penicillin acylase in the industrial production of β-lactam antibiotics. *Org.Proc.Res.Dev.* **2**: 128-133.

Chaubal, M.V., G.F. Payne, C.H. Reynolds, and R.L. Albright (1995). Equilibria for the adsorption of antibiotics onto neutral polymeric sorbents: experimental and modelling studies. *Biotechnol.Bioeng.* **47**: 215-226.

Cheetham, P.S.J. (2000). Case studies in the application of biocatalysts for the production of (bio)chemicals. In: *Applied Biocatalysis,* 2nd Edition (A.J.J. Straathof and P. Adlercreutz, eds.), pp. 93-152. Harwood Academic Publishers, Amsterdam, The Netherlands.

Clausen, K. (1995). US patent, Method for the preparation of certain β-lactam antibiotics. Assigned to: Gist-brocades, The Netherlands. Patent number 5,470,717.

Cornish-Bowden, A. (1995). How to derive steady-state equations. In: *Fundamentals of enzyme kinetics.* London: Portland Press, pp. 73-92.

De Vroom, E. (1997). An improved immobilized penicillin G acylase. International Patent Application WO 97/04086 to Gist-brocades.

Diender, M.B., A.J.J. Straathof, L.A.M. van der Wielen, C. Ras, J.J. Heijnen (1998) *J.Mol.Cat. B: Enz.* **5**: 249-253.

Duggleby,, H.J., S.P. Tolley, C.P. Hill, E.J. Dodson, G. Dodson, and P.C.E. Moody (1995). Penicillin acylase has a single-amino-acid catalytic centre. *Nature* **373**: 264-268.

Guisan, J.M., G. Alvaro, C. M. Rosell, and R. Fernandez-Lafuente (1994). Industrial design of enzymic processes catalyzed by very active immobilized derivatives: utilization of diffusional limitations (gradients of pH) as a profitable tool in enzyme engineering. *Biotechnol.Appl.Biochem.* **20**: 357-369.

Hernandez-Justiz, O., R. Fernandez-Lafuente, M. Terrini, and J. M. Guisan (1998). Use of aqueous two-phase systems for in situ extraction of water soluble antibiotics during their synthesis by enzymes immobilized on porous support. *Biotechnol.Bioeng.* **59**: 73-79.

Jovetic, S., Tramper, J., Marinelli, F. (1998). Biotransformation of the lipoglycopeptide antibiotic A40926 with immobilized cells of *Actinoplanes teichomyceticus* - in situ supply of nutrients. *Enzyme Microb.Technol.* **22**: 117-121.

Kaasgaard, S.G., and U. Veitland (1992). Process for preparation of β-lactams. International Patent Application to DSM N.V. The Netherlands.

Kasche, V. (1985). Ampicillin- and Cephalexin-synthesis catalyzed by E.Coli penicillin amidase. Yield increase due to substrate recycling. *Biotechnol.Lett.* **7**: 877-882.

Kasche, V. (1986). Mechanism and yields in enzyme catalyzed equilibrium and kinetically controlled synthesis of β-lactam antibiotics, peptides and other condensation products. *Enzyme Microb.Technol.* **8**: 4-16.

Kemperman, G.J., R. de Gelder, F.J. Dommerholt, P.C. Raemaekers-Franken, A.J.H Klunder, and B. Zwanenburg (1999). Clathrate-type complexation of cephalosporins with β-naftol. *Chem.Eur.J.* **5**: 2163-2168.

Michielsen, M.J.F. (1999). Modelling solid-to-solid biocatalysis. Ph.D. thesis, Wageningen University.

Mita, G.D. (2000) Biocatalytic membranes operating under non-isothermal conditions: fundamentals and applications. In: *Research advances in Biotechnology and Bioengineering*, Trivandrum, Kerole, India: Global research Net Work, p 1-16

Mohy Eldin, M.S., M. Santucci, S. Rossi, U. Bencivenga, P. Canciglia, F.S. Gaeta, J. Tramper, A.E.M. Janssen, C.G.P.H. Schroën, and G.D. Mita (2000). Non-isothermal Cephalexin hydrolysis by penicillin G acylase immobilized on grafted nylon membranes. *J Mol.Catal. B-Enzym.* **8**: 221-232.

Mohy Eldin, M.S., A.E.M. Janssen, D.G. Mita, J. Tramper (2001). Immobilized enzymes. Strategies for overcoming the diffusion-limitation problem. Submitted for publication.

Nam, D.H., C. Kim and D.D.Y. Ryu (1985). Reaction kinetics of Cephalexin synthesizing enzyme from Xanthomonas citri. *Biotechnol.Bioeng.* **27**: 953-960.

Nierstrasz, V.A., C.G.P.H. Schroën, R. Bosma, P.J. Kroon, H.H. Beeftink, A.E.M. Janssen, and J. Tramper (1999) Thermodymically controlled synthesis of cefamandole. *Biocat.Biotransf.* **17**: 209-223.

Schroën, C.G.P.H., V.A. Nierstrasz, P.J. Kroon, R. Bosma, A.E.M. Janssen, H.H. Beeftink and J. Tramper (1999) Thermodymically controlled synthesis of β-lactam antibiotics. Equilibrium concentration and side-chain properties. *Enzyme Microb.Technol.* **24**: 498-506.

Schroën, C.G.P.H., S. van de Wiel, P.J. Kroon, E. de Vroom, A.E.M. Janssen and J. Tramper, J. (2000). Equilibrium position, kinetics, and reactor concepts for the adipyl-7-ADCA hydrolysis process. *Biotechnol.Bioeng.* **70**: 654-661.

Schroën, C.G.P.H., C.B. Fretz, V.H. de Bruin, W. Berendsen, H.M. Moody, E.C. Roos, J.L. van Roon, P.J. Kroon, M. Strubel, A.E.M. Janssen, J. Tramper (2001[a]). Modelling of the enzymatic kinetic synthesis of Cephalexin -Influence of diffusion limitation. Submitted for publication.

Schroën, C.G.P.H., P.J. Kroon, J.M. van der Laan, A.E.M. Janssen, and J. Tramper (2001[b]). Enhancement of enzymatic adipyl-7-ADCA hydrolysis. Submitted for publication.

Schroën, C.G.P.H., M.S. Mohy Eldin, A.E.M. Janssen, G.D. Mita, and J.Tramper (2001[c]). Cephalexin synthesis by immobilized penicillin G acylase under non-isothermal conditions: reduction of diffusion limitation. Submitted for publication.

Schroën, C.G.P.H., V.A. Nierstrasz, R. Bosma, G.J. Kemperman, H.H. Beeftink, and J. Tramper (2001[d]). *In situ* product removal during enzymatic Cephalexin synthesis using complexation. Submitted for publication.

Schroën, C.G.P.H., V.A. Nierstrasz, R. Bosma, P.J. Kroon, P.S. Tjeerdsma, E. de Vroom, J.M. van der Laan, H.M. Moody, H.H. Beeftink, A.E.M. Janssen, and J. Tramper (2001[e]). Integrated reactor concepts for the enzymatic kinetic synthesis of Cephalexin. Submitted for publication.

Schroën, C.G.P.H., V.A. Nierstrasz, H. Moody, M.J. Hoogschagen, P.J. Kroon, R. Bosma, H.H. Beeftink, A.E.M. Janssen and J. Tramper (2001[f]). Modelling of the enzymatic kinetic synthesis of Cephalexin. Influence of substrate concentration and temperature. *Biotechnol.Bioeng.*, in press.

Schroën, C.G.P.H., V.A. Nierstrasz, R. Bosma, Z.J. Dijkstra, E. de Vroom, H.H. Beeftink and J. Tramper (2001[g]). Process design for enzymatic adipyl-7-ADCA hydrolysis. Submitted for publication.

Spieß, A., R.-C. Schlothauer, J. Hinrichs, B. Scheidat and V. Kasche (1999). *pH* gradients in immobilized amidases and their influence on rates and yields of β-lactam hydrolysis. *Biotechnol.Bioeng.* **62**: 267-277.

Svedas, V.K., A.L. Margolin, and I.V. Berezin (1980). Enzymatic synthesis of β-lactam antibiotics: a thermodynamic background. *Enzyme Microb.Technol.* **2**: 138-144.

Tramper, J. (1996). Chemical versus biochemical conversion: when and how to use biocatalysts? *Biotechnol.Bioeng.* **52**: 290-295.

Tramper, J., S.A.G.F. Angelino, F. Müller and H.C. VanderPlas (1979). Kinetics and stability of immobilized chicken-liver xanthine dehydrogenase. *Biotechnol.Bioeng.* **21**: .

Ulijn, R.V., A.E.M. Janssen, B.D. Moore, P.J. Halling (2001). Predicting when precipitation driven synthesis is feasible: application to biocatalysis. Submitted for publication.

Van 't Riet, K., and J. Tramper (1991). *Basic bioreactor design.* Marcel Dekker Inc., New York, USA.

Van de Sandt, E.J.A.X., De Vroom, E. (2000). Innovations in cephalosporin and penicillin production: painting the antibiotics industry green. *Chimica Oggi*, **18** (2001), 25.

Verweij, J., and E. de Vroom (1993). Industrial transformations of penicillins and cephalosporins. *Recl. Trav. Chim. Pays-Bas.* **112**: 66-81.
Weast, R.C. (1971). *Handbook of Chemistry and Physics.* The Chemical Rubber Company,Cleveland, USA.
Yamana, T., and A. Tsuji (1976). Comparative stability of cephalosporins in aqueous solution: kinetics and mechanisms of degradation. *J.Pharm.Sci.* **65**: 1563-1574.
Yamana, T., A. Tsuji, K. Kanayama, and O. Nakano (1974). Comparative stabilities of cephalosporins in aqueous solution. *J.Antibiotics* **27**: 1000-1001.

Chapter VI Engineering enzymes for the synthesis of semi-synthetic antibiotics *251*

§1 Introduction ***251***

§2 Penicillin acylases ***252***

§3 *E. coli* acylase ***253***
3.1 Substrate specificity of penicillin acylase *255*
3.2 The 3-D structure *256*

§4 Application of *E. coli* acylase in synthesis of semi-synthetic β-lactam antibiotics ***258***
4.1 Enzyme-substrate interactions in penicillin acylase *259*
4.2 Kinetics *261*
4.3 Protein engineering *265*

§5 Prospects ***269***
5.1 Optimization of existing biocatalyst *270*
5.2 New penicillin acylases *270*
5.3 α-Amino acid ester hydrolases *271*
5.4 Acyl transferases *272*
5.5 Fermentative production of semi-synthetic penicillins and cephalosporins *273*

§6 Acknowledgments ***275***

§7 References ***275***

Chapter VI Engineering enzymes for the synthesis of semi-synthetic antibiotics

WYNAND B. L. ALKEMA‡, ERIK J. DE VRIES‡, CHARLES M. H. HENSGENS§, JOLANDA J. POLDERMAN-TIJMES‡, BAUKE W. DIJKSTRA§ AND DICK B. JANSSEN‡ (‡DEPARTMENT OF BIOCHEMISTRY AND §LABORATORY OF BIOPHYSICAL CHEMISTRY, BIOSON RESEARCH INSTITUTE, GRONINGEN BIOMOLECULAR SCIENCES AND BIOTECHNOLOGY INSTITUTE, UNIVERSITY OF GRONINGEN)

§1 Introduction

Enzymatic reactions offer promising alternatives to chemical processes for the industrial synthesis of β-lactam antibiotics. Whereas chemical methods may produce environmentally harmful side products, or require high temperatures, pressure and the use of organic solvents, enzymatic reactions often are more selective and can take place in an aqueous environment at neutral pH.

Chemical processes that have been substituted by enzymatic reactions or may be replaced in the future are the hydrolysis of penicillin G to produce 6-aminopenicillanic acid (6-APA), the transformation of 6-APA to 7-amino-deacetoxycephalosporanic acid (7-ADCA), or the condensation of synthetic acyl donors to β-lactam nuclei to produce new β-lactam antibiotics. Alternatively, new enzymatic steps may be introduced in or added to biosynthetic routes, in order to produce semi-synthetic antibiotics by fermentation from sugars rather than by using additional derivatisation of fermentation products.

In this Chapter enzymes that may be used in the preparation of semi-synthetic β-lactam antibiotics are discussed. In particular, attention is given to the properties of penicillin acylase of *Escherichia coli*, an enzyme that can be used for coupling of synthetic acyl donors to the penicillin G derived β-lactam nucleus. Recent structural and biochemical studies have provided insight into the catalytic mechanism and factors that determine substrate specificity of this enzyme and have opened up new ways to improve the enzyme for biocatalytic applications. Furthermore, characteristics of some novel enzymes that may be used for in vitro or fermentative synthesis, will be presented.

§2 Penicillin acylases

Penicillin acylases catalyze the hydrolysis of an amide bond between a carboxylic acid and a β-lactam nucleus while leaving the β-lactam ring intact (Fig. VI.1).

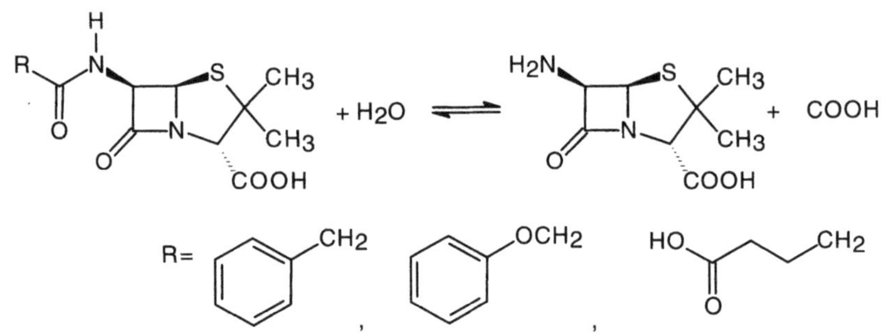

Fig. VI.1 Reaction catalyzed by penicillin G acylases (R = benzyl), penicillin V acylases (R = phenoxymethyl) and glutaryl acylases (R = carboxypropyl). Also cephalosporin nuclei can be accepted in place of 6-APA.

The penicillin acylases display different substrate specificities for the acid moiety of the substrate and are divided accordingly into penicillin G acylases, penicillin V acylases and glutaryl acylases. Penicillin acylases are used in the industry for the production of 6-APA by hydrolysis of penicillin G obtained by fermentation. Another application of penicillin acylase is found in the production of semi-synthetic antibiotics in which the enzyme catalyzes the condensation of 6-APA or 7-ADCA to a synthetic acyl group such as phenylglycine. The enzymes are usually applied in an immobilized form, which allows for easy separation and recycling.
Microbial conversion of penicillin G to phenylacetic acid and a substance called penicin was reported in 1950 by Sakaguchi and Murao[1]. The occurrence of penicillin acylases in *Streptomyces lavendulae, Nocardia, Proteus rettgeri, E. coli, Bacillus megaterium* and *A. faecalis* was subsequently reported in the 1960's[1]. The genes coding for several proteins with acylase activity have been cloned[2-6] including those for new penicillin acylases from *Kluyvera citrophila*[7] and *Arthrobacter viscosus*[8]. Recent data obtained from genome sequencing projects indicate the existence of putative penicillin acylases in the thermophiles *Archeoglobus fulgidus, Thermoplasma acidophilum* and *Aeropyrum pernix*, three organisms belonging to the Archae, and in the actinomycete *Streptomyces coelicolor* as well as in the cyanobacterium *Synechocystis* sp.[9-12]. An overview of the distribution and the relationship between penicillin acylases and putative penicillin acylases is given in figure VI.2. In this tree, the penicillin G acylases (*E. coli, K. citrophila, A. faecalis, P. rettgeri, B. megaterium* and *A. viscosus*) and cephalosporin acylases (*Pseudomonas* sp. SE83, *Pseudomonas* sp. 130 and *Pseudomonas diminuta*) form distinct sub-groups. The acylase of *Actinoplanes utahensis* even accepts palmitoyl and linoleoyl acyl groups as a substrate[13]. The level of identity of the enzymes with

known substrate specificity with the amino acid sequence of putative acylases is too low to allow accurate prediction of the substrate specificity of these (putative) enzymes. Cloning and further characterization should place the enzymes in the appropriate sub-class or provide evidence for a new class of acylases with yet another substrate specificity.

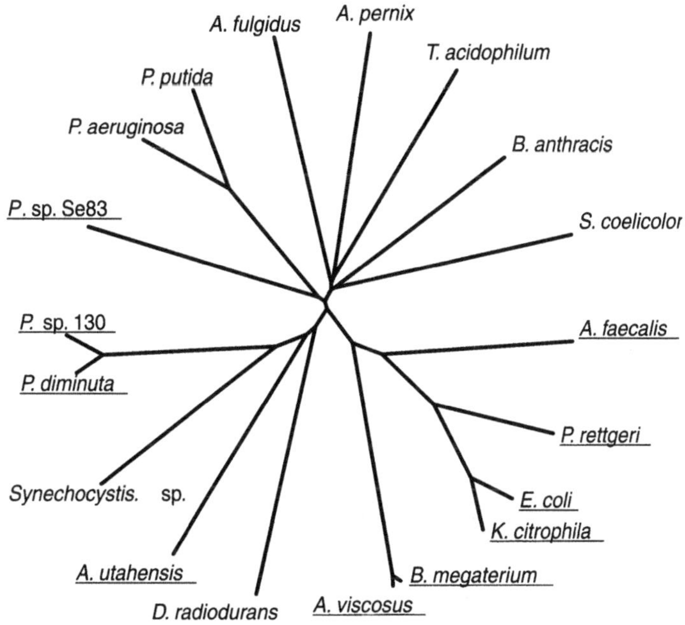

Fig. VI.2 Evolutionary tree of penicillin acylase and putative penicillin acylases. The penicillin acylases shown in this tree were obtained by doing a BLAST search[103] using the amino acid sequence of *E. coli* acylase as a query. Proteins with known acylase activity are underlined. The tree was generated using the program Treeview (available on the WWW).

The physiological function of penicillin acylases in these organisms is not known. Expression of penicillin acylase is regulated by glucose and phenylacetic acid as well as by cAMP, indicating that penicillin acylase is involved in assimilation of aromatic carbon sources[3,14-20]. This hypothesis has been supported by the finding that the penicillin acylase gene in *E. coli* is localized in an operon containing genes coding for proteins involved in the breakdown of aromatic compounds[21].

§3 *E. coli* acylase

Penicillin acylase from *E. coli* is the best-studied enzyme with respect to synthesis of semi-synthetic antibiotics. The enzyme is a heterodimer with a small α-subunit of 23 kD and a large β-subunit of 63 kD. Both subunits are produced from a common pre-proprotein in which the mature peptides are separated by a spacer peptide, and are preceded by a leader sequence that directs the protein to the periplasm. During or after translocation to the periplasm the leader peptide is removed, the protein folds and the spacer peptide is cut out to produce the mature αβ-enzyme. Biochemical and structural studies have indicated that this maturation process is probably an autocatalytic cleavage process[4,22-25]. Mutations of penicillin acylase that affect the catalytic activity lead to impaired processing as concluded from the accumulation of unprocessed protein in the periplasm[26] (Fig. VI.3).

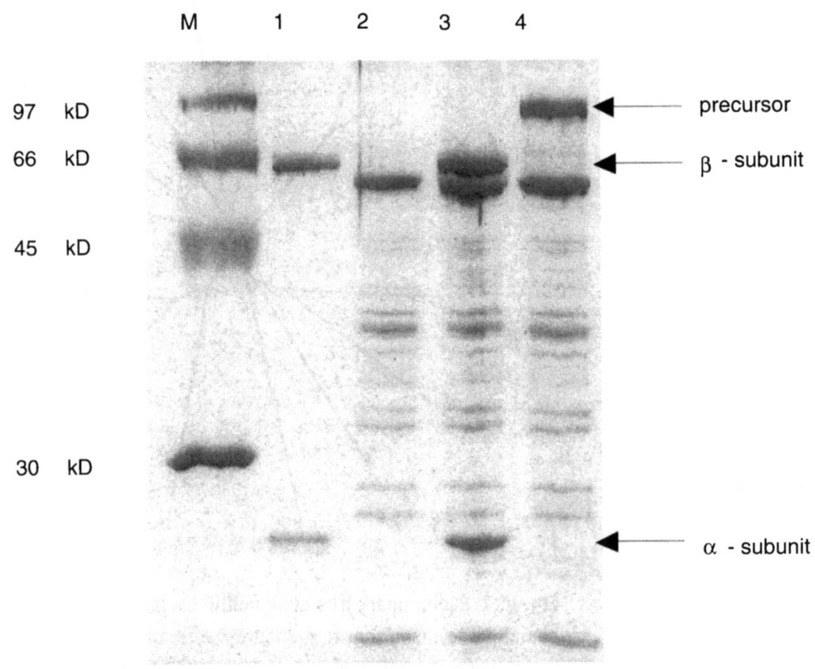

Fig. VI.3 SDS-PAGE analysis of periplasmic extracts of *E. coli* HB101 expressing wild-type and mutant *E. coli* penicillin acylase[26]. Normal processing leads to production of an α-subunit of 23 kD and a β-subunit of 63 kD. Protein bands with an apparent Mw of 90 kD are indicative of impaired processing. Lanes: M) molecular weight marker, masses are indicated on the left side; 1) Purified *E. coli* penicillin acylase; 2) Periplasmic extract of *E. coli* without penicillin acylase; 3) Periplasmic extract of *E. coli* with wild-type penicillin acylase; 4) Periplasmic extract of *E. coli* with a mutant penicillin acylase impaired in processing. The position of the subunits and the precursor protein are indicated on the right side.

The structure of the precursor protein shows that the spacer peptide is located in the active site of penicillin acylase. From the configuration of the spacer peptide in the active site it is apparent that processing involves an autocatalytic hydrolytic cleavage to liberate the N-terminal serine.

The complicated processing mechanism has hindered the development of high-level overexpression systems for penicillin acylase. Several reports, however, have appeared in which improved expression has been achieved by changing the carbon sources[14], by co-expression of penicillin acylase with chaperone proteins[27], by increasing the transcriptional and translational efficiency[28] or by using more suitable combinations of host/vector systems[29,30].

3.1 Substrate specificity of penicillin acylase

Even before the structure of penicillin acylase became available, biochemical studies identified the βSer1 as the essential catalytic residue and proposed that it was involved in formation of an acyl-enzyme intermediate[31]. Based on substrate specificity studies, Svedas proposed a model for the binding pockets of penicillin acylase (Fig. VI.4)[32].

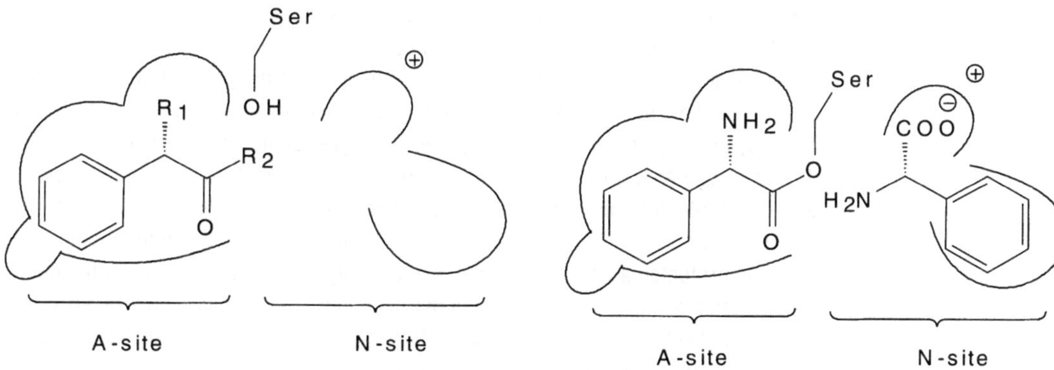

Fig. VI.4 Schematic active-site model of *E. coli* penicillin acylase based on the observed substrate specificity[32].

According to this model, the active site can be divided into two parts: one part where the acyl moiety binds (A-site) and one site where the leaving group binds (N-site), which is also the place where the nucleophile binds. The acyl binding pocket is hydrophobic and binds apolar aromatic molecules very well. There is some space to accommodate a substituent (R_1) on the α-carbon atom and NH_2, OH and CH_3 groups are accepted. The enantioselectivity of this subsite is such that only moderate e.e. values are observed during hydrolysis of racemates (R_2 is OCH_3 or NH_2) with a mild preference for the (R)-configuration[33,34]. Furthermore, a small group is accepted on the 4-position of the phenyl ring, thus giving access to the 4-OH-substituted antibiotics Amoxicillin and Cefadroxil[35].

The N-site is able to accept a large variety of groups, ranging from the large β-lactam nuclei to much smaller groups such as amines[36], alcohols, (β)-amino acids[37] and amino phosphonic acids[32,38] (Fig. VI.5).

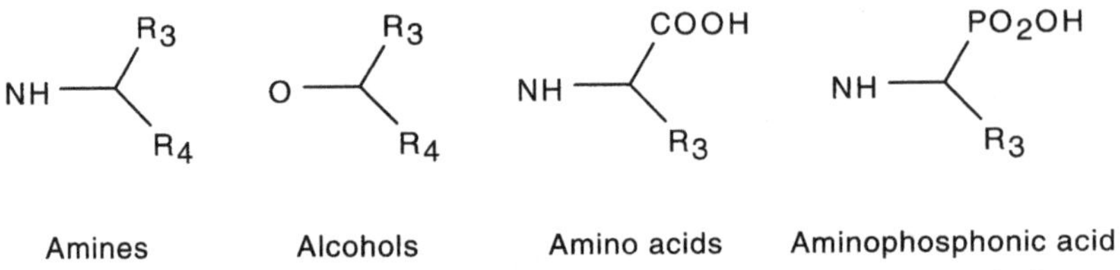

Fig. VI.5 A selection of the groups that bind to the N-site of penicillin acylase. In the case of chiral groups, the enzyme generally displays a high enantioselectivity for the (S)-enantiomer.

The N-site has a high preference for the (S)-configuration of the leaving group or nucleophile and this selectivity can be used in hydrolytic processes as well as in synthetic processes. Several reports have appeared in the literature that describe the kinetic resolution of phenylacetylated chiral compounds into products with high *e.e.* Not only hydrolytic reactions can be used in a kinetic resolution. The fact that (S)-phenylglycine acts as a nucleophile can be used in the production of optically pure (R)-phenylglycine[39] or in the production of diketopiperazines[34]. In both cases only (S)-phenylglycine(amide) deacylates the acyl-enzyme intermediate.

The observed enantioselectivity in the hydrolysis of phenylacetylated amino acids and aminophosphonic acids is much lower when the negative charge in the substrate is not present. This is the case when the carboxylate or phosphonic acid group is protected as its ester. These results imply that there is an ionic interaction between this negative charge in the substrate and a positive charge on the enzyme. The most likely candidate furnishing this positive charge is αArg145 as was concluded from structural work (see below).

The low substrate specificity of the N-site has major ramifications for antibiotic synthesis. In a synthesis reaction both the β-lactam nucleus and (S)-phenylglycine amide can bind in the N-site and deacylate the acyl-enzyme (Fig. VI.4). This will lead to the concomitant production of the desired antibiotic and (S)-phenylglycine-(S)-phenylglycine amide or (R)-phenylglycine-(S)-phenylglycine amide dipeptides or even tripeptides. The occurrence of this coupling reaction is the reason why in an enzymatic synthesis process for semi-synthetic antibiotics, enantiomerically pure (R)-phenylglycine derivatives must be used.

3.2 The 3-D structure

The 3-D structure of penicillin acylase was solved in 1995, and has provided insight into the catalytic mechanism of penicillin acylase. It appeared that the enzyme contains a single catalytic residue, a serine positioned at the N-terminal end of the β-subunit[40]. Recently, other enzymes have been characterized which share the same fold around the active site, and also contain a catalytic cysteine, serine or threonine at the N-terminal position[22]. This Ntn (N-terminal nucleophile)-hydrolase family comprises an increasing number of enzymes of markedly varying size and complexity, with a diverse range of catalytic activities and operating in very different biological contexts. Members of this family of Ntn-hydrolases include the proteolytic components of the 20S proteasome[41], penicillin V acylase of *Bacillus sphaericus*[42] and penicillin acylase of *E. coli* and *P. rettgeri*[43], an aminopeptidase of *Ochrobactrum anthropi*[44], aspartylglucosylaminase[45], glutaryl acylase from *Pseudomonas sp. GK16*[46], and cephalosporin acylase from *Pseudomonas diminuta*[47]. All these enzymes share no or little sequence similarity but they have in common that they are synthesized as a pro-enzyme which is autocatalytically cleaved to produce the active form[22,40,42] The catalytic mechanism of penicillin acylase resembles the well known mechanism found in serine proteases, involving nucleophilic attack by the serine after activation by a catalytic base, and stabilization of the transition state by interactions of the negative charge developing on the carbonyl oxygen by oxyanion hole residues (Fig. VI.6).

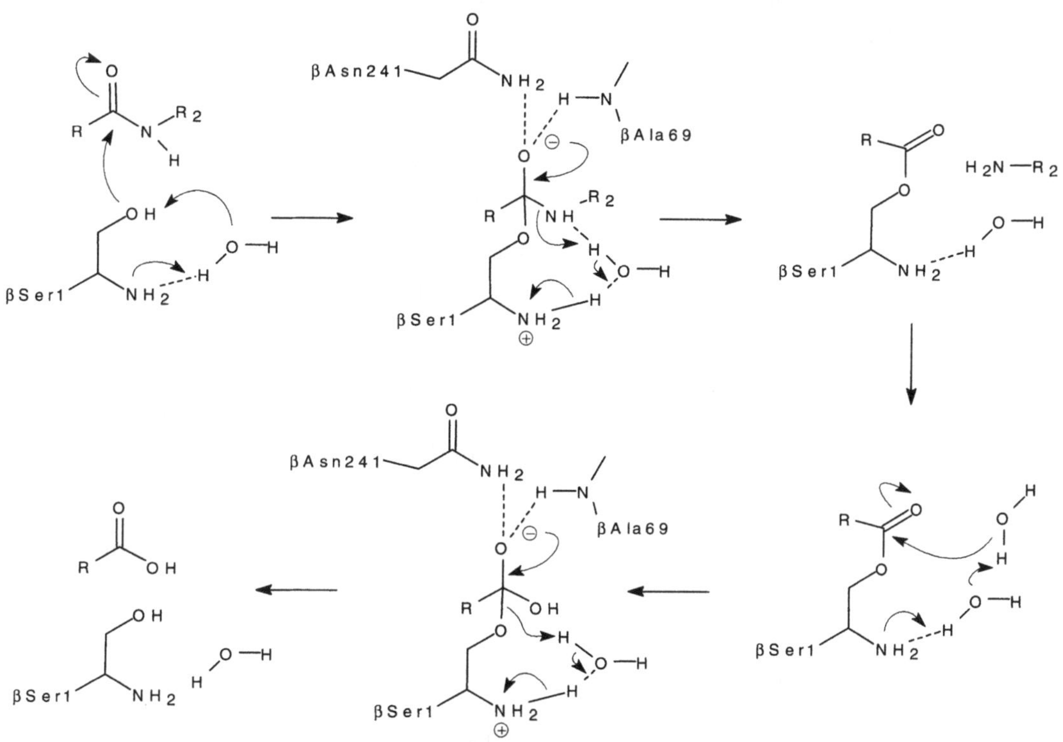

Fig. VI.6 Catalytic mechanism of penicillin acylase as proposed by Duggleby *et al.*[40]. The serine is activated by its own amino group, after which the carbonyl carbon of the substrate is attacked. In the transition state, a negatively charged oxyanion is formed that is stabilized by hydrogen bonds to the ND1 of βAsn241 and the backbone amide of βAla69. Collapse of the tetrahedral intermediate under expulsion of the leaving group gives an acyl-enzyme intermediate, which can be cleaved by water or a different nucleophile yielding the acid or the condensation product, respectively.

In serine proteases the catalytic center consists of three residues, the catalytic triad Ser-His-Asp. In this triad, the histidine residue is the base that activates the nucleophilic serine. The aspartic acid is hydrogen bonded to the histidine and serves to stabilize the positive charge and position the imidazole ring in such a way that proton transfer between the serine and the histidine is facilitated[48]. In penicillin acylase the serine is activated by its own free N-terminal amino group, possibly by a bridging H_2O molecule[40]. The oxyanion in the tetrahedral intermediate, that is formed after attack of the nucleophile on the carbonyl carbon of the substrate, is stabilized by hydrogen bonds provided by protons of an asparagine side chain (βAsn241) and a backbone amide (βAla69). Collapse of this unstable intermediate yields an acyl-enzyme, which can be deacylated by H_2O to yield the acid of the acyl group and active enzyme[40]. The nucleophilicity of the deacylating nucleophile is likely to be enhanced by the free NH_2 of the catalytic serine.

§4 Application of *E. coli* acylase in synthesis of semi-synthetic β-lactam antibiotics

In the synthesis of β-lactam antibiotics, a synthetic acyl donor, which in general is an amide or a methyl ester of a phenylacetic acid derivative, acylates the enzyme under expulsion of ammonia or methanol. The resulting acyl-enzyme can be aminolysed by a β-lactam nucleophile, *e.g.* 6-APA or 7-ADCA, yielding a semi-synthetic penicillin or cephalosporin, respectively. Since this is a kinetically controlled synthesis, the kinetic properties of the enzyme will determine the suitability of the application of penicillin acylase in the synthesis of β-lactam antibiotics.

Both the β-lactam nucleus and H_2O are nucleophiles that compete for the acyl-enzyme. The enzyme displays a moderate affinity towards β-lactam nucleophiles, with binding constants to the free enzyme ranging from 10-100 mM[49]. In principle, the ester bond in the acyl-enzyme can be rapidly hydrolyzed by H_2O as indicated by the high rate of acyl-enzyme cleavage in the hydrolytic reaction. This combination between the low affinity for β-lactam nucleophiles and the accessibility of the acyl-enzyme to water may cause significant hydrolysis of the acyl-enzyme by H_2O thereby reducing the yields in the synthesis of semi-synthetic antibiotics (see below).

Another important property of penicillin acylase is that its affinity for phenylacetic acid, which is a product of penicillin G hydrolysis, is up to 100-fold higher than the affinity for synthetic acyl donors such as phenylglycine amide and *p*-hydroxyphenylglycine amide. Therefore, trace amounts of phenylacetic acid in preparations of 6-APA can severely inhibit enzyme activity even at high concentrations of the acyl donor. A further problem in large scale application is that significant product hydrolysis may occur. The rate of product hydrolysis (V_Q) versus the rate of hydrolysis of the acyl donor (V_{AD}) at a certain concentration of acyl donor and product is given by[50]

$$\frac{V_Q}{V_{AD}} = \alpha \cdot \frac{[\mathrm{Q}]}{[\mathrm{AD}]} \qquad [1]$$

in which α is defined as

$$\alpha = \frac{k_{catQ}/K_{mQ}}{k_{catAD}/K_{mAD}} \qquad [2]$$

The subscripts AD and Q refer to the steady-state kinetic parameters for the conversion of the acyl donor and the product, respectively. The specificity constant (k_{cat}/K_m) of penicillin acylase for the antibiotic is in general higher than the specificity constant for the corresponding acyl donor, with values of α ranging from 10 to 50 for the synthesis of Ampicillin, Cephalexin, Amoxicillin and Cefadroxil, depending on the type of acyl donor. This high value for α leads to significant rates of product hydrolysis even at a relatively high concentration of the acyl donor.
In order to improve the biocatalytic properties of the enzyme with respect to application in the synthesis of semi-synthetic antibiotics, several approaches can be followed. Much attention has been given in the past to improving the synthesis process by altering reaction conditions such as optimization of temperature and pH, addition of an organic cosolvent and changing the concentrations of the reactants[51-54]. Alteration of the biocatalyst itself has mostly been done by various immobilization techniques[55-60]. Only few studies have attempted to alter the biocatalyst active-site by protein engineering[31,61,62]. To improve *E. coli* penicillin acylase by protein engineering, knowledge of the structure of the enzyme and the interactions with its substrates as well as an understanding of the kinetics of wild-type and mutant enzymes are of paramount importance.

4.1 Enzyme-substrate interactions in penicillin acylase

The elucidation of the 3-D structure of penicillin acylase complexed with phenylacetic acid (Protein Data Bank Accession number: 1PNL.PDB) and with the covalent inhibitor PMSF (1PNM.PDB) has provided information about the substrate binding site[40]. The enzyme has a well-defined binding site for the acyl moiety of the substrate. It is formed by the hydrophobic side chains of αPhe146, αMet142, βPhe24, βVal56 and βPhe57, rendering the enzyme highly specific for hydrophobic phenylacetylated substrates. Based on the 3-D structure and the deduced kinetic mechanism, an inactive mutant was made by removing the side chain of βAsn241, that stabilizes the transition state[62,63]. This mutant, βAsn241Ala, was impaired in catalysis but was still able to bind substrates. Soaking the crystals of this mutant with penicillin G revealed the β-lactam binding site of penicillin acylase (Fig. VI.7, 1FXV.PDB).

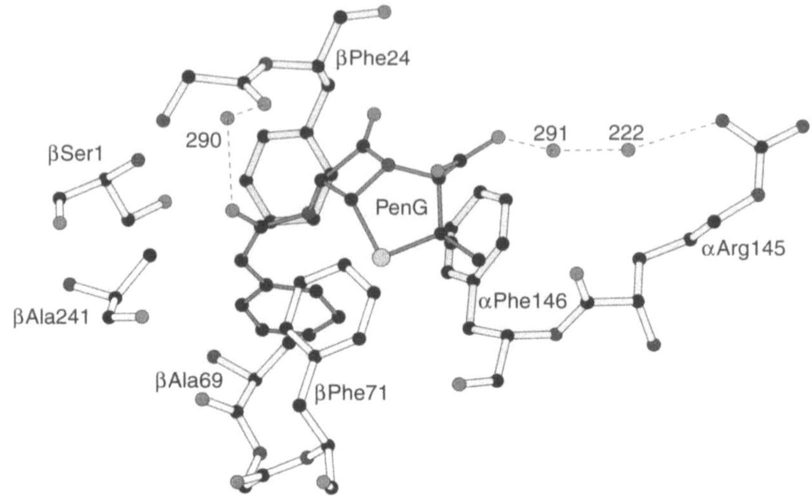

Fig. VI.7 Structure of βAsn241Ala complexed with penicillin G (1FXV.PDB). The enzyme is in the open conformation allowing binding of the β-lactam nucleus. The β-lactam moiety is bound in the active site by van der Waals interactions with the side chains of βPhe146 and βPhe71 and by hydrogen bonds between the carboxylate group of the substrate and the guanidinium group of αArg145. In the acyl binding site the hydrophobic phenylacetyl group interacts with βAla69 and with βSer1. The entrance to this site is through a narrow bottleneck which is formed by βPhe71 and by βPhe24. The catalytic water molecule (H_2O_{290}) is bound between the carbonyl oxygens of the substrate and the backbone of βGln23. Distances between the enzyme substrate complex and the water molecules are: αArg145(Nη$_1$)-H_2O_{222}: 2.7 Å; H_2O_{222}-H_2O_{291}: 2.9 Å; H_2O_{291}-PenG(O_3): 2.9 Å; βGln23-H_2O_{290}: 2.6 Å; H_2O_{290}-PenG(O_4): 2.6 Å[62].

It appeared that upon binding of penicillin G, a conformational change takes place, in agreement with an induced fit mechanism. During this change, the side chains of αArg145 and αPhe146 move away from the active site, and in this process create a β-lactam binding site. The β-lactam nucleus is bound by hydrophobic interactions between the 2β-methyl group of the thiazolidine ring and the side chain of αPhe146 and between the β-lactam ring and the aromatic ring of βPhe71. The carboxylate group of the thiazolidine ring is H-bonded to the positively charged side chain of αArg145, the H-bond being transferred via two water molecules. In this structure also the putative catalytic H_2O could be identified, which is bound to the backbone carbonyl oxygen of βGln23. A similar conformational change as the one that occurs upon binding of penicillin G has been observed in crystals of wild-type penicillin acylase that had been soaked with substituted phenylacetic acids[64]. This suggests that when binding occurs of substrates that do not precisely fit into the active site, the enzyme switches from a "closed" conformation to a new low energy "open" conformation in which the substrate can be accommodated.

4.2 Kinetics

Kinetically controlled synthesis and hydrolysis of β-lactam antibiotics catalyzed by penicillin acylase may be described by a general scheme for protease-catalyzed peptide synthesis and hydrolysis[50,65] (Fig. VI.8).

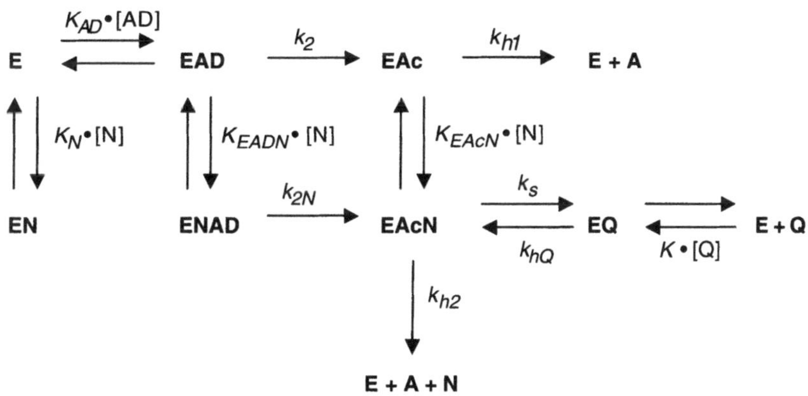

Fig. VI.8 Kinetic scheme describing the synthesis and hydrolysis catalyzed by penicillin acylase, adapted from a similar scheme for chymotrypsin-catalysed reactions[50]. In this scheme AD is the acyl donor, N the β-lactam nucleophile, Q the product of aminolysis (*i.e.* the antibiotic) and A the product of hydrolysis (*i.e.* the acid of the acyl donor). EAD, EN, ENAD and EQ are the enzymes with substrates non-covalently bound, whereas EAc and EAcN represent the acyl-enzyme intermediate with and without bound nucleophile. k_2, k_{2N} and k_{hQ} are the rate constants for the acylation of the enzyme and k_s, k_{h1}, and k_{h2} are the rate constants for deacylation. K_N, K_{EADN}, and K_{EAcN} are the binding constants of the nucleophile to the various forms of the enzyme and K_Q and K_{AD} are the binding constants of the substrates to the free enzyme.

Using this scheme, rate equations for the production of antibiotic (Q) and free acid (A) and for the depletion of nucleophile (N) and acyl donor (AD) can be derived. Assuming rapid binding, and applying a steady-state assumption for EAc (acyl-enzyme) and EAcN (acyl-enzyme with nucleophile bound) the following expressions for the rate of production of antibiotic ($d[Q]/dt$) and the rate of formation of hydrolysis product A ($d[A]/dt$) may be obtained

$$\frac{d[\mathrm{Q}]}{dt} = [\mathrm{E}_0]\cdot\frac{\dfrac{\dfrac{[\mathrm{N}]\cdot[\mathrm{AD}]\cdot k_{2N}}{K_{AD}\cdot K_{EADN}}+\dfrac{[\mathrm{AD}]\cdot k_2}{K_{AD}}+\dfrac{[\mathrm{Q}]\cdot k_{hQ}}{K_Q}}{[\mathrm{N}]\cdot(k_s+k_{h2})+k_{h1}\cdot K_{EAcN}}\cdot[\mathrm{N}]\cdot k_s-\dfrac{[\mathrm{Q}]}{K_Q}\cdot k_{hQ}}{Z} \qquad [3]$$

and

$$\frac{d[\mathrm{A}]}{dt}=[\mathrm{E}_0]\cdot\frac{\dfrac{\dfrac{[\mathrm{N}]\cdot[\mathrm{AD}]\cdot k_{2N}}{K_{AD}\cdot K_{EADN}}+\dfrac{[\mathrm{AD}]\cdot k_2}{K_{AD}}+\dfrac{[\mathrm{Q}]\cdot k_{hQ}}{K_Q}}{[\mathrm{N}]\cdot(k_s+k_{h2})+k_{h1}\cdot K_{EAcN}}\cdot\left(K_{EAcN}\cdot k_{h1}+k_{h2}\cdot[\mathrm{N}]\right)}{Z} \qquad [4]$$

in which

$$Z=1+\frac{[\mathrm{AD}]}{K_{AD}}+\frac{[\mathrm{N}]}{K_N}+\frac{[\mathrm{N}]\cdot[\mathrm{AD}]}{K_{AD}\cdot K_{EADN}}+\frac{[\mathrm{Q}]}{K_Q}+\frac{\left(\dfrac{[\mathrm{N}]\cdot[\mathrm{AD}]\cdot k_{2N}}{K_{AD}\cdot K_{EADN}}+\dfrac{[\mathrm{AD}]\cdot k_2}{K_{AD}}+\dfrac{[\mathrm{Q}]\cdot k_{hQ}}{K_Q}\right)\cdot\left(K_{EAcN}+[\mathrm{N}]\right)}{[\mathrm{N}]\cdot(k_s+k_{h2})+k_{h1}\cdot K_{EAcN}}$$

Analyzing equation [3] for the case that

$$\frac{d[\mathrm{Q}]}{dt}=0 \qquad [5]$$

an expression for the yield of the desired product, *i.e.* the maximum product accumulation $[\mathrm{Q}]_{max}$, can be obtained

$$[\mathrm{Q}]_{\max}=\frac{K_Q}{k_{hQ}}\cdot\frac{\dfrac{k_{2N}\cdot[\mathrm{N}]}{K_{EADN}}+k_2}{K_{AD}}\cdot\frac{V_s}{V_h}\cdot[\mathrm{AD}] \qquad [6]$$

in which [AD] and [N] represent the concentrations of acyl donor and nucleophile at this point, and

$$\frac{V_s}{V_h}=[\mathrm{N}]\cdot\frac{k_s}{k_{h1}\cdot K_{EAcN}+k_{h2}\cdot[\mathrm{N}]} \qquad [7]$$

In equation [7], V_s/V_h represents the ratio between the intrinsic rate constants of aminolysis versus hydrolysis of the acyl enzyme. This ratio is only dependent on the kinetic parameters for deacylation and [N]. Equations [6] and [7] show that for increased yields in synthesis reactions, all kinetic parameters are targets for improvement by protein engineering rather than only the V_s/V_h ratio.

4.2.1 The V_s/V_h ratio

The microscopic rate constants that describe the deacylation of the enzyme k_s, k_{h1} and k_{h2} and the binding constant of the nucleophile to the acyl-enzyme, K_{EAcN}, determine the V_s/V_h ratio of the enzyme. The value of these constants can be determined only if a significant amount of acyl-enzyme accumulates during the reaction, which happens when acylation of the enzyme by either the antibiotic or the acyl donor is faster than the hydrolysis or aminolysis of the acyl-enzyme. These constants can then in principle be determined by rapid reaction techniques. However, stopped-flow experiments with penicillin acylase of *K. citrophila* and *E. coli* have revealed that formation of the covalent intermediate is the rate-limiting step in the reaction[66-69]. Analysis of the conversion of chromogenic penicillin G analogues has shown that the hydrolysis of the phenylacetylated acyl-enzyme proceeds with a rate constant k_{h1}, of at least 1000 s^{-1}. Given the V_s/V_h ratios of well above 1 which can be obtained for the synthesis of penicillin G, it can be concluded that the value for $k_s > 1000\ s^{-1}$. Since the k_{cat} values for phenylacetylated substrates are <100 s^{-1} for most substrates, one can conclude that the steady-state kinetic parameters and thereby the important factor α (equation [2]) are determined by the individual rate constants for *acylation* of the enzyme.

4.2.2 Steady-state parameters

The steady-state parameters of penicillin acylase, and thereby the rate constants for acylation, can in principle be determined with several methods. The 6-APA produced during hydrolysis of penicillin G can be detected by using HPLC methods, by fluorescence measurements after reaction with fluorescamine[70], or by colorimetric measurements after derivatization with 4-(dimethylamino)-benzaldehyde[49]. The released phenylacetic acid can also be monitored by acid/base titration[71].

These assays are complicated by the chemical instability of the β-lactam compounds. Furthermore, the low sensitivity of some of these methods complicates determination of the low K_m values typical for phenylacetylated substrates. The classical method of measuring initial rates is error prone because of the strong product inhibition by phenylacetic acid. One method that overcomes these difficulties, is to measure the conversion of penicillin G in the presence of a second substrate, from which the conversion can easily be followed and the kinetics are well known. Compounds that can be used as reference substrate are the chromogenic substrates 2-nitro-5-[(phenylacetyl)amino]-benzoic acid (NIPAB) and 2-nitro-5-[(phenylglycyl)amino]-benzoic acid (NIPGB) (Fig. VI.9).

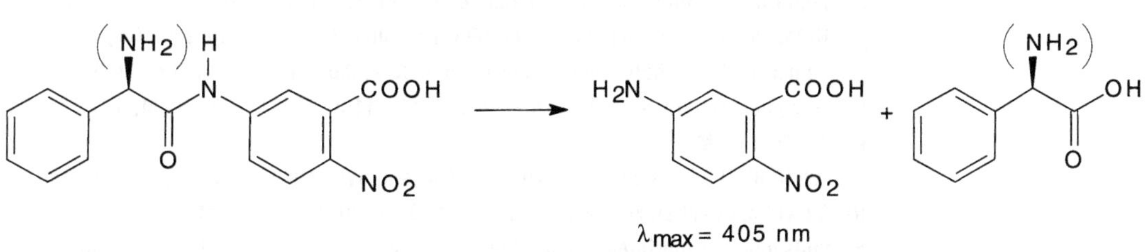

Fig. VI.9 Chromogenic substrates used for kinetic studies of penicillin acylase. Hydrolysis of NIPAB (NIPGB) releases 2-amino-5-nitrobenzoic acid. The Δε for the conversion of these substrates is 9090 $M^{-1}\ cm^{-1}$ at 405 nm at pH 7.

An example of such an experiment is shown in figure VI.10 in which the conversion of penicillin G and Cephalexin was studied in combination with the hydrolysis of the chromogenic substrates NIPAB or NIPGB[72]. The steady-state parameters for the hydrolysis of the second substrate can be calculated, using the equations

$$\frac{d[\mathrm{NIPAB}]}{dt} = -\frac{k_{cat}^{NIPAB}\cdot[\mathrm{E_0}]\cdot[\mathrm{NIPAB}]}{K_m^{NIPAB}\cdot\left(1+\frac{[\mathrm{PenG}]}{K_m^{PenG}}+\frac{[\mathrm{PAA}]}{K_i^{PAA}}+\frac{[\mathrm{APA}]}{K_i^{APA}}\right)+[\mathrm{NIPAB}]} \quad [8]$$

and

$$\frac{d[\mathrm{PenG}]}{dt} = -\frac{k_{cat}^{PenG}\cdot[\mathrm{E_0}]\cdot[\mathrm{PenG}]}{K_m^{PenG}\cdot\left(1+\frac{[\mathrm{NIPAB}]}{K_m^{NIPAB}}+\frac{[\mathrm{PAA}]}{K_i^{PAA}}+\frac{[\mathrm{APA}]}{K_i^{APA}}\right)+[\mathrm{PenG}]} \quad [9]$$

which describe the rate of conversion of both substrates including inhibition by both products.

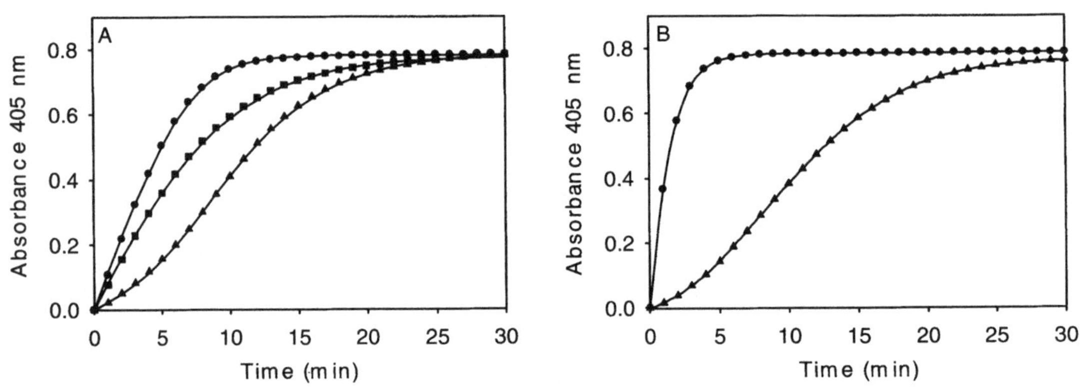

Fig. VI.10 A) Progress curves of NIPAB hydrolysis in the presence of different concentrations of phenylacetic acid and penicillin G[72]. Symbols: (●) Hydrolysis of 86 μM NIPAB without a second substrate; (■) with 200 μM phenylacetic acid; (▲) with 200 μM penicillin G. Symbols denote experimental data, the line shows the fit with the following parameters: k_{cat}^{NIPAB} = 16.2 s^{-1}, K_m^{NIPAB} = 11.3 μM, K_i^{PAA}= 57.3 μM, k_{cat}^{PenG} = 42 s^{-1}, K_m^{PenG} = 6 μM.

Fig. VI.10 B) Progress curves of NIPGB hydrolysis. Symbols: (●) hydrolysis of 86 μM NIPGB without Cephalexin; (▲) hydrolysis of 86 μM NIPGB with 25 mM Cephalexin. Symbols denote experimental data, the line shows the fit with the following parameters: k_{cat}^{NIPGB} = 11 s^{-1}, K_m^{NIPGB} = 1 mM, $k_{cat}^{cephalexin}$ = 57 s^{-1}, $K_m^{cephalexin}$ = 1.25 mM, K_i^{ADCA} = 7.5 mM. Inhibition by phenylglycine and 6-APA by the concentrations used in these experiments is negligible and K_i for these compounds were not included in the fitting procedure.

4.2.3 Progress curves for product accumulation

Using the rate equations derived from the kinetic scheme (Fig. VI.8) and the kinetic constants derived from steady-state and pre-steady-state experiments, synthesis of Ampicillin by wild-type *E. coli* penicillin acylase can be modeled satisfactorily (Fig. VI.11). These models and kinetic descriptions may be used to characterize and describe the properties of mutant penicillin acylases.

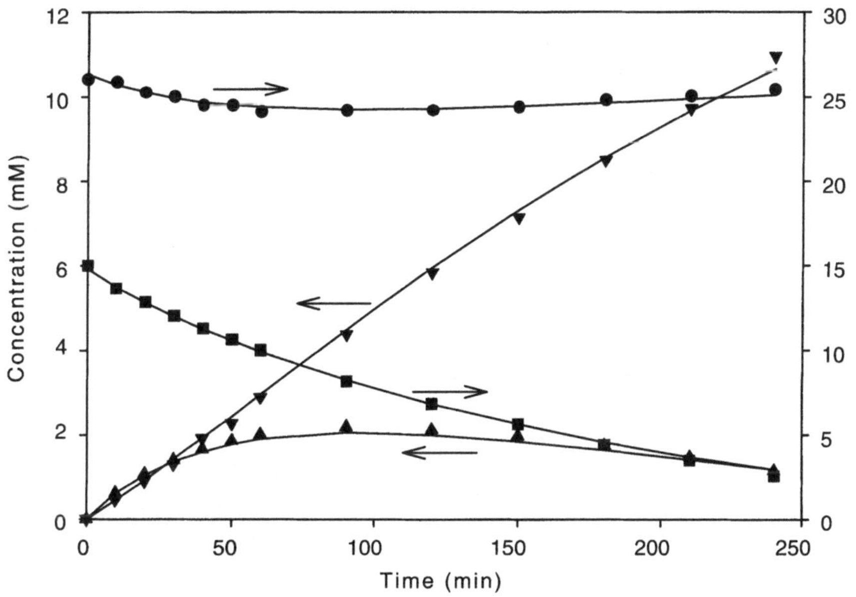

Fig. VI.11 Synthesis of Ampicillin (▲) and phenylglycine (▼) from 15 mM PGA (■) and 26 mM 6-APA (●) using penicillin acylase from *E. coli*. Symbols represent experimental data, the lines are simulations using rate equations derived from the scheme in Fig. VI.8. The kinetic parameters used to describe the course of the reaction were obtained from steady-state and pre-steady state experiments: $k_2 = 30\ s^{-1}$, $k_{2N} = 30\ s^{-1}$, $k_s = 6000\ s^{-1}$, $k_{h1} = 1000\ s^{-1}$, $k_{h2} = 2000\ s^{-1}$, $k_{hQ} = 21\ s^{-1}$, $K_N = 40$ mM, $K_{EADN} = 39.8$ mM, $K_{EAcN} = 34$ mM, $K_Q = 2.5$ mM, $K_{AD} = 40$ mM. k_s, k_{h1} and k_{h2} can be any value as long as the values are >>k_2 and the ratio between the three constants is the same.

4.3 Protein engineering

4.3.1 Site-directed versus random mutagenesis

As mentioned above, the formation of semi-synthetic antibiotics catalyzed by penicillin acylase is a kinetically controlled process; the yield is determined by intrinsic rate and binding constants at the active site of the enzyme. Using protein engineering these properties may be altered and optimized for use in synthesis reactions. In general, two approaches may be followed in protein engineering of an enzyme: site directed mutagenesis (design) or directed evolution (random).

For site-directed mutagenesis, the 3-D structure of an enzyme is preferably used as a starting point. From the structure, residues are identified which are important for catalysis and substrate specificity. When a 3-D structure is not available, functions of certain amino acid residues may be obtained by comparing the amino acid sequence to the sequence of a protein of which the structure is known. Alternatively, alignment with sequences from a family of homologues may reveal conserved catalytic residues or regions. Mutation of putative catalytic residues or residues that may be involved in substrate binding and subsequent kinetic analysis of the mutants may give insight into the structure-function relationships of selected residues and may yield mutant enzymes with improved catalytic properties.

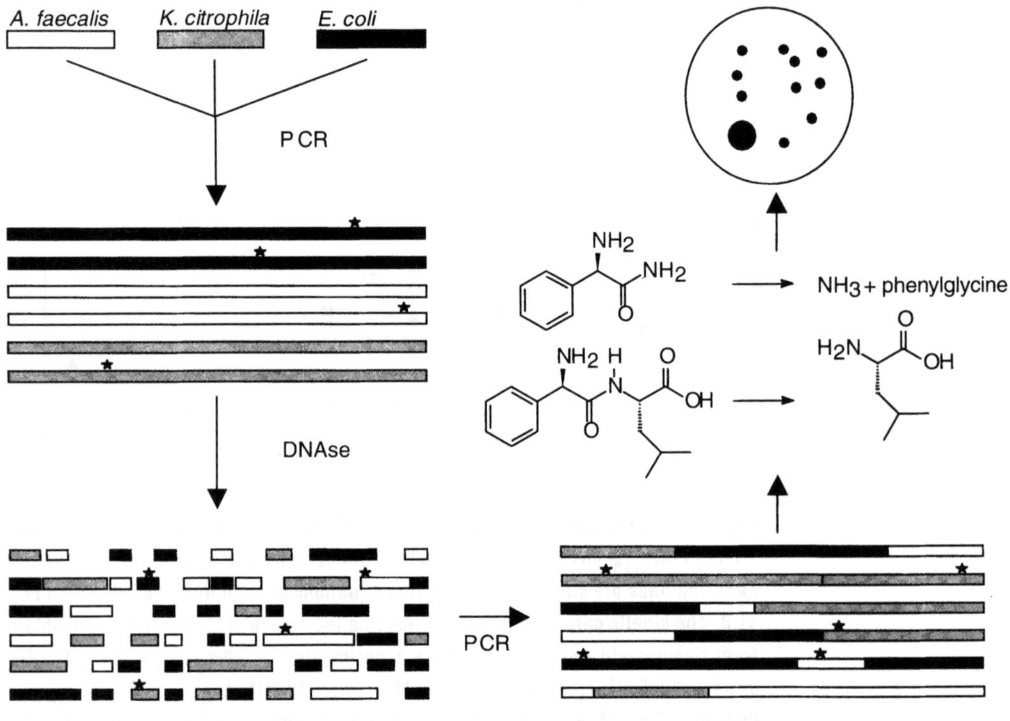

Fig. VI.12 Family shuffling[77] of penicillin acylase from *E. coli*, *A. faecalis* and *K. citrophila*. Homologous sequences are amplified using error-proof or error-prone PCR. The obtained pool of genes is subjected to DNAse treatment, generating small DNA fragments. In a second PCR the full-length genes are formed back, thereby allowing recombination between DNA regions with high similarity. After cloning the full-length PCR products in an expression vector, transformants may be screened for desired properties. In the case of acylases, fast growing colonies are selected using amides or leucine derivatives of acyl donors (see text).

A method that does not require any knowledge of the 3-D structure is directed evolution. This method involves the mutagenesis of the DNA that encodes the protein by error prone PCR or chemical methods and subsequent screening of the expressed mutant proteins for improved properties of interest. A method that has become popular in recent years is DNA shuffling[73]; it has been successfully used to optimize enzymes for industrial processes[74-76]. DNA shuffling involves not only mutagenesis by error prone PCR, but also a step in which the DNA molecules carrying different mutations are recombined, allowing combinations of several mutations to occur. This technique has therefore also been called "sexual PCR". A variant of this method is family shuffling in which homologous genes with a high level of DNA sequence identity are recombined[77] (Fig. VI.12).

4.3.2 Screening

Random mutagenesis with subsequent identification of improved mutants requires a robust screening or selection method. In general, selection on a growth substrate or resistance to a toxic substrate is preferred, since this method is relatively easy and large numbers of mutants may be screened. When such a method is not available, screening of increased activity of transformants using chromogenic substrates may be used.

In the case of penicillin acylase several methods are available. The enzyme displays a relatively low selectivity for the leaving group of the substrate and is capable of hydrolysing a variety of phenylacetyl derivatives. With a leucine auxotrophic mutant of *E. coli* as a host organism, one can select for transformants with penicillin acylase activity using phenylacetyl-L-leucine as the sole source of leucine. This strategy has been used to select penicillin acylase of *A. faecalis* from a genomic library[2] and to obtain mutants of *E. coli* penicillin acylase[78,79] and *K. citrophila* penicillin acylase with improved activity for glutaryl-L-leucine[80]. Instead of a leucine derivative, the amides of acyl donors may be used for selection of transformants which can use the substrate of interest as a nitrogen source[81]. An advantage of the latter method is that it does not require the use of an auxotrophic host for expression, and that screening is more directed towards specificity for the acyl donor and for selecting mutants with a decreased α (equation [2]).

Screening for a property such as an improved V_s/V_h ratio is more difficult. The antibiotic that is formed by transformants with improved synthetic capacities can be detected by chemical means[70,82] or by using an overlay culture with an indicator strain which is sensitive to the antibiotic produced. The nature of these assays requires that the transformants are replica plated, making the method elaborate and limiting the number of colonies that can be screened. Another factor that complicates this screening method is the sensitivity of indicator cultures to intrinsically toxic substrates used for synthesis such as 6-APA and 7-ADCA. Furthermore the [Q]/[A] ratio has a kinetic character, *i.e.* it changes during the reaction, making an endpoint assay less suitable.

4.3.3 Protein engineering of E. coli penicillin acylase

In order to improve penicillin acylase for application in the preparation of semi-synthetic antibiotics, several properties may be optimized. From equation [2] it follows that a decrease of the relative specificity of the enzyme towards the product will result in a decrease in α, which will give higher yields during synthesis. Several studies have appeared in which the substrate specificity of penicillin acylases of *E. coli*[26,31,61,62,78,79,83-85] and *K. citrophila*[80,86-88] has been changed. DSM, in particular,

has filed a number of patents identifying the residues of crucial importance for activity and specificity of these β-lactam acylases (see DSM patents WO 9116435 and WO 9605318). In these studies both a directed approach as well as a random approach has been used.
The second parameter which may be optimized is the V_s/V_h ratio. The V_s/V_h ratio is determined by the individual rate constants for deacylation of the enzyme (equation [7]). The competition between 6-APA or 7-ADCA and H_2O may be influenced by small differences in position of the nucleophile when bound to the acyl enzyme. Structures of these complexes are not available and difficult to obtain. Furthermore, the changes in the values of the rate constants that are affected by mutations cannot be measured directly, and it is therefore difficult to establish detailed structure-function relationships for residues that influence the V_s/V_h ratio. Consequently, only a few studies are reported in which the V_s/V_h ratio of an enzyme was altered by site-directed mutagenesis. Wells and coworkers replaced the active-site serine in subtilisin with a cysteine, thereby creating subtiligase. Subtiligase shows a much higher V_s/V_h ratio, probably because of the higher reactivity of thioesters towards amine nucleophiles compared to water or alcohols[89]. Attempts to replace βSer1 by Cys in penicillin acylase resulted in βSer1Cys mutants which were correctly processed but were inactive[31,83]. Elliott *et al.* were able to change protease B of *Streptomyces griseus* into an effective ligase by replacing the active-site serine by an alanine[90]. In this case, the histidine, which normally serves as the general acid/ base, became the nucleophile and catalysis proceeded via an acyl-imidazole intermediate.

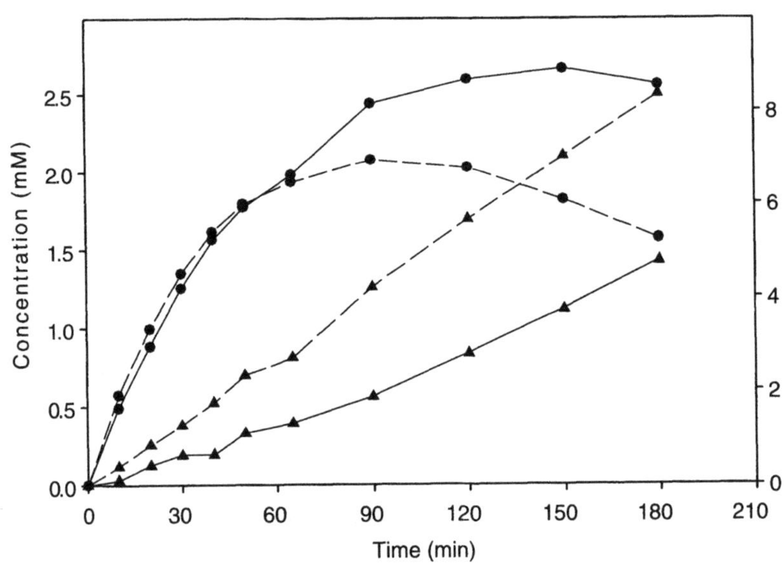

Fig. VI.13 Synthetic capacities of a mutant of *E. coli* penicillin acylase generated using random mutagenesis. Ampicillin synthesis from PGA and 6-APA catalyzed by the *E. coli-K. citrophila* hybrid enzyme and *E. coli* wild-type penicillin acylase. Symbols: (▲) phenylglycine, (●) Ampicillin. The dashed lines represent data for the wild type and the solid lines the data for the mutant.

An overview of the results obtained with protein engineering of penicillin acylase using both site-directed mutagenesis and random techniques is shown in table VI.1.

Table VI.1. Function of the catalytic residues of *E. coli* penicillin acylase and properties of some mutants on these positions.

Mutant	Location / function	Properties
βSer1Cys	Catalytic center. Performs nucleophilic attack on the substrate.	Correct processing, but no hydrolytic or synthetic activity.
βAsn241Ala	Catalytic center. Stabilizes the negative charge on the carbonyl oxygen in the transition state.	Impaired processing and inactive, but able to bind substrates. Allows crystallization with intact β-lactam substrates and acyl donors.
αPhe146Leu	Acyl and β-lactam binding site.	5-fold increased V_s/V_h ratio for penicillin G.
αPhe146Tyr	Hydrophobic interactions with the 2β-methyl group of the thiazolidine ring.	10-fold increased specificity for PGA.
βPhe71X	β-lactam binding site. Van der Waals interactions with thiazolidine moiety of the β-lactam nucleus.	Decreased V_s/V_h ratio for all mutants tested. For efficient aminolysis a Phe is probably needed on this position.
E. coli-K. citrophila hybrid		Improved V_s/V_h ratio and yield in Ampicillin synthesis compared to *E. coli* wild type (Fig. VI.13).

From these results one can conclude that protein engineering is a fruitful approach for improving *E. coli* penicillin acylase for synthetic applications.

§5 Prospects

Penicillin acylase has high potential for application in industrial preparation of semi-synthetic antibiotics (see above). Further optimization may be achieved by protein engineering of this enzyme to improve the properties for biocatalytic use. Another strategy is to search in nature for new acylases which have more beneficial characteristics. These new acylases may involve variants of *E. coli* penicillin acylase with improved properties or totally new enzymes capable of synthesizing β-lactam antibiotics from a β-lactam nucleus and a synthetic acyl donor.

5.1 Optimization of existing biocatalyst

The kinetic properties of an ideal catalyst include an infinitely high V_s/V_h ratio and a high specificity (k_{cat}/K_m) for the acyl donor relative to the product and no inhibition by phenylacetic acid.

There is no theoretical limit for the ratio between aminolysis and hydrolysis. Suppressing k_{h1} and k_{h2}, *i.e.* removing the factors that stabilize the transition state of hydrolysis of the acyl-enzyme, but leaving intact the factors that stabilize the transition state for aminolysis, would result in an enzyme which can only synthesize antibiotics. With such an enzyme, there is no need to perform a kinetically controlled reaction since the yield of the produced antibiotic will be dependent on the thermodynamic equilibrium between the reactants. Given the structure of the enzyme and the scant knowledge of the factors that influence the deacylation, it will be hard to construct such an enzyme by knowledge based protein engineering. However, smaller changes in V_s/V_h ratio can be obtained by site-directed mutagenesis as shown in table VI.1.

The fact that penicillin acylase seems to be optimized in evolution for the hydrolysis of a wide range of phenylacetylated compounds indicates that there is room for improvement with respect to synthesis of phenylglycylated compounds using 6-APA as a nucleophile. In other words, it is highly unlikely that an enzyme which probably has a hydrolytic function in nature is optimized for a non-natural synthetic reaction.

The possibility to screen large amounts of mutants with high throughput screening and the availability of the X-ray structures of penicillin acylase complexed with phenylglycine, Ampicillin and other β-lactams form the toolkit to improve penicillin acylase by both site directed and random mutagenesis. Using these techniques it is also possible to engineer properties which are less influenced by kinetic factors but are nevertheless important such as high stability or resistance to organic solvents.

5.2 New penicillin acylases

New acylases have traditionally been obtained by screening organisms for penicillin acylase activity. Large scale cultivation of the organism and isolation of the protein followed by N-terminal sequencing and PCR cloning into expression vectors is often necessary to improve enzyme production for kinetic and structural studies and, more important, for industrial application. However, this approach is elaborate and new penicillin acylases may be acquired from other, more readily available, sources. Genome sequencing projects have revealed the existence of several putative penicillin acylases[9-12] (Fig. VI.2). As new genomes continue being sequenced, this source of new penicillin acylases becomes increasingly important. Since the DNA sequence of these genes is known, they can be amplified by PCR. Subsequent cloning into a suitable expression vector permits rapid evaluation of the applicability of these enzymes as a biocatalyst. Investigation of the kinetic properties of these enzymes may give more insight into the function of certain amino acid residues or regions in the penicillin acylase gene family.

Another method that circumvents the necessity of cultivating the organisms which contain the penicillin acylase genes or knowledge of the DNA sequence is screening of environmental DNA expression libraries. In this method, DNA is isolated from soil samples and cloned into suitable expression vectors. The transformants may be subsequently screened for the desired properties using high-throughput screening methods.

5.3 α-Amino acid ester hydrolases

The α-amino acid ester hydrolases (AEHs) catalyze the transfer of an acyl group from an amino acid ester to H_2O. However, several organisms containing α-amino acid ester hydrolases capable of catalyzing acyltransfer to 7-aminocephem and 6-aminopenam compounds have been identified[91] (Table VI.2).

Table VI.2: Proposed quaternary structure of α-amino acid ester hydrolases capable of synthesizing β-lactam antibiotics.

Organism	Structure	Mw
Xanthomonas citri[93] IFO 3835	$(\alpha)_4$	72 kD
Acetobacter turbidans[91] ATCC 9325	$(\alpha)_2(\beta)_2$	70/72 kD
Pseudomonas melanogenum[102] IFO 12020	$(\alpha)_2$	70 kD

In contrast to most of the penicillin and cephalosporin acylases, which are heterodimers composed of a small and a large subunit, the AEHs consist of two subunits of almost identical size. These enzymes possess several characteristics that make them useful for application in the synthesis of semi-synthetic penicillins and cephalosporins. AEHs require a Cα-amino group on the acyl moiety of the substrate, and are in general not inhibited by phenylacetic acid[92,93]. Moreover, AEHs have a high specificity towards esters as compared to amides, and it is therefore likely that the hydrolysis of the product (an amide) is diminished. Furthermore, these enzymes have an optimum pH of around 6 compared to the optimum of around pH 7.0 for penicillin acylase, which can be advantageous for biocatalytic synthesis[92,94]. It has been suggested that the α-amino acid ester hydrolases are able to accept the charged forms of their substrates[95], which extends the pH range in which the biocatalyst can be used to lower pH values. Together with the structural differences, these properties of the AEHs suggest a catalytic mechanism that differs from that of the penicillin acylases. Little information is available, however, on the biochemistry and genetics of these enzymes and overexpressing by recombinant strains has not been reported in the literature.
The genes encoding the AEHs of *Xanthomonas citri*, *Acetobacter pasteurianus* and *Acetobacter turbidans* have been cloned and are being investigated further in our

lab. Amino acid sequence comparison showed 60% identity among these proteins. No homology of these sequences with any other known penicillin or cephalosporin acylase has been found, indicating that the α-amino acid ester hydrolases form a new family of acylases. Further studies are aimed at unraveling the catalytic mechanism of the enzymes.

5.4 Acyl transferases

Another group of enzymes which also have a potential use in the production of semi-synthetic β-lactam antibiotics are the enzymes of the penicillin and cephalosporin biosynthetic routes. Acyl CoA:isopenicillin N acyltransferase (AT), the last enzyme in the penicillin biosynthetic route, exchanges the polar α-aminoadipic acid side chain in isopenicillin N with the hydrophobic phenylacetic acid group. It uses the CoA-activated derivative of phenylacetic acid to make penicillin G, the end product of the penicillin biosynthetic route. This promiscuity in the recognition of fundamentally different side chains makes AT a suitable candidate for semi-synthetic β-lactam biosynthesis, because it may be employed in the exchange of α-aminoadipic acid for phenylglycine to make Ampicillin.

Figure VI.14 reviews the main enzymatic activities of AT. Isopenicillin N may be directly transacylated to penicillin G (reaction 3), or via hydrolysis of isopenicillin N (reaction 1) and a subsequent acylation of 6-APA (reaction 2). Relevant unproductive side reactions are the hydrolysis of penicillin G and phenylacetyl-CoA. In the absence of acyl-CoA derivatives the hydrolysis of isopenicillin N is slowed down considerably, which suggests some kind of allosteric control. AT accepts a wide range of hydrophobic and hydrophilic acyl-CoA derivatives. It also accepts various non-CoA thioesters. For the reaction mechanism an ordered Bi-Bi-Ping-Pong mechanism has been proposed, with competitive inhibition by phenylacetyl-CoA.

Mature AT is a heterodimeric protein with subunits of 11 and 29 kD. It is produced as a 40 kD precursor, which is autocatalytically processed by cleavage of the bond between Gly102 and Cys103[96]. The two peptide chains remain associated in the heterodimer.

AT does not show any homology with other enzymes or enzyme families. This and the lack of direct structural information makes a rational redesign by site-directed mutations a difficult task. The 3-D X-ray structure elucidation of AT would tremendously facilitate the modification of AT for the biosynthetic or *in vitro* production of semi-synthetic β-lactams, as well as indicate amino acid residues important in catalysis and clarify the reaction mechanism. The importance of X-ray structural information has already been demonstrated for isopenicillin N synthase and deacetoxy cephalosporin C synthase, which are involved in the penicillin and cephalosporin biosynthetic routes, respectively[97,98].

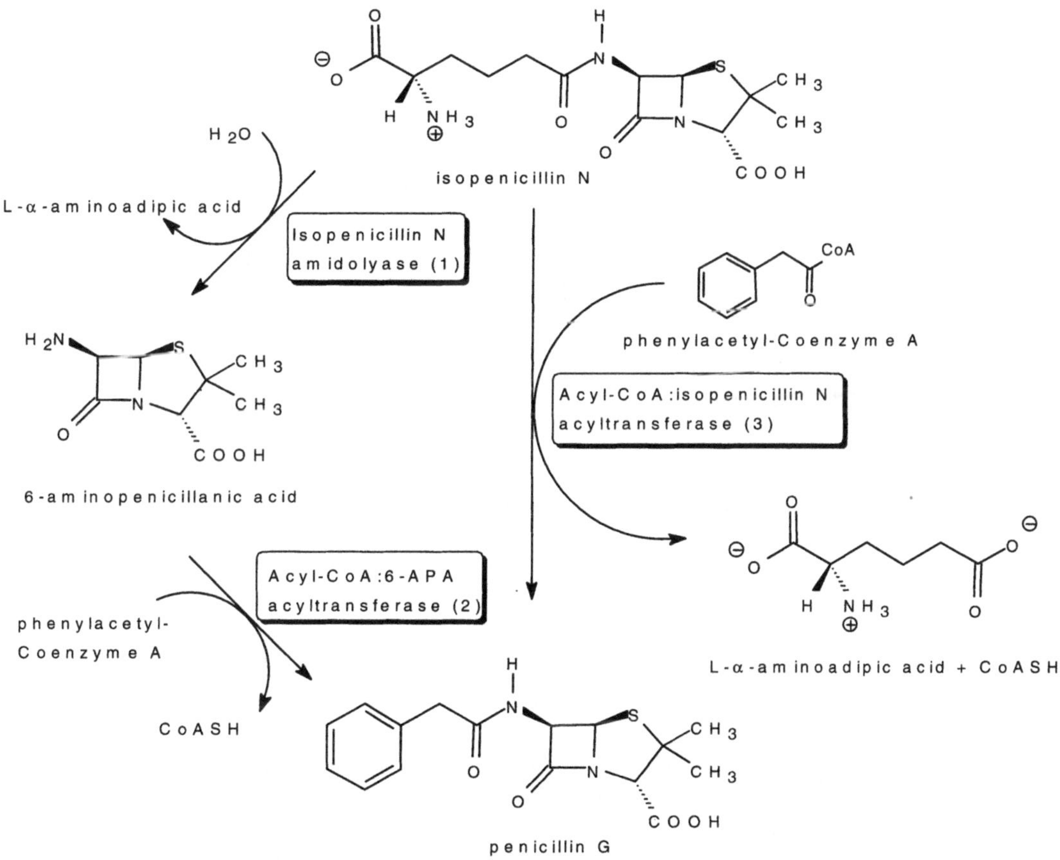

Fig. VI.14 Reaction catalyzed by CoA:isopenicillin N acyltransferase. The enzyme converts isopenicillin N into penicillin G, either by a direct transacylation (3), or by a hydrolysis reaction (1) followed by an acylation reaction (2).

To accommodate the desire for structural information as a basis for site-directed mutations, X-ray structural studies on AT have been initiated and successfully concluded within the framework of the cluster project. Although attempts to solve the structure of the mature AT were not successful until now, the structure of the precursor of AT has been elucidated to 1.85 Å resolution (Hensgens and Dijkstra, unpublished results) and forms the basis for a protein engineering project in which AT may be modified for the production of β-lactam antibiotics other than penicillin G.

5.5 Fermentative production of semi-synthetic penicillins and cephalosporins

Ideally, the modification of natural penicillins to produce semi-synthetic β-lactam antibiotics would be carried out *in vivo*, rather than by isolation of a natural antibiotic followed by post-fermentative modification.
The initial steps in penicillin and cephalosporin biosynthesis are identical and diverge only after the synthesis of isopenicillin N (Fig. VI.15).

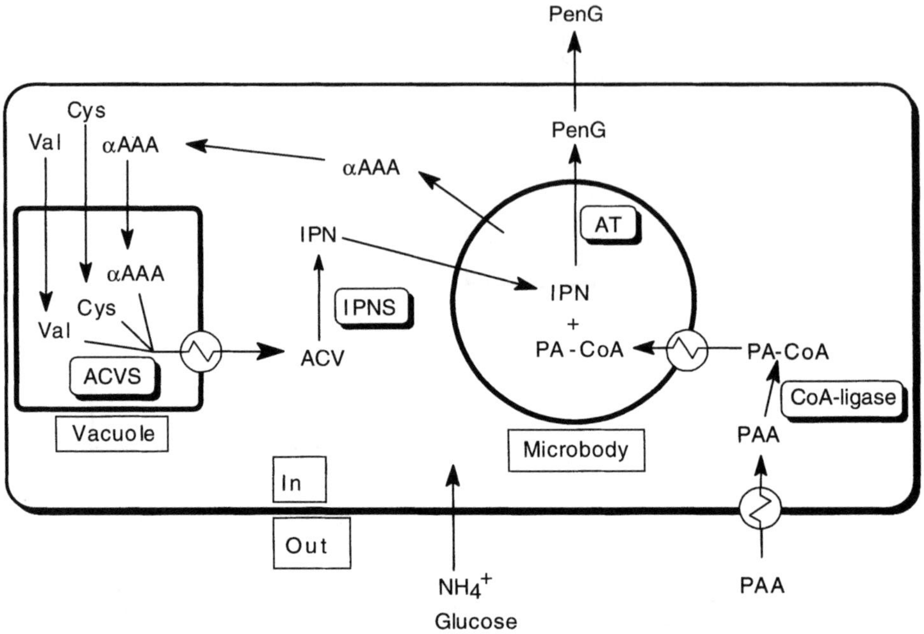

Fig. VI.15 Biosynthesis of penicillin G from Val, Cys and αAAA by *Penicillium chrysogenum*. See text for details.

The first step is the production of δ-(L-α-aminoadipoyl)-L-cysteinyl-D-valine (ACV) by ACV synthetase. ACV is then converted by isopenicillin N synthase (IPNS) into the first functional penicillin, isopenicillin N (IPN). In *Penicillium chrysogenum* and related organisms, the polar IPN is transported to the microbody and converted by acyl CoA:isopenicillin N acyltransferase (AT) into the hydrophobic penicillin G, which is subsequently excreted. In organisms that produce cephalosporins such as the fungus *Acremonium chrysogenum* (formerly known as *Cephalosporium acremonium*) and the bacterium *Streptomyces clavilugerus*, IPN is converted to penicillin N by an isomerase and then funneled into the cephalosporin biosynthetic route. The first enzyme in this route is deacetoxy cephalosporin C synthase (expandase), which expands the five-membered thiazolidine ring of penicillins to the six-membered dihydrothiazine ring of cephalosporins. Additional enzymes such as deacetyl cephalosporin C synthase are active further down the pathway, resulting in the ultimate formation of cephalosporin C. In contrast to the production of penicillin G, cephalosporin is synthesized in the cytoplasm.

If one considers the natural penicillin biosynthetic pathway (Fig. VI.15), it is evident that a major cell-engineering and protein-engineering effort is required for the fermentative production of semi-synthetic antibiotics[99]. Current knowledge of the diversity, specificity and engineering potential of the acyl precursor and β-lactam biosynthetic pathways is insufficient to judge the feasibility of such an approach.

Even though complete fermentative production may only be feasible in the far future, significant improvements in the productivity and product spectrum have been achieved already. In fact, classical strain improvement and use of phenylacetic acid during fermentation already have switched fermentative production from penicillin V, which carries a phenoxyacetyl group, to penicillin G.

Strain productivity has also been improved by introducing a bacterial phenylacetyl CoA ligase gene, which was obtained from a *Pseudomonas sp.* growing on phenylacetic acid, into *Penicillium chrysogenum*[100]. A two-fold increased accumulation of penicillin G was observed, indicating that phenylacetyl CoA levels limit the rate of antibiotic production.

Furthermore, it is possible to perform *in vivo* deacetoxy cephalosporin C production. This was achieved by introducing the expandase gene (*cefE*) from *Streptomyces clavuligerus* into *Acremonium chrysogenum* from which the expandase/hydroxylase activity was removed. Subsequent treatment with D-amino acid oxidase and glutaryl acylase leads to 7-ADCA. In this way, the environmentally unfriendly chemical ring expansion method is no longer needed[101].

§6 Acknowledgments

The authors would like to thank Jan-Metske van der Laan, Theo Sonke and Herman Slijkhuis for their collaboration and fruitful discussions, which greatly helped the authors in preparing this manuscript.

§7 References

1) Huber, F. M., Chauvette, R. R. and Jackson, B. G. Preparative methods for 7-ACA and 6-APA; Flynn, E. H., Ed.; Academic Press: New York, **1972**.
2) Verhaert, R. M., Riemens, A. M., van der Laan, J. M., van Duin, J. and Quax, W. J. *Appl Environ Microbiol* **1997**, *63*, 3412-8.
3) Oh, S. J., Kim, Y. C., Park, Y. W., Min, S. Y., Kim, I. S. and Kang, H. S. *Gene* **1987**, *56*, 87-97.
4) Schumacher, G., Sizmann, D., Haug, H., Buckel, P. and Bock, A. *Nucleic Acids Res* **1986**, *14*, 5713-27.
5) Martin, L., Prieto, M. A., Cortes, E. and Garcia, J. L. *FEMS Microbiol Lett* **1995**, *125*, 287-92.
6) Ljubijankic, G., Konstantinovic, M. and Glisin, V. *DNA Seq* **1992**, *3*, 195-200.
7) Barbero, J. L., Buesa, J. M., Gonzalez de Buitrago, G., Mendez, E., Pez-Aranda, A. and Garcia, J. L. *Gene* **1986**, *49*, 69-80.
8) Ohashi, H., Katsuta, Y., Hashizume, T., Abe, S. N., Kajiura, H., Hattori, H., Kamei, T. and Yano, M. *Appl Environ Microbiol* **1988**, *54*, 2603-7.
9) Klenk, H. P., Clayton, R. A., Tomb, J. F., White, O., Nelson, K. E., Ketchum, K. A., Dodson, R. J., Gwinn, M., Hickey, E. K., Peterson, J. D., Richardson, D. L., Kerlavage, A. R., Graham, D. E., Kyrpides, N. C., Fleischmann, R. D., Quackenbush, J., Lee, N. H., Sutton, G. G., Gill, S., Kirkness, E. F., Dougherty, B. A., McKenney, K., Adams, M. D., Loftus, B., Venter, J. C. *et al.* *Nature* **1997**, *390*, 364-70.

10) Kawarabayasi, Y., Hino, Y., Horikawa, H., Yamazaki, S., Haikawa, Y., Jin-no, K., Takahashi, M., Sekine, M., Baba, S., Ankai, A., Kosugi, H., Hosoyama, A., Fukui, S., Nagai, Y., Nishijima, K., Nakazawa, H., Takamiya, M., Masuda, S., Funahashi, T., Tanaka, T., Kudoh, Y., Yamazaki, J., Kushida, N., Oguchi, A., Kikuchi, H. *et al. DNA Res* **1999**, *6*, 83-101, 145-52.

11) Kaneko, T. and Tabata, S. *Plant Cell Physiol* **1997**, *38*, 1171-6.

12) Ruepp, A., Graml, W., Santos-Martinez, M. L., Koretke, K. K., Volker, C., Mewes, H. W., Frishman, D., Stocker, S., Lupas, A. N. and Baumeister, W. *Nature* **2000**, *407*, 508-13.

13) Inokoshi, J., Takeshima, H., Ikeda, H. and Omura, S. *Gene* **1992**, *119*, 29-35.

14) Chou, C. P., Tseng, J. H., Lin, M. I., Lin, H. K. and Yu, C. C. *J Biotechnol* **1999**, *69*, 27-38.

15) Daumy, G. O., McColl, A. S. and Apostolakos, D. *J Bacteriol* **1982**, *152*, 104 10.

16) Gang, D. M. and Shaikh, K. *Biochim Biophys Acta* **1976**, *425*, 110-4.

17) Jiang, Q., Wu, R. and Yang, S. *Chin J Biotechnol* **1992**, *8*, 153-8.

18) Merino, E., Balbas, P., Recillas, F., Becerril, B., Valle, F. and Bolivar, F. *Mol Microbiol* **1992**, *6*, 2175-82.

19) Roa, A. and Garcia, J. L. *FEMS Microbiol Lett* **1997**, *151*, 9-16.

20) Roa, A. and Garcia, J. L. *FEMS Microbiol Lett* **1999**, *177*, 7-14.

21) Prieto, M. A., Diaz, E. and Garcia, J. L. *J Bacteriol* **1996**, *178*, 111-20.

22) Brannigan, J. A., Dodson, G., Duggleby, H. J., Moody, P. C., Smith, J. L., Tomchick, D. R. and Murzin, A. G. *Nature* **1995**, *378*, 416-9.

23) Hewitt, L., Kasche, V., Lummer, K., Lewis, R. J., Murshudov, G. N., Verma, C. S., Dodson, G. G. and Wilson, K. S. *J Mol Biol* **2000**, *302*, 887-98.

24) Kasche, V., Lummer, K., Nurk, A., Piotraschke, E., Rieks, A., Stoeva, S. and Voelter, W. *Biochim Biophys Acta* **1999**, *1433*, 76-86.

25) Sizmann, D., Keilmann, C. and Bock, A. *Eur J Biochem* **1990**, *192*, 143-51.

26) Alkema, W. B. L., Prins, A. P., De Vries, E. J. and Janssen, D. B. *Manuscript in prep.*

27) Chou, C. P., Tseng, J. H., Kuo, B. Y., Lai, K. M., Lin, M. I. and Lin, H. K. *Biotechnol Prog* **1999**, *15*, 439-45.

28) Chou, C. P., Yu, C. C., Tseng, J. H., Lin, M. I. and Lin, H. K. *Biotechnol Bioeng* **1999**, *63*, 263-72.

29) Gumpert, J., Cron, H., Plapp, R., Niersbach, H. and Hoischen, C. *J Basic Microbiol* **1996**, *36*, 89-98.

30) Chou, C. F., Kuo, B. Y. and Lin, W. J. *J Biosc Bioeng* **1999**, *88*, 160-7.

31) Slade, A., Horrocks, A. J., Lindsay, C. D., Dunbar, B. and Virden, R. *Eur J Biochem* **1991**, *197*, 75-80.

32) Svedas, V. K., Savchenko, M. V., Beltser, A. I. and Guranda, D. F. *Ann N Y Acad Sci* **1996**, *799*, 659-69.

33) Fuganti, C., Rosell, C. M., Servi, S., Tagliani, A. and Terreni, M. *Tet Asymm* **1992**, *3*, 383-5.

34) van Langen, L. M., van Rantwijk, F., Svedas, V. K. and Sheldon, R. A. *Tet Asymm* **2000**, *11*, 1077-83.

35) Guy, A., Dumant, A. and Sziraky, P. *Bioorg Med Chem Lett* **1993**, *3*, 1041-4.
36) Rossi, D., Calcagni, A. and Romeo, A. *J Org Chem* **1979**, *44*, 2222-5.
37) Cardillo, C., Tolomelli, A. and Tomasini, C. *J Org Chem* **1996**, *61*, 8651-4.
38) Solodenko, V. A., Kasheva, T. N., Kukhar, V. P., Kozlova, E. V., Mironenko, D. A. and Svedas, V. K. *Tetrahedron* **1991**, *47*, 3989-98.
39) Bossi, A., Cretich, M. and Righetti Pier, G. *Biotechnol Bioeng* **1998**, *60*, 454-61.
40) Duggleby, H. J., Tolley, S. P., Hill, C. P., Dodson, E. J., Dodson, G. and Moody, P. C. *Nature* **1995**, *373*, 264-8.
41) Seemuller, E., Lupas, A. and Baumeister, W. *Nature* **1996**, *382*, 468-71.
42) Suresh, C. G., Pundle, A. V., SivaRaman, H., Rao, K. N., Brannigan, J. A., McVey, C. E., Verma, C. S., Dauter, Z., Dodson, E. J. and Dodson, G. G. *Nat Struct Biol* **1999**, *6*, 414-6.
43) McDonough, M. A., Klei, H. E. and Kelly, J. A. *Protein Sci* **1999**, *8*, 1971-81.
44) Bompard-Gilles, C., Villeret, V., Davies, G. J., Fanuel, L., Joris, B., Frere, J. M. and Van Beeumen, J. *Structure Fold Des* **2000**, *8*, 153-62.
45) Liu, Y., Guan, C. and Aronson, N. N., Jr. *J Biol Chem* **1998**, *273*, 9688-94.
46) Lee, Y. S., Kim, H. W. and Park, S. S. *J Biol Chem* **2000**, *275*, 39200-6.
47) Kim, Y., Yoon, K., Khang, Y., Turley, S. and Hol, W. G. *Structure Fold Des* **2000**, *8*, 1059-68.
48) Polgar, L. *Mechanism of protease action*; CRC Press Inc.: Boca Raton, Florida, **1989**.
49) Balasingham, K., Warburton, D., Dunnill, P. and Lilly, M. D. *Biochim Biophys Acta* **1972**, *276*, 250-6.
50) Gololobov, M. Y., Borisov, I. L. and Svedas, V. K. *J Theor Biol* **1989**, *140*, 193-204.
51) Fernandez-Lafuente, R., Rosell, C. M. and Guisan, J. M. *Biotechnol Appl Biochem* **1996**, *24*, 139-43.
52) Lummer, K., Rieks, A., Galunsky, B. and Kasche, V. *Biochim Biophys Acta* **1999**, *1433*, 327-34.
53) Youshko M. I., van Langen L. M., de Vroom, E., Moody H. M., van Rantwijk, F., Sheldon R. A. and Svedas, V. K. *J Mol Cat: B Enz* **2000**, *10*, 509-15.
54) Park C., B., Lee S., B. and Ryu D., D. Y. *J Mol Cat: B Enz* **2000**, *9*, 275-81.
55) Alvaro, G., Fernandez-Lafuente, R., Blanco, R. M. and Guisan, J. M. *Appl Biochem Biotechnol* **1990**, *26*, 181-95.
56) Prabhune, A. and Sivaraman, H. *Appl Biochem Biotechnol* **1991**, *30*, 265-72.
57) Boccu, E., Carenza, M., Lora, S., Palma, G. and Veronese, F. M. *Appl Biochem Biotechnol* **1987**, *15*, 1-10.
58) Bryjak, J. and Noworyta, A. *J Chem Technol Biotechnol* **1993**, *57*, 79-85.
59) Fernandez-Lafuente, R., Rosell, C. M. and Guisan, J. M. *Enzyme Microb Technol* **1991**, *13*, 898-905.
60) Fonseca, L. P., Cardoso, J. P. and Cabral, J. M. *J Chem Technol Biotechnol* **1993**, *58*, 27-37.

61) You, L., Usher, J. J., White, B. J. and Novotny, J. *Mutant penicillin G acylases patent WO 98/20120.*
62) Alkema, W. B. L., Hensgens, C. M. H., de Vries, E. J., Floris, R., Kroezinga, E., Dijkstra, B. W. and Janssen, D. B. *Prot Eng* **2000**, *13*, 857-863.
63) Brannigan, J. A., Dodson, G. C., Done, S. H., Hewitt, L., McVey, C. E. and Wilson, K. S. *Applied Biochem Biotech* **2000**, *88*, 313-9.
64) Done, S. H., Brannigan, J. A., Moody, P. C. and Hubbard, R. E. *J Mol Biol* **1998**, *284*, 463-75.
65) Kasche, V., Haufler, U. and Zollner, R. *Hoppe Seylers Z Physiol Chem* **1984**, *365*, 1435-43.
66) Morillas, M., Goble, M. L. and Virden, R. *Biochem J* **1999**, *338*, 235-9.
67) Roa, A., Goble, M. L., Garcia, J. L., Acebal, C. and Virden, R. *Biochem J* **1996**, *316*, 409-12.
68) Roa, A., Castillon, M. P., Goble, M. L., Virden, R. and Garcia, J. L. *Biochem Biophys Res Commun* **1995**, *206*, 629-36.
69) Alkema, W. B. L., Floris, R., de Vries, E. J. and Janssen, D. B. *Manuscript in prep.*
70) Baker, W. L. *Antimicrob Agents Chemother* **1983**, *23*, 26-30.
71) Svedas, V., Guranda, D., van Langen, L., van Rantwijk, F. and Sheldon, R. *FEBS Lett* **1997**, *417*, 414-8.
72) Alkema, W. B. L., Floris, R. and Janssen, D. B. *Anal Biochem* **1999**, *275*, 47-53.
73) Stemmer, W. P. *Nature* **1994**, *370*, 389-91.
74) Zhao, H. and Arnold, F. H. *Protein Eng* **1999**, *12*, 47-53.
75) Arnold, F. H. and Moore, J. C. *Adv Biochem Eng Biotechnol* **1997**, *58*, 1-14.
76) Giver, L., Gershenson, A., Freskgard, P. O. and Arnold, F. H. *Proc Natl Acad Sci USA* **1998**, *95*, 12809-13.
77) Crameri, A., Raillard, S. A., Bermudez, E. and Stemmer, W. P. *Nature* **1998**, *391*, 288-91.
78) Forney, L. J., Wong, D. C. and Ferber, D. M. *Appl Environ Microbiol* **1989**, *55*, 2550-5.
79) Forney, L. J. and Wong, D. C. *Appl Environ Microbiol* **1989**, *55*, 2556-60.
80) Roa, A., Garcia, J. L., Salto, F. and Cortes, E. *Biochem J* **1994**, *303*, 869-75.
81) Daumy, G. O., Danley, D., McColl, A. S., Apostolakos, D. and Vinick, F. J. *J Bacteriol* **1985**, *163*, 925-32.
82) Fujii, T., Matsumoto, K. and Watanabe, T. *Process Biochem* **1976**, *11*, 21-4.
83) Choi, K. S., Kim, J. A. and Kang, H. S. *J Bacteriol* **1992**, *174*, 6270-6.
84) Niersbach, H., Kuhne, A., Tischer, W., Weber, M., Wedekind, F. and Plapp, R. *Appl Microbiol Biotechnol* **1995**, *43*, 679-84.
85) Alkema, W. B. L., Dijkhuis, A.-J., De Vries, E. J. and Janssen, D. B. *Manuscript in prep .*
86) Prieto, I., Rodriguez, M. C., Marquez, G., Perez-Aranda, A. and Barbero, J. L. *Appl Microbiol Biotechnol* **1992**, *36*, 659-62.
87) Martin, J., Prieto, I., Barbero, J. L., Perez-Gil, J., Mancheno, J. M. and Arche, R. *Biochim Biophys Acta* **1990**, *1037*, 133-9.

88) Martin, J., Slade, A., Aitken, A., Arche, R. and Virden, R. *Biochem J* **1991**, *280*, 659-62.
89) Abrahmsen, L., Tom, J., Burnier, J., Butcher, K. A., Kossiakoff, A. and Wells, J. A. *Biochemistry* **1991**, *30*, 4151-9.
90) Elliott, R. J., Bennet, A. J., Braun, C. A., MacLeod, A. M. and Borgford, T. J. *Chem Biol* **2000**, *7*, 163-71.
91) Takahashi, T., Yamazaki, Y., Kato, K. and Isona, M. *J Am Chem Soc* **1972**, *94*, 4035-7.
92) Takahashi, T., Yamazaki, Y. and Kato, K. *Biochem J* **1974**, *137*, 497-503.
93) Kato, K., Kawahara, K., Takahashi, T. and Kakinuma, A. *Agric Biol Chem* **1980**, *44*, 1075-1081.
94) Ryu, Y. W. and Ryu, D. D. Y. *Enz Microb Technol* **1988**, *10*, 239-245.
95) Blinkovsky, A. M. and Markaryan, A. N. *Enz Microb Technol* **1993**, *15*, 965-973.
96) Aplin, R. T., Baldwin, J. E., Cole, S. C., Sutherland, J. D. and Tobin, M. B. *FEBS Lett* **1993**, *319*, 166-70.
97) Roach, P. L., Clifton, I. J., Fulop, V., Harlos, K., Barton, G. J., Hajdu, J., Andersson, I., Schofield, C. J. and Baldwin, J. E. *Nature* **1995**, *375*, 700-4.
98) Valegard, K., van Scheltinga, A. C., Lloyd, M. D., Hara, T., Ramaswamy, S., Perrakis, A., Thompson, A., Lee, H. J., Baldwin, J. E., Schofield, C. J., Hajdu, J. and Andersson, I. *Nature* **1998**, *394*, 805-9.
99) Van de Kamp, M., Driessen, A. J. and Konings, W. N. *Antonie Van Leeuwenhoek* **1999**, *75*, 41-78.
100) Minambres, B., Martinez-Blanco, H., Olivera, E. R., Garcia, B., Diez, B., Barredo, J. L., Moreno, M. A., Schleissner, C., Salto, F. and Luengo, J. M. *J Biol Chem* **1996**, *271*, 33531-8.
101) Velasco, J., Luis Adrio, J., Angel Moreno, M., Diez, B., Soler, G. and Barredo, J. L. *Nat Biotechnol* **2000**, *18*, 857-61.
102) Kim, D. J. and Byun, S. M. *Biochim Biophys Acta* **1990**, *1040*, 12-18.
103) Altschul, S. F., Madden, T. L., Schaffer, A. A., Zhang, J., Zhang, Z., Miller, W. and Lipman, D. J. *Nucleic Acids Res* **1997**, *25*, 3389-402.

Chapter VII **Modeling the metabolism of Penicillin-G formation** ***283***

§1 Introduction ***283***
1.1 The "cell factory" concept *283*
1.2 Comparison of current fine chemistry practice and "cell factories" *283*
1.3 The *Penicillium chrysogenum* Cell Factory *286*
1.4 The cell factory, chances problems and a "rational intuitive approach" *286*
1.5 Key problems in metabolic engineering *289*

§2 Possible location of metabolic bottlenecks for Penicillin-G biosynthesis ***290***

§3 Development of a stoichiometric model for growth and Penicillin-G production in *Penicillium chrysogenum* ***294***
3.1 Anaplerotic pathways *294*
3.2 Regeneration of cytosolic NADPH *294*
3.3 Biosynthesis of cysteine *295*
3.4 Intracellular compartmentation *295*
3.5 Solute transport across the plasmalemma *296*
3.6 Intracellular transport *297*

§4 Energetics of growth and Penicillin production in a high producing strain of *Penicillium chrysogenum* ***298***
4.1 Estimation of maximum yields *298*
4.2 Model for the ATP stoichiometry of the metabolic network *299*
4.3 Estimation of ATP stoichiometry parameters *301*
4.4 Maximum yield and maintenance parameters on substrate *305*
4.5 Maximum theoretical penicillin yield *306*
4.6 Maximum yield and maintenance parameters on oxygen *306*
4.7 Validation of estimated additional energy needs for penicillin production *307*
4.8 Expressions for the maximum yield and maintenance coefficients as a function of the ATP stoichiometry parameters *309*

§5 Application of metabolic flux analysis for the identification of metabolic bottlenecks in the biosynthesis of Penicillin-G ***311***
5.1 Identification of principal nodes for Penicillin production *311*
5.2 Glucose, ethanol and acetate limited chemostat cultures *313*
5.3 Balancing of steady state data *313*
5.4 Penicillin production as a function of the growth rate *315*
5.5 Changes in flux partitioning around principal nodes *317*
5.6 Supply of cytosolic NADPH *318*

§6 Application of ^{13}C-NMR for the estimation of intracellular fluxes ***319***
6.1 Problems in steady state metabolic flux analysis *319*
6.2 Solutions to overcome inobservability of fluxes *321*
6.3 Using ^{13}C-NMR for metabolic flux analysis *323*
6.4 Results of ^{13}C-labeling studies *325*

§7 A first attempt to model the regulation of Penicillin-G production in *Penicillium chrysogenum* ***325***
7.1 Control networks and product formation *325*
7.2 Simple dynamic model for penicillin production *326*

§8 Conclusions and outlook ***328***

§9 References ***329***

Chapter VII Modeling the metabolism of Penicillin-G formation

W.M. VAN GULIK, W.A. VAN WINDEN AND J.J. HEIJNEN (TECHNICAL UNIVERSITY DELFT, KLUYVER LABORATORY FOR BIOTECHNOLOGY)

§1 Introduction

1.1 The "cell factory" concept

Micro-organisms have been used since many decades for the production of valuable chemicals for food, pharmaceutical and bulk industries. Examples are amino acids, vitamins, antibiotics or alcohols and organic acids (Table VII.1). Improvement of the production properties has been achieved using random classical mutation techniques. The development of recombinant-DNA techniques, the unraveling of complete genomes and genome wide information measurement (DNA chips) have recently opened the possibility of precise modifications in microbial metabolism. The goals of such "rational metabolic engineering" are “de novo” or improved production of desirable chemical compounds. Rational metabolic engineering opens the possibility to use micro-organisms for the production of a wider scope of bulk and fine chemicals. This is called the "cell factory concept".

It is now useful to:

- compare current fine chemistry practice with microbial performance
- present an intuitive approach to rational metabolic engineering
- discuss the problem area's to metabolic engineering

Table VII.1 Microbial production processes

Product	Market volume (tons/yr.)
Lysine	100.000
Glutamic acid	1000.000
Functional proteins	10.000
β-lactam antibiotics	45.000
Ethanol	25.000.000
Lactic acid	50.000

1.2 Comparison of current fine chemistry practice and "cell factories"

Current processes for the production of fine chemicals are typically cascades of batch steps to achieve a multi-step synthesis (Figure VII.1). Each step starts with high substrate concentration and ends with high intermediate product concentration.

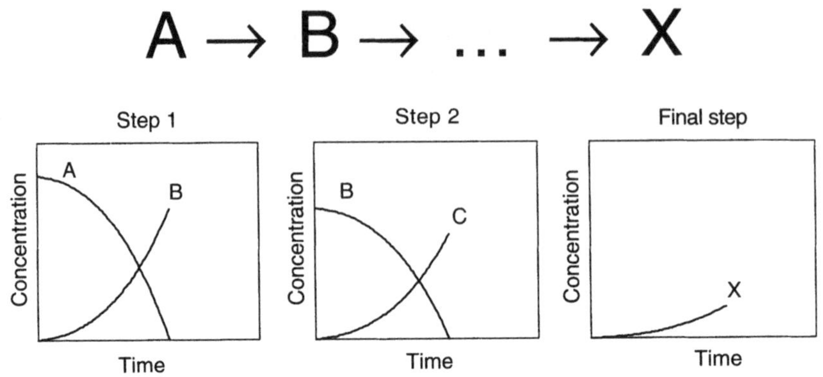

Fig. VII.1 Cascade of batch steps in fine chemistry.

Due to the high concentrations (typically Molar) frequently the compounds (A, B, etc.) are subject to spontaneous, undesirable side reactions (hydrolysis, decarboxylations etc.).

In addition there is stoichiometric consumption of:

- activating groups to provide thermodynamic driving force
- protecting groups to provide selectivity
- oxidation/reduction agents to provide the desired redox status of the compounds.

Their consumption leads to the stoïchiometric accumulation of the spent groups (e.g. ammonia due to amide usage etc.), which can again lead to undesirable kinetic interferences. All this leads to complex reaction mixtures, necessitating costly and (for their development) time-consuming downstream processes. A final point of concern is that most fine chemical processes use non-renewable feed stocks (derived from fossil carbon sources as oil, coal, gas).

In conclusion current fine chemical processes use non-renewable feed stocks, consume stoichiometric amounts of auxiliary chemicals, result in low yields, require complex downstream processing, produce large amounts of waste leading to high costs and long development times.

This current fine chemistry practice can be compared with the fine chemistry approach found in micro-organisms (cell factories) (Figure VII.2).

Micro-organisms use nearly always glucose, which is a renewable feedstock. The feedstock (A) enters the cell factory by controlled feeding, using specific transport proteins in the cell membrane. This results in a low substrate concentration (A) at the start of the multi-step reaction sequence, where each step is enzyme catalyzed.

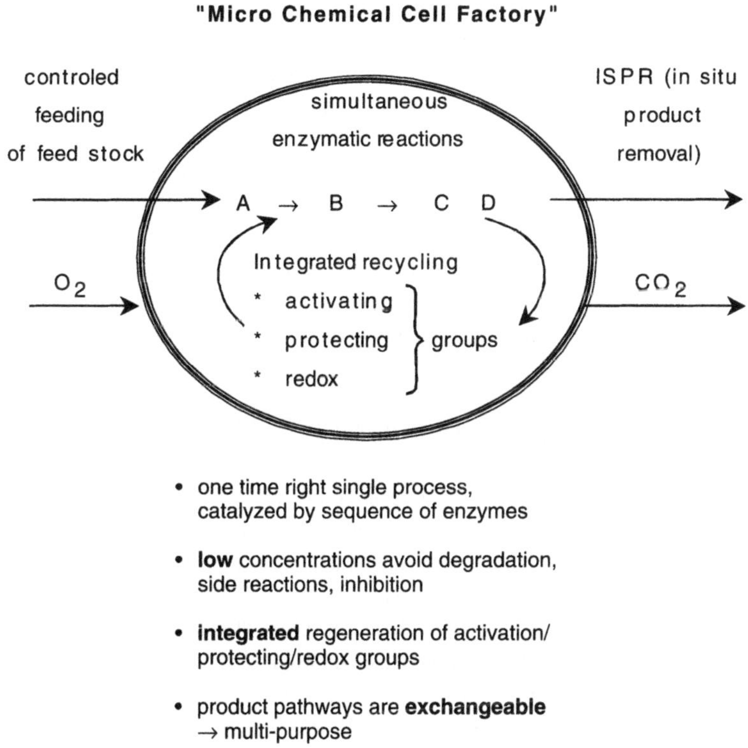

Fig. VII.2 Fine chemistry approach in micro-organisms.

Because all reactions occur simultaneously the concentrations of the intermediates between initial substrate and final product remains very low (order 10^{-3} to 10^{-6} M). Consequently, there is hardly any spontaneous degradation and also a-specific inhibition phenomena are absent. Most importantly: micro-organisms use activating groups (e.g. ATP, Co-enzyme A), redox groups (e.g. NAD(P)H) and protecting groups (e.g. N-acetyl) which are regenerated continuously at the expense of energy. Hence only catalytic amounts of these groups (called cofactors) need to be present. The regeneration of all these groups and the starting molecule (A) of the product-forming pathway (A → D) all are realized in the so-called primary metabolism of micro-organisms at the expense of energy, derived from the oxidation of the renewable feedstock glucose with oxygen.
To avoid intracellular accumulation the product is exported (in situ product removal, ISPR) from the micro-organisms using membrane embedded transport proteins,

again at the expense of energy (the external product concentration is usually much higher than the intracellular concentration). Finally it should be realized that in principle many different product pathways can be served by the available primary metabolism. Micro-organisms can therefore be considered as multi-purpose factories.

In summary, the fine chemistry of micro-organisms is completely integrated with respect to material and energy, the simultaneous execution of multi-step enzyme catalyzed synthesis and the recycling of spent cofactors. The only waste materials are CO_2, H_2O and spent biomass. The main used feedstock's are glucose and O_2. It should be expected that, due to the ever-increasing developments in life sciences, these cell factories could become the main production technology of future chemicals.

1.3 The *Penicillium chrysogenum* Cell Factory

Penicillin has been produced for many decades in the filamentous fungus *P. chrysogenum* with ever increasing yield. This has been achieved based on random mutation and subsequent selection of high producing mutants. However, current production strains are suspected to have such high productivity's that primary metabolism might become limiting. Therefore metabolic engineering of *P. chrysogenum* based on rational concepts is highly relevant.

This is even more so because of the recently introduced 7-ADCA production process (Chapter I), which heavily relies on insight in the metabolism of *P. chrysogenum*.

1.4 The cell factory, chances problems and a "rational intuitive approach"

Figure VII.3shows the essential features of microbial metabolism (cell factory).

The Cell Factory

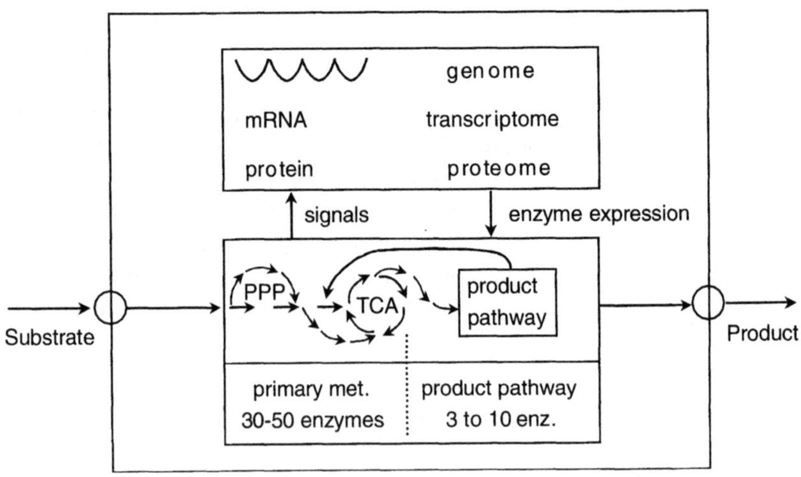

Fig. VII.3 Schematic overview of the essential features of microbial metabolism.

The reaction network can be distinguished as containing a product pathway (typically 3-10 enzymes in a linear sequence) and primary metabolism (containing typically 30-50 enzymes in a network of linear sequences, cycles, parallel routes). The product pathway has the product as its final compound. The starting compound of the product pathway is delivered by primary metabolism, which also provides and regenerates all energy, redox and other cofactors involved in product formation. Primary metabolism is fed with the substrate. Above the reaction network there exists a multi layered control network (genome, transcriptome and proteome), which responds to the metabolite levels (= signals) in the reaction network and controls the levels of the enzymes. E.g. when a metabolite level drops, the multi-layered control network reacts to increase the level of several enzymes leading to an increased synthesis of the dropping metabolite, restoring the original metabolite level (homeostasis concept).
The current explosive growth of genomic information and the fast increase in molecular biology tools provides the chance for the precise design of such cell factories for a multitude of desirable products.
Problems to be solved in order to apply this approach are:

- a lack of quantitative understanding of the complex interactions in the cellular metabolic network
- the need for studying the whole cell (in vivo behavior) and not its parts (in vitro behavior).

These problems have led to an approach where the focus is on improving the product pathway capacity by increasing the involved enzyme levels. The successes have been limited, however. Recently it has been recognized that this might be due to negligence of the effect of these increased capacities on the intracellular metabolites. Especially the importance of the product export may have been underestimated. Knowing that feedback inhibition of the final product on the rate of the first product pathway step is a general property of many product pathways; it is essential to keep this product concentration low. The following "rational intuitive approach" without detailed knowledge therefore seems appropriate. This approach is based on pull on the product and not on push on the substrate (Figure VII.4). First the export of product is increased by increasing the export capacity. The lower intracellular product concentration will achieve:

- less inhibition of the product on the first step of the product pathway, which leads to increased flux of product
- increased levels of enzymes in the network because of genetic control mechanisms

The cell factory

Intuitive approach of metabolic reprogramming

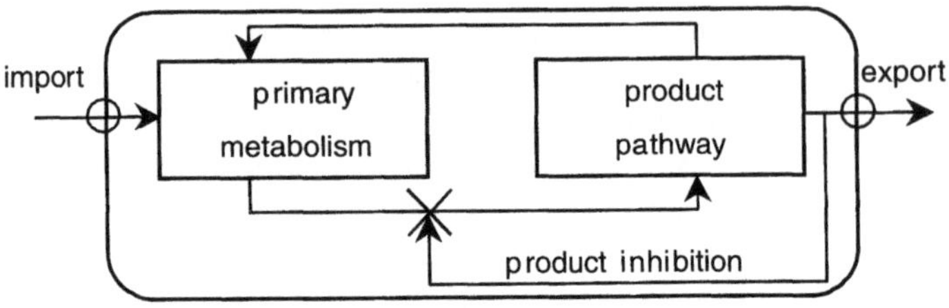

* increase of product export keeps intermediate concentrations low, minimal adverse effects from high concentrations

* decrease product inhibition (opens the flow between primary metabolism → product pathway)

* increase enzyme levels in product pathway

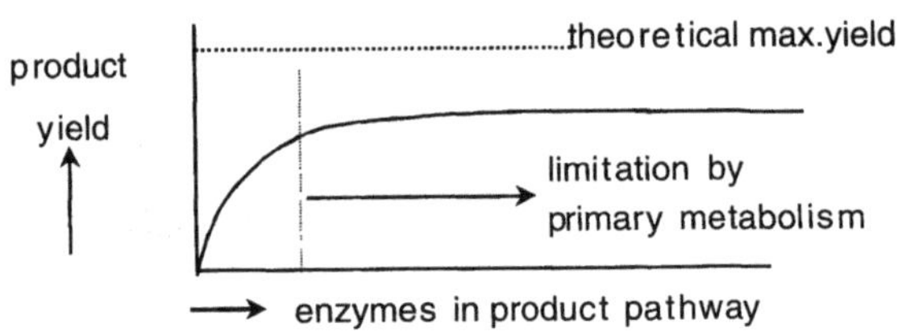

* increase substrate import

Fig. VII.4 Metabolic reprogramming of cell factories.

Secondly, the product inhibition of the first enzyme is alleviated (using e.g. gene shuffling or rational protein engineering). Usually an increased flux is not easily observed because (in absence of sufficient export capacity) the increased intracellular product concentration counteracts the increased reaction capacity. However, having first increased the export capacity, allows the full benefit of decreased feedback inhibition.

Thirdly, although the enzyme levels of the product pathway will have been increased, following increased export and decreased product inhibition by existing genetically based control mechanisms, the enzymes of the product pathway are simultaneous increased in level to achieve a further flux increase.

However, the inevitable point will be reached where changes in the product pathway have no effect anymore.

At this stage product formation is not limited anymore by the product pathway. The limitations reside then in primary metabolism in:

- supply of the starting compound of the product pathway
- regeneration or supply of spent energy, redox, protecting group cofactors (ATP, NAD(P)H, Ac-CoA etc.)

Improving primary metabolism is a much more demanding task, where the above intuition fails due to the many interactions.
The only intuition here is to focus on the so-called anaplerotic reactions of the TCA-cycle if the product pathway takes a TCA-cycle intermediate as a starting compound to increase substrate import capacity.
Practice has however shown that changing primary metabolism has only met limited success.

1.5 Key problems in metabolic engineering

Given the above situation the following key problems can be defined in designing cell factories.

Export of product.
This has been studied far too little. Much more knowledge is required on the genetics of active transporters of important classes of products.

What is the maximal product yield on substrate?
Good insight in the maximal product yield is highly valuable to compare this value with the current yield. The difference quantifies the maximal possible improvement. As indicated, improvement of only the product pathway leads to saturation in yield. The yield may only be improved subsequently by improving primary metabolism.
The reward of such an endeavour can be judged by comparing the maximal theoretical yield with the yield at saturation product pathway capacity. Also, methods to calculate and experimentally determine the maximal yield as function of network changes are highly valuable for a priori quantification of possible improvement.
In paragraph 4, for the penicillin fermentation, a new method to determine this maximal yield will be presented, with surprising results.

Is the product pathway or primary metabolism limiting?
A practical method to decide whether the product pathway is still limiting the production rate would be most useful. In paragraph 5 a new simple in-vivo method will be presented, using the penicillin fermentation as a model system, to answer this question.

How to determine where to change primary metabolism using rec-DNA methods

Improving product pathway capacity always will lead to the situation that primary metabolism becomes limiting.

Improving primary metabolism requires:

- Knowledge of the reactions employed and their rates. Here many uncertainties exist in e.g. *P. chrysogenum* with respect to NADPH metabolism. A promising method to elucidate such problems is use of labelled ^{13}C. Its application to P. chrysogenum will be discussed in paragraph 6.
- Knowledge of the in-vivo kinetics of primary metabolism. Here, especially for *P. chrysogenum*, a large "black box" area exists , which needs to be resolved in the coming years. Most important is then the development of :
  - ⇒ Intracellular metabolite measurements, including compartmentation
  - ⇒ Measurement of intracellular fluxes
  - ⇒ Measurement of enzyme activities for the whole primary metabolism

A quantitative understanding of enzyme expression

The mechanisms of enzyme expression are very complex. Detailed models of repression/induction exist and also global concepts as cybernetic models are proposed. In paragraph 7 it will be shown that the expression of the penicillin pathway can, surprisingly simple, be described by an induction/repression model. However, with the development of DNA chips and e.g. 2D-gel electrophoresis a much more detailed study of the expression/status of enzymes will be possible and must be undertaken in future.

§2 Possible location of metabolic bottlenecks for Penicillin-G biosynthesis

With the current methods for genetic modification the possibility to introduce specific changes in the microbial genome became available. The problem is, however, to identify at which locations in the metabolic network bottlenecks occur, in other words to identify the targets for rational genetic engineering. Obvious targets are the genes coding for the enzymes in the product pathway. Metabolic control analysis of the biosynthetic pathway of penicillin-V has been applied to identify the rate-limiting step in a high producing strain of *P. chrysogenum* grown in fed batch culture (Nielsen and Jørgensen, 1995[1]; Pissara et al. 1996[2]). Their analysis indicated that in the strain of *P. chrysogenum* they used, the flux through the product pathway was determined by both ACV synthethase (ACVS) and isopenicillin-N synthethase (IPNS). However, from a recent analysis performed with the same *P. chrysogenum* strain, it was concluded that the flux control totally resides at the IPNS (Theilgaard and Nielsen, 1999[3]).

In almost all cases production of a secondary metabolite requires carbon precursors from primary metabolism, energy in the form of ATP and reduction equivalents, either NADH or NADPH. An important aspect which has received relatively little attention so far is the fact that in microbial strains producing high levels of a secondary metabolite, e.g. penicillin, often competition exists between growth and product formation for common precursors, cofactors and energy which all have to be delivered by primary metabolism. Increasing the rate of product formation by increasing the levels of the enzymes in the product pathway will at some point inevitably lead to the occurrence of metabolic bottlenecks for product formation in primary metabolism (Figure VII.5).

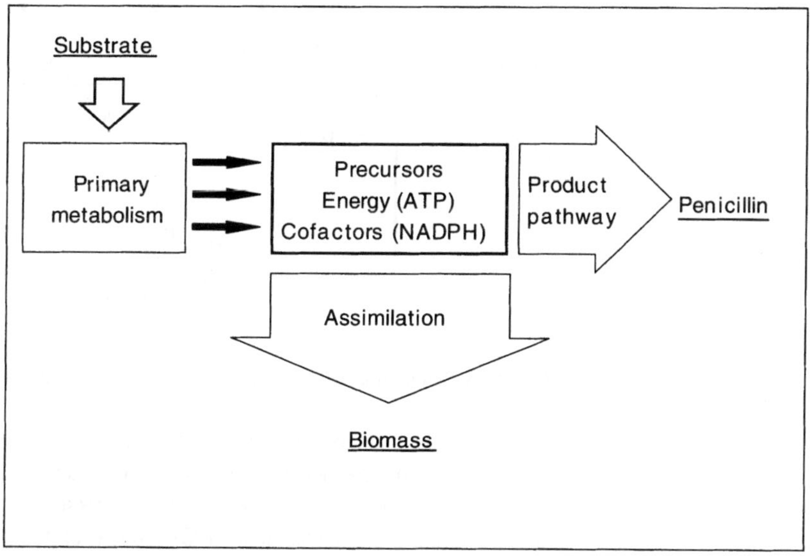

Fig. VII.5 Formation of a substantial amount of a secondary product, like e.g. penicillin, requires significant amounts of precursors, energy and cofactors from primary metabolism, leading to competition between growth and product formation.

In case of penicillin production (Figure VII.6) product formation starts with the condensation of the intermediate of the lysine pathway, L-α-amino-adipic acid, and the two amino acids L-cysteine and L-valine to the tripeptide L-α-aminoadipyl-L-cysteinyl-D-valine (LLD-ACV). In the second step, where ACV is converted to isopenicillin-N, the characteristic bicyclic penam ring structure is formed. In the final step the side chain exchange is carried out. Feeding of phenylacetic acid (PAA) leads to the replacement of the α-amino-adipic acid moiety by PAA thus forming penicillin-G.

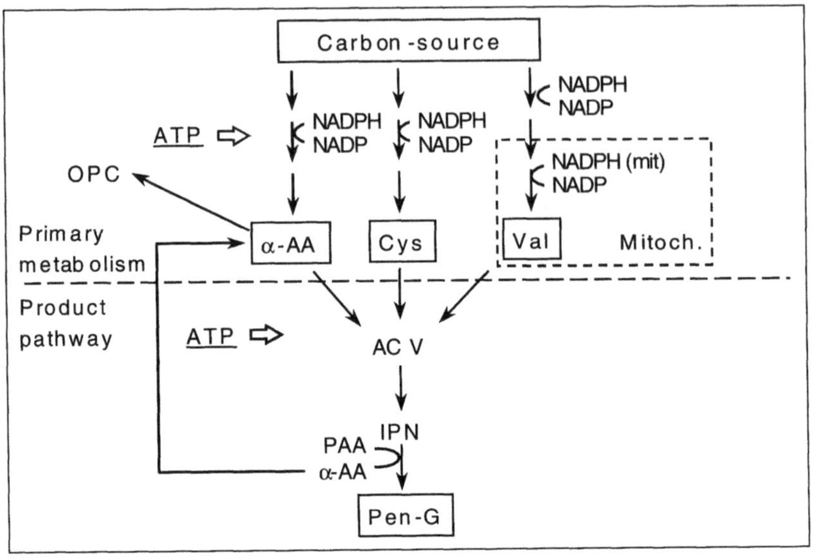

Fig. VII.6 Schematic representation of the penicillin-G biosynthesis pathway.

However, side chain exchange requires the activation of PAA to a CoA-thioester. The greatest part of the released α-amino-adipic acid is recycled, a minor part is converted by a yet unknown mechanism (Henriksen et al., 1998[4]) to a cyclic form, 6-oxopiperide-2-carboxylic-acid (OPC) and is excreted by the cells. Precursors and cofactors for penicillin formation which have to be delivered by the primary metabolic pathways are therefore i) the amino acids cysteine and valine which form the carbon skeleton of penicillin ii) reducing equivalents in the form of cytosolic and mitochondrial NADPH and iii) energy (ATP).

In previous studies on metabolic flux calculations in *P. chrysogenum*, using data obtained from fed-batch cultures or chemostats, the fluxes through the central metabolic pathways as well as through the product pathway have been estimated (Jørgensen et al., 1995[5]; Henriksen et al., 1996[6]) using a detailed stoichiometric model of the metabolism of *P. chrysogenum*. However, so far no attempts were undertaken to identify potential bottlenecks in the primary metabolic network of *P. chrysogenum* by flux manipulation.

The theoretical concepts for identifying metabolic bottlenecks for metabolite overproduction have been developed by Stephanopoulos and Vallino (1991[7]). According to their analysis product yield typically depends on the flux partitioning that occurs only at a small subset of all possible nodes that comprise the metabolic network. Identification of these nodes, referred to as principal nodes, is accomplished by searching for those nodes in the metabolic network where significant changes in flux partitioning must occur as a function of the rate of product formation. The (differences in the) kinetic parameters of the enzymes participating in each branch point of such a principal node, as well as the complete control structure of the node, determines whether it can be considered as flexible, weakly rigid or strongly rigid. The flux partitioning around a rigid node is dominated by the kinetics of one of its branches. Such nodes may form potential bottlenecks since they are resistant against changes in flux partitioning necessary to meet the metabolic demands for product formation.

The authors propose different ways to investigate the rigidity of principal nodes such as: i) attenuation of enzyme activity by the addition of a specific inhibitor; ii) amplification or attenuation of enzyme activity by genetic modification; iii) deregulation of a certain metabolite to increase metabolic burden; and iv) environmental perturbation such as a change in carbon source. In subsequent work they applied these methods to lysine overproduction in *Corynebacterium glutamicum* (Vallino and Stephanopoulos, 1994a[8]; 1994b[9]). Although especially the induction of flux changes by changing the environmental conditions seems an attractive method, certainly for micro-organisms which are genetically less accessible, like e.g. *P. chrysogenum*, this method has to our knowledge never been fully exploited.

In paragraph 3 the development of a detailed stoichiometric model for growth and Penicillin-G formation in *P. chrysogenum* is described. In this model intracellular compartmentation is taken into account, i.e. cytosol, mitochondrion and peroxisome are distinguished as separate compartments and material fluxes between compartments occur via specific transporters.

Paragraph 4 of this chapter focuses on the energetics of penicillin-G production. Several workers used stoichiometric models to calculate maximum theoretical yields of penicillin on glucose (Cooney and Acevedo, 1977[10]; Hersbach et al., 1984[11]; Jørgensen et al., 1995)[5]. In these studies the energetics of penicillin formation has been based on the well known biochemistry of the biosynthesis pathway. However, until now no attempts have been made to estimate possible extra energetic costs associated with penicillin formation, in the same way as has been done for biomass formation (Henriksen et al., 1996[6]; Jørgensen et al., 1995[5]; vanGulik and Heijnen, 1995[12]; Vanrolleghem et al., 1996[13]; Verduyn et al., 1991[14], 1992[15]). Estimation of the total in-vivo energetic costs of penicillin production and its impact on maximum product yield will be addressed in paragraph 4.

Paragraph 5 of this chapter describes the results of a study on the interaction between primary metabolism and penicillin-G production with respect to the supply of carbon precursors and reduction equivalents (NADPH). The main question thereby is whether product formation is limited by primary metabolism or by the product pathway. We applied metabolic flux analysis for the theoretical identification of principal nodes in primary metabolism, i.e. branch points at which significant changes in flux partitioning occur at increasing rates of product formation. In principle these branch points may form potential metabolic bottlenecks for penicillin production. The flux partitioning around these nodes was manipulated experimentally by cultivation of the cells at different growth rates in carbon limited chemostat cultures on different carbon and nitrogen sources. Metabolic flux analysis, based on measured input and output fluxes, has then been used to investigate the effect of drastic changes in flux partitioning around these nodes on penicillin-G production.

§3 Development of a stoichiometric model for growth and penicillin-G production in *Penicillium chrysogenum*

A stoichiometric model for growth and product formation of *P. chrysogenum* was developed, based on a previously published stoichiometric model for yeast (VanGulik and Heijnen, 1995[12]) and a review on the biochemistry of the catabolic, anabolic and product pathways of *P. chrysogenum* (Nielsen, 1995[16]). In the model three cellular compartments are distinguished, namely cytosol, mitochondrion and peroxisome. For some parts of the biochemical reaction network assumptions had to be made with respect to the pathways applied and/or their stoichiometry as well as intracellular compartmentation. These assumptions will be discussed below. A listing of the complete model can be found in vanGulik et al., 2000[17].

3.1 Anaplerotic pathways

Part of the TCA-cycle intermediates are withdrawn for anabolic purposes and therefore the cycle has to be replenished. This is brought about by the anaplerotic pathways. For our model we assume that the most important anaplerotic route for growth on glucose and xylose is pyruvate carboxylase, converting pyruvate to oxaloacetate. In most fungi the location of this enzyme is exclusively cytosolic (Osmani and Scrutton, 1983[18]; Osmani and Scrutton, 1985[19]; Pronk et al. 1996[20] and references therein). However, in some Aspergillus species also mitochondrial pyruvate carboxylase activity has been detected (Bercovitz et al., 1990[21]). Considering the above we assumed for our model that pyruvate carboxylase is present only in the cytosol. The oxaloacetate formed in the cytosol cannot cross the mitochondrial membrane. It is first converted into malate via cytosolic malate dehydrogenase. Malate can be readily transported into the mitochondrion. For growth on C-2 carbon sources the most important anaplerotic route is the glyoxylate cycle. In case of growth on glucose cytosolic AcetylCoA, which is mainly needed for lipid biosynthesis, is assumed to be formed via ATP citrate lyase converting citrate into AcetylCoA and oxaloacetate. In case of growth on ethanol and acetate cytosolic AcetylCoA is assumed to be synthesized from acetate via cytosolic AcetylCoA synthethase.

3.2 Regeneration of cytosolic NADPH

For growth on glucose the pentose phosphate pathway is assumed to be the major source of cytosolic NADPH (Nielsen, 1995[16] and references therein). It is most likely that this also holds for growth on acetate. However, when the cells are grown on ethanol there exists another, energetically cheaper, way of NADPH regeneration namely through NADP dependent acetaldehyde dehydrogenase. It has been found for *Candida utilis* grown on ethanol that both NAD- and NADP-dependent acetaldehyde dehydrogenases are present. It has been found that the NADP-dependent isoenzyme is the most important one because it is responsible for more than 80% of total acetaldehyde dehydrogenase activity (Bruinenberg et al., 1983[22]).

For this model it is therefore assumed that when *P. chrysogenum* is grown on ethanol cytosolic NADPH is generated mainly through NADP dependent acetaldehyde dehydrogenase and that the PP-pathway has only an anabolic function.

3.3 Biosynthesis of cysteine

In fungi two routes for cysteine biosynthesis exist. The simplest one is the direct sulfhydrylation pathway in which reduced sulfur is incorporated by O-Acetylserine sulfhydrylase into O-acetyl-L-serine to give cysteine. In the second more complicated route, the transsulfuration pathway, reduced sulfur is incorporated into O-acetyl-homoserine by the enzyme O-Acetyl-homoserine sulfhydrylase yielding homocysteine. In the next reaction homocysteine and L-serine are combined to cystathionine. In the last step, catalyzed by cystathionine β-lyase, the homoserine moiety is cleaved off as α-ketobutyric acid and ammonia. The main difference of this route with the direct sulfhydrylation pathway is that per molecule of cysteine synthesized one molecule of aspartate, which is the precursor of homoserine, is converted into a-ketobutyrate. The fate of α-ketobutyrate in *P. chrysogenum* is not known. Some authors claimed that in *P. chrysogenum* only the transsulfuration route is operative, however, no evidence was presented (Treichler et al., 1979[23]; Döbeli and Nüesch, 1980[24]). Recently Østergaard et al. (1998[25]) demonstrated that P. chrysogenum posesses O-acetyl-L-serine sulphydrylase, the last enzyme of the direct sulfhydrylation pathway. For reasons of simplicity and to avoid the incorporation of questionable routes for the re-cycling of α-ketobutyrate in the metabolic model, only the direct sulfhydrylation pathway was assumed to be operative.

3.4 Intracellular compartmentation

The compartmentation of the majority of the biochemical pathways of primary metabolism in fungi is known and can be found in textbooks. In the metabolic model for *P. chrysogenum* three compartments are distinguished. The cytosol, containing the reactions of the glycolysis, gluconeogenesis, the pentose phosphate pathway and the majority of the anabolic pathways. The mitochondrion, containing the TCA-cycle enzymes, the biosynthetic pathways of the amino acids isoleucine, valine, arginine, part of the biosynthetic pathway of leucine and oxidative phosphorylation. The peroxisome, which contains the glyoxylate cycle enzymes isocitrate lyase and malate synthase which are necessary for growth on C-2 carbon sources such as ethanol and acetate.

The biosynthesis of valine is catalyzed by the same set of four enzymes which catalyses also the biosynthesis of isoleucine. These enzymes form a complex, which is located in the mitochondrial matrix (Bender, 1985[26]; Smith and Berry, 1976[27]). For the biosynthesis of both amino acids one mitochondrial NADPH per amino acid is required in the second step in the biosynthesis pathway (Ketol-acid reductoisomerase). Because NADPH can not pass the mitochondrial membrane it

must be generated in the mitochondrial compartment trough NADP dependent isocitrate dehydrogenase. A second NADPH is needed for the biosynthesis of glutamate, which is used in the fourth step i.e. the transamination reaction where the amino group of glutamate is transferred to α-ketoisovalerate to form valine. In *Aspergillus nidulans* both NAD and NADP dependent isoenzymes of glutamate dehydrogenase are present exclusively in the cytosol. The activity of the NADP dependent isoenzyme was found to be almost 20 times higher than of the NAD dependent one (Osmani et al., 1983[18]). For our model we therefore assume that the biosynthesis of glutamate occurs in the cytosol through NADP dependent glutamate dehydrogenase and that the glutamate needed in the transamination reaction is transported from the cytosol to the mitochondrion.
Penicillin-G biosynthesis is assumed to be compartmentalized according to the model of Lendenfeld at al. (1993[28]) which is partly based on the subcellular localization studies of Müller et al. (1991[29]). According to this model the formation of the tripeptide α-aminoadipyl-cysteinyl-valine (ACV) by α-aminoadipyl-cysteinyl-valine synthethase (ACVS) and the conversion of ACV to isopenicillin-N (IPN) by isopenicillin-N synthethase (IPNS) occurs in the cytosol. Lendenfeld et al. thereby assume that the precursor amino acids α-aminoadipic acid, cysteine and valine are drained from the vacuole. Because the present metabolic model does not contain a vacuolar compartment the three amino acids are assumed to be taken up from the cytosol. The activation of PAA to PAA-CoA and the exchange of the α-aminoadipic acid moiety of IPN by PAA by the action of acetyltransferase (AT) occurs in the peroxisome.

3.5 Solute transport across the plasmalemma

Unfortunately, detailed information on transport in *P. chrysogenum* is scarce. In case that exact information is lacking, available information on other fungi, e.g *S. cerevisiae*, has been used.
In principle three modes of transport of metabolites across the plasma membrane are possible: simple diffusion, facilitated diffusion and active transport. Active transport of solutes occurs in most cases by direct proton symport. The proton motive force is generated by a plasma membrane H^+ -ATPase with an H^+/ATP stoichiometry of 1 (Serrano, 1988[30]). The exact nature of the transport systems for glucose and xylose in *P. chrysogenum* is not known. However, many fungi possess proton symport systems for glucose uptake (Nielsen, 1995[16]; Boles and Hollenberg, 1997[31]), while xylose utilizing yeasts have been found to contain similar transport systems for xylose (Does and Bisson, 1989[32]; Kilian and VanUden, 1988[33]). For the present model it is assumed that in *P. chrysogenum* glucose and xylose are accumulated by proton symport with an H^+/glucose stoichiometry of 1. Ethanol is assumed to enter the cells by passive diffusion. Transport of short-chain monocarboxylates (e.g. acetate) into yeast cells has been reported to occur via a proton symport with an H^+/monocarboxylate stoichiometry of 1 (Cássio et al., 1987[34]; Leao and VanUden, 1986[35]). Sulfate transport across the plasma membrane

of *P. chrysogenum* has been the subject of a study by Hillenga et al. (1996[36]). It was reported that the sulfate transport system catalyzes the symport of two protons with one sulfate anion.

Less is known about the uptake of ammonia, nitrate and phosphate by the cells. Although it has been shown that ammonia uptake by *P. chrysogenum* is an active process (Hackette et al., 1970[37]), little is known about its energetics. According to the model for ammonium transport in *S. cerevisiae*, proposed by Roon et al. (1977[38]), ammonium is supposed to be taken up by proton symport with an H^+/ammonium stoichiometry of 1. Also phosphate is taken up by a proton symport mechanism in yeasts (Eddy, 1982[39]) with an H^+/phosphate stoichiometry of 2 or 3 depending on the extracellular phosphate concentration. Here the H^+/phosphate stoichiometry of the phosphate transporter in *P. chrysogenum* is assumed to be 2. With respect to nitrate, it is assumed that uptake is similar to nitrate uptake in *Neurospora crassa* i.e. active transport through proton symport with a stoichiometry of 2 H^+ per NO_3^- anion transported across the membrane (Blatt et al., 1997[40]). Transport of phenylacetic acid in *P. chrysogenum* has been reported to occur through passive diffusion (Hillenga et al., 1995[36]).

3.6 Intracellular transport

Transport of metabolites across the mitochondrial membrane is brought about by specific carriers and is driven by the electrochemical proton gradient across the membrane. It has been assumed that the stoichiometry of mitochondrial transport in *P. chrysogenum* is similar as described in the reviews of Walker et al. (1992[41]) and Palmieri (1994[42]).

Less is known about transport of solutes across the peroxisomal membrane. It has been found that an electrochemical proton gradient exists across the membranes of peroxisomes isolated from the yeast *Hansenula polymorpha* (Nicolay et al., 1987[43]) and that these peroxisomes contain a membrane bound proton translocating ATP-ase which pumps protons into the peroxisomal matrix (Douma et al., 1987[44]). Transport of metabolites across the peroxisomal membrane is therefore also likely to be driven by an electrochemical proton gradient. Localization of the glyoxylate cycle enzymes isocitrate lyase and malate synthase in the peroxisome requires the transport of isocitrate and acetyl coenzyme-A from the cytosol to the peroxisome and the transport of succinate and malate in the opposite direction. It is assumed that isocitrate is exchanged for succinate and that malate is transported via a proton symport carrier analogous to transport across the mitochondrial membrane. Acetyl coenzyme-A is assumed to enter the peroxisomes via a carnithine shuttle (Elgersma et al., 1995[45]).

§4 Energetics of growth and penicillin production in a high producing strain of *Penicillium chrysogenum*

4.1 Estimation of maximum yields

Reliable estimation of maximum theoretical yields of product on substrate is of great value in industrial strain improvement programs. If properly done, such estimations may provide a good indication for the available room for further improvement. A prerequisite is that sufficient information is available on the metabolism of the particular micro-organism, not only with respect to product biosynthesis but also with respect to the central metabolic pathways. An essential aspect for the proper estimation of biomass and/or product yields is sufficient information on the ATP-stoichiometry and thermodynamic/biochemical irreversibility constraints. From this information a stoichiometric model can be constructed allowing a straightforward calculation of the maximum yields of biomass and product on substrate and the maintenance coefficient (Roels, 1983[46]; vanGulik and Heijnen, 1995[12]; Vanrolleghem et al., 1996[13]) i.e. the parameters and m_S of the well known linear equation for substrate consumption:

$$q_S = \frac{\mu}{Y_{SX}^{\max}} + \frac{q_P}{Y_{SP}^{\max}} + m_S \tag{1}$$

A general problem in calculating these maximum yields is that the ATP-stoichiometry of some metabolic processes is not or insufficiently known. Examples of such processes are intracellular transport (Andre, 1995[47]), turnover of macromolecules and organelles, enzyme regulation by phosphorylation / de-phosphorylation (Hinnebusch and Liebman, 1991[48]), proton leakage (Rigoulet et al., 1998[49]) etc. To drive these processes unknown amounts of energy (ATP) are required. A practical solution to this problem is the assumption that an unknown amount of additional energy is required for growth related and non-growth related maintenance leading to 2 unknown ATP stoichiometry parameters (K_X, m_{ATP}) on the consumption side. On the ATP production side we have a third parameter, namely the ratio between oxygen consumption in oxidative phosphorylation and ATP formation, i.e. the P/O-ratio. This approach has been applied in the metabolic modeling of the growth of yeast cultures on a large number of different carbon sources (vanGulik and Heijnen, 1995[12]) and growth on glucose/ethanol mixtures (Vanrolleghem et al., 1996[13]). It has been shown that experimental data for growth on at least two different substrates at different growth rates allows the estimation of the unknown ATP-stoichiometry parameters P/O, K_X and m_{ATP} (vanGulik and Heijnen, 1995[12]; Vanrolleghem et al., 1996[13]). Here we want to extend this approach to microbial growth with product formation i.e. the production of penicillin-G in *Penicillium chrysogenum*.

Although the energy needed for the biosynthesis of a molecule of penicillin-G can be calculated straightforward from the well-known biosynthesis pathway, it is evident that additional energy is required e.g. for intracellular transport and for product excretion. Recently it has been found that the energy demand for the formation of the tripeptide ACV, which under optimal conditions needs the hydrolysis of 3 mol

ATP to AMP per mol of tripeptide formed, may require more than 20 mol of ATP under unfavourable conditions (Kallow, et al, 1998[50]). The additional energy consumed was reported to be caused by the hydrolytic loss of activated intermediates.
To account for additional energy consumption (mol ATP/mol penicillin) associated with penicillin production, the parameter K_P is introduced. The value of this parameter represents an additional amount of ATP needed for penicillin-G biosynthesis, which comes on top of the amount dictated by the known ATP-stoichiometry of the biosynthesis pathway specified in the metabolic model. The result is that the metabolic network model contains 4 unknown ATP-stoichiometry parameters, which have to be estimated from experimental data.
The results of carbon limited steady state chemostat cultures of *P. chrysogenum* at different dilution rates on three different carbon sources, i.e. glucose, ethanol and acetate have been used to estimate the 4 unknown ATP-stoichiometry parameters. Subsequently, from these estimates theoretical maximum yields of biomass and product on substrate and oxygen for the three substrates were calculated. Finally the validity of the estimated extra energy associated with penicillin-G biosynthesis, i.e. the estimated value of the parameter K_P, has been assessed in independent experiments.

4.2 Model for the ATP stoichiometry of the metabolic network

With the four unknown ATP-stoichiometry parameters the ATP-balance can be expressed as:

$$P/O \cdot q_{2e} - \sum q_i^{ATP} - K_X \cdot \mu - K_P \cdot q_{Pen} - m_{ATP} = 0 \qquad (2)$$

where the first term represents the production of ATP in oxidative phosphorylation, $\sum q_i^{ATP}$ represents the net rate of ATP consumption in the part of the metabolic network model of which the ATP stoichiometry is known (i.e. the result of all stoichiometrically fixed ATP usage, as well as production in substrate level phosphorylation), μ is the specific growth rate and q_{Pen} is the specific β-lactam production rate (i.e. the sum of penicillin-G (Pen-G), penicilloic acid (PIO), 6-aminopenicillanic acid (6-APA), 8-hydroxypenillic acid (8-HPA) and isopenicillin-N (IPN) production). The parameters K_X and m_{ATP} are operational values for growth associated maintenance, and non-growth associated maintenance respectively (Van Gulik et al., 1995[12]; Vanrolleghem et al., 1996[13]). The parameter K_P represents additional ATP requirements associated with the production of β-lactam compounds.
It should be realized that the P/O-ratio as defined in Eq. 2 can not be considered as a fixed parameter. The reason is that this ratio is determined by the division of the electron flux over the different proton translocating complexes (I, III and IV) of the respiratory chain which have different $H^+/2e$ stoichiometries (Verduyn et al. , 1991[14]). The growth conditions, e.g. the carbon substrate used, the growth rate, the rate of product formation etc., determine the relative amounts of mitochondrial (NADH and FADH) and cytosolic (NADH) electron fluxes. Because the electrons

derived from mitochondrial NADH pass the complete respiratory chain whereas the electrons derived from FADH and cytosolic NADH generally pass only complexes III and IV the division of the electron flux over the different complexes is a function of the growth conditions. Therefore the operational P/O-ratio will be a function of the growth conditions as well.

If the metabolic model is sufficiently detailed, e.g. intracellular compartmentation is included, the origin of the electrons generated in microbial catabolism is known and thus the relative contributions of complexes I, II and IV of the respiratory chain to oxidative phosphorylation. To include this the first term of Eq. 2 has to be replaced by:

$$P/O \cdot q_{2e} = \delta \cdot (q_{2e}^{NADH:mit} + \alpha \cdot q_{2e}^{NADH:cyt} + \beta \cdot q_{2e}^{FADH}) \qquad (3)$$

Where α and β represent the relative contributions to proton translocation of electrons delivered by cytosolic NADH and FADH respectively. The values of these parameters depend on the configuration of the electron transport chain. Unfortunately no details are available on the configuration of the electron transport chain of *P. chrysogenum*. However, information on the electron transport chain of strongly related fungi (various *Aspergillus* species and *Neurospora crassa*) indicates that all the elements of the electron transport chain are available (references in Nielsen, 1995[16]).

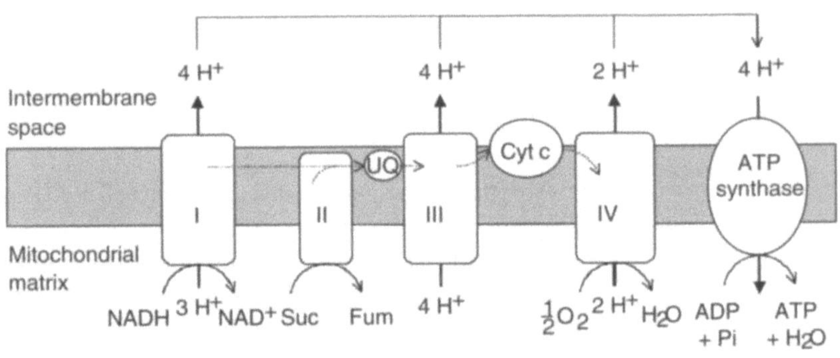

Fig. VII.7 Schematic representation of the electron transport chain of eukaryotic cells.

Because no specific information on the H+/2e stoichiometry of proton translocation in the respiratory chain of *P. chrysogenum* is available, literature values have been used. The H+/2e stoichiometry of the complexes I and III is assumed to be equal to 4 where complex IV has a lower stoichiometry of 2 (Senior, 1988[51]). The stoichiometry of proton translocation and ATP synthesis used in our model is summarized in Figure VII.7.

Because NADH can not pass the mitochondrial membrane some mechanism must exist to transport the electrons derived from cytosolic NADH to the respiratory chain. In principle there are two ways for the oxidation of cytosolic NADH, 1) via a NADH dehydrogenase which faces the outer surface of the inner mitochondrial membrane and 2) via a shuttle system e.g. the dihydroxyacetone phosphate /

glycerol-3-phosphate shuttle (Zubay, 1988[52]). In both cases the electrons are delivered to ubiquinone and thus pass only complexes III and IV. From the above it can be concluded that per 2 electrons delivered by mitochondrial NADH 10 protons are translocated (4, 4 and 2 for complexes I, III and IV respectively) and that per 2 electrons delivered by $FADH_2$ and cytosolic NADH 6 protons are translocated (4 and 2 for complexes III and IV respectively) which results in $\alpha = \beta = 0.6$.

From studies on isolated yeast mitochondria it has been found that the operational P/O ratio is a function of the electron flux through the respiratory chain (Rigoulet et al., 1998[49]). The authors found that at relatively low flux values (i.e. low respiration rates) the P/O ratio decreased when the flux increased but remained constant when the flux was increased further. The nature of the mechanism involved was reported to be slipping at the level of the respiratory chain rather than a change in the H^+/ATP stoichiometry of the ATP synthase. Although it can be inferred from these findings that also in *P. chrysogenum* the P/O ratio might be variable at low respiration rates (i.e. at low specific growth rates) we will, as a first assumption, consider the H^+/2e stoichiometry in proton translocation and the H^+/ATP stoichiometry of the ATP-synthase to be independent of the growth rate. Furthermore we have no indications that the stoichiometry of proton translocation and/or the ATP-synthase is different among the substrates used in this study (glucose, ethanol and acetate). Under these assumptions the parameter δ is independent of the growth conditions as well. Note that by its definition the parameter δ represents the P/O-ratio for the oxidation of mitochondrial NADH, i.e. when the electrons pass the complete electron transport chain.

Combination of Eq. 2 and Eq. 3 with $\alpha = \beta = 0.6$ yields the ATP-balance as it has been used for the estimation of the four ATP stoichiometry parameters (δ, K_X, K_P and m_{ATP}):

$$\delta \cdot (q_{2e}^{NADH:mit} + 0.6 \cdot q_{2e}^{NADH:cyt} + 0.6 \cdot q_{2e}^{FADH}) - \sum q_i^{ATP} - K_X \cdot \mu - K_P \cdot q_{Pen} - m_{ATP} = 0 \qquad (4)$$

4.3 Estimation of ATP stoichiometry parameters

For each chemostat experiment a linear equation of the form of Eq. 4 was derived. The (reconciled) specific fluxes, i.e. $q_{2e}^{NADH:mit}$, $q_{2e}^{NADH:cyt}$, q_{2e}^{FADH}, $\Sigma\, q_i^{ATP}$, (and q_{Pen}, have been obtained for each experiment from metabolic flux balancing using the measured uptake rates of substrate, oxygen, ammonia and PAA and the production rates of biomass, penicillin, carbon dioxide and seven different by-products (6-APA, 8-HPA, 6-oxopiperidine-2-carboxylic acid (OPC), PIO, ortho-hydroxyphenylacetic acid (o-OH-PAA), excreted polysaccharides and excreted peptides). From a chi-square test performed for each data set it appeared that in all cases the fit of the experimental data to the stoichiometric model was satisfactory (confidence level for errors in data and/or model less than 75%) for further details we refer to vanGulik et al., 2000[17] and vanGulik et al., 2001[53].

With the assumption that the ATP stoichiometry parameters can be considered as constants (see Theoretical Aspects) and thus independent of the growth rate and the substrate used, their values can be estimated by linear regression. This was done using experimental data from 20 steady state glucose, ethanol and acetate limited chemostat cultures performed at different dilution rates. The results of the estimations are shown in Table VII.2.

Table VII.2 Estimated values of the ATP-stoichiometry parameters with their 95% confidence intervals.

Parameter	Value (measurement unit)	
δ	1.84 ± 0.08	mol ATP / mol O
K_X	0.38 ± 0.11	mol ATP / Cmol biomass
K_P	73 ± 20	mol ATP / mol penicillin
m_{ATP}	0.033 ± 0.012	mol ATP / Cmol biomass / h

With respect to the P/O ratio little information is available with respect to in-vivo estimations of the overall stoichiometry of oxidative phosphorylation of eukaryotic cells in which the complete respiratory chain is active. In this case the maximum P/O ratio for the oxidation of mitochondrial NADH equals 2.5. The only example we know of, has been reported by vanGulik and Heijnen (1995[12]) who estimated a value of 1.53 for the P/O-ratio of the yeast *Candida utilis*. In this yeast the complete respiratory chain is operative (Light and Garland, 1971[54]; Aiking et al., 1977[55]). However, vanGulik and Heijnen used a stoichiometric model in which intracellular compartmentation was not taken into account. This results in an underestimation of the P/O-ratio because both the electrons derived from mitochondrial and cytosolic NADH are assumed to pass the complete electron transport chain (Verduyn et al., 1991[14]).

Several workers estimated the in-vivo P/O ratio of the yeast *Saccharomyces cerevisiae* of which it is known that only two phosphorylation sites are active (Onishi, 1973[56]). In this case the maximum P/O ratio for the oxidation of mitochondrial NADH equals 1.5 (Verduyn et al., 1991[14]). From chemostat experiments on glucose/ethanol mixtures Bonnet et al., (1980[57]) estimated a P/O ratio of 1.18 which is close to the value of 1.09 estimated by Vanrolleghem et al. (1996[13]) from the same type of experiments. The fact that, due to imperfect coupling between respiration and ATP formation, the operational P/O ratio in a functioning organism is significantly lower than the maximum theoretical value is a well accepted phenomenon (Bonnet et al., 1980[57], Roels, 1983[46]; Ferguson, 1986[58]). Therefore our estimation of the P/O-ratio for oxidation of mitochondrial NADH, δ, of 1.84 fits well with the presence of three phosphorylation sites.

The estimated value for non-growth associated maintenance energy requirement, m_{ATP} (on a dry weight basis approx. 1.2 mmol/g/h) is somewhat lower than previously estimated values for *P. chrysogenum* of between 2.1 and 3.4 mmol/g/h (Henriksen et al., 1998[59]; Jørgensen et al., 199[55]) but comparable with values estimated for other micro-organisms, e.g. *Saccharomyces cerevisiae* (Roels, 1983[46]).

From the stoichiometric model, using the estimated ATP-stoichiometry, we calculated the total ATP needs of biomass synthesis from glucose, Y_{xATP}, to be 61 mmol ATP/g dwt. This figure is slightly lower than the values estimated by Henriksen et al. (1998[59]) and Jørgensen et al. (1995[5]) for their strain of *P. chrysogenum*. Assuming an operational P/O ratio of 1.5 Henriksen estimated from chemostat experiments that Y_{xATP} = 76 mmol ATP/g dwt whereas Jørgensen, assuming a P/O ratio for NADH oxidation of 2.6, estimated a higher value of 87 mmol / g dwt from fed batch cultures.
With respect to the growth associated maintenance, K_X, the estimated value for our *P. chrysogenum* strain of 14 mmol/g biomass is significantly lower than previously reported values of 57 mmol/g biomass estimated from fed-batch experiments of *P. chrysogenum* by Jørgensen et al. (1995[5]) and of 36 mmol/g biomass estimated from chemostat experiments. The latter value can be calculated from the difference between the theoretical ATP requirement for growth of 40 mmol/g biomass (Henriksen et al., (1996[6]) and the operational ATP requirement of 76 mmol/g biomass (Henriksen et al., 1998[59]). The high value estimated by Jørgensen et al. (1995[5]) is probably caused by the high value for the P/O-ratio they assumed in their metabolic model. A too high value of the P/O-ratio results in a too high ATP synthesis rate and consequently in a too high flux of excess ATP in the corresponding metabolic flux calculations. Because Jørgensen et al. estimated the growth dependent maintenance from this flux of excess ATP they probably overestimated the ATP costs for growth dependent maintenance. In spite of these differences our estimation of K_X is still on the low side. However, it cannot be excluded that different high producing *P. chrysogenum* strains, which have been obtained after a long period of repeated mutagenesis and selection of high producing variants, display different maintenance energy requirements.
Also when compared with values estimated for *S. cerevisiae* and *Candida utilis* of 22 - 30 mmol/g (vanGulik and Heijnen, 1995[12]; Vanrolleghem et al., 1996[13]; Verduyn et al., 1991[14]) our estimation of K_X is relatively low. Another reason for the low value estimated in the present study is that a rather detailed compartmentalized stoichiometric model has been used in which available information on the energetics of intracellular transport has been incorporated. This results in a smaller amount of ATP consumption which has to be attributed to unspecified processes and thus a lower value of K_X.
Clearly a very surprising result is the high value of K_P of 73 mol ATP per mol of penicillin. If this value is realistic it has important consequences for the maximum theoretical yield of penicillin on C-source, which will be much lower than previously calculated values based on the present knowledge of the ATP stoïchiometry of the penicillin biosynthesis pathway (Cooney and Acevedo, 1977[10]; Hersbach et al., 1984[11]; Jørgensen et al., 1995[5]).
Substitution of the estimated ATP stoichiometry parameters into the metabolic model allows the calculation of operational biomass yields on substrate and oxygen for given experimental conditions i.e. substrate, growth rate and rates of product and by-product formation. This was done for all chemostat experiments to compare calculated and measured operational biomass yields on substrate and oxygen. The results are presented as parity plots in Figures. VII.8 and VII.9.

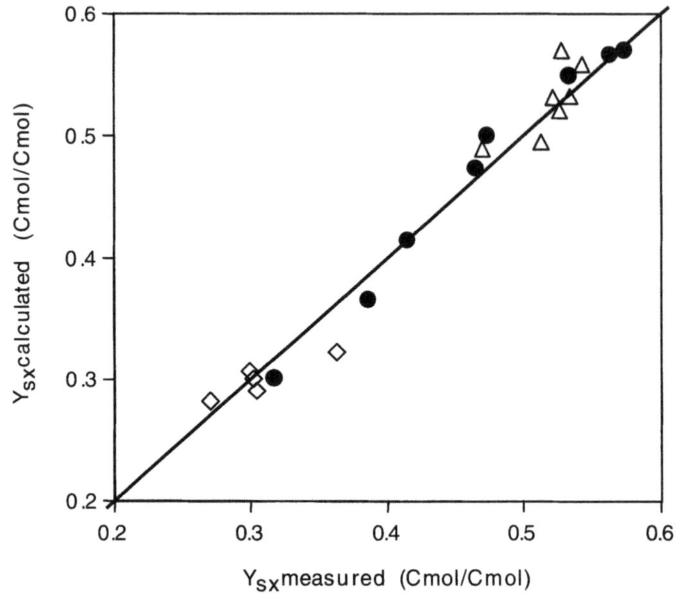

Fig. VII.8 Calculated versus measured biomass yields of *P. chrysogenum* on carbon source in carbon limited chemostat culture, (●) glucose limited cultures, (Δ) ethanol limited cultures, (◇) acetate limited cultures.

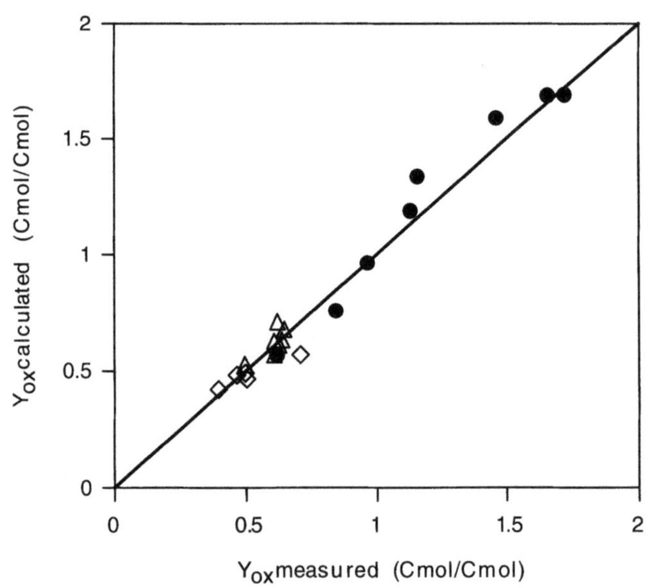

Fig. VII.9 Calculated versus measured biomass yields of *P. chrysogenum* on oxygen in carbon limited chemostat culture, (●) glucose limited cultures, (Δ) ethanol limited cultures, (◇) acetate limited cultures.

From these plots it can be seen that the experimental biomass yields on substrate and oxygen are close to the calculated values for all different growth conditions applied. From this good agreement between calculated and observed yields it can be inferred that the assumption we used for the estimation of the ATP-stoichiometry coefficients, i.e. that these coefficients are independent of the growth rate and the substrate used, leads to accurate yield calculations for all growth conditions used (substrates and growth rates).

4.4 Maximum yield and maintenance parameters on substrate

From the metabolic model algebraic expressions can be derived for the maximum yield and maintenance coefficients on substrate as a function of the estimated ATP stoichiometry parameters (see Table VII.8). By substituting the estimated values of these parameters (Table VII.2) in the obtained expressions (Table VII.8) the stoichiometric coefficients of the linear equation (Eq. 1) for substrate consumption for glucose, ethanol and acetate were calculated. The results are shown in Table VII.3. Our estimations are compared in tables VII.4 and VII.5 with literature values for *P. chrysogenum*.

Table VII.3 Calculated yield and maintenance parameters of penicillin and biomass on carbon source with their 95% confidence intervals.

C-source	Y_{SX}^{max} (Cmol/Cmol)	Y_{SP}^{max} (mol/Cmol)	m_S (Cmol/Cmol/h)
Glucose	0.663 ± 0.013	0.029 ± 0.004	0.0088 ± 0.0032
Ethanol	0.721 ± 0.015	0.034 ± 0.005	0.0071 ± 0.0026
Acetate	0.425 ± 0.010	0.020 ± 0.003	0.0117 ± 0.0042

With respect to the maximum biomass yield and maintenance coefficients on glucose a number of estimated values have been reported in literature. A comparison with our estimations (Table VII.4) shows that our value for the maximum biomass yield is higher, but the maintenance coefficient is lower than the values reported in literature.

Table VII.4 Estimated maximum biomass yields (Y_{SX}^{max}) and maintenance (m_S) coefficients for growth of *P. chrysogenum* on glucose.

Reference	Y_{SX}^{max} (g/g)	m_S (g/g/h)
This work	0.61	0.010
Christensen et al.,1995[63]	0.51	0.028
Ryu and Hospodka, 1980[67]	0.45	0.024
Mason and Righelato, 1976[60]	0.48	0.023
Righelato et al., 1968[62]	0.45	0.022

The most probable reason for this is that in previous studies the authors estimated these parameters in a different way, i.e. by determining the slope and intercept from a plot of specific glucose consumption against dilution rate whereby penicillin production as well as by-product formation were neglected. This leads to a lower value of the maximum biomass yield parameter and/or a higher value of the maintenance coefficient.

To our knowledge no published information is available on the yield and maintenance coefficients of *P. chrysogenum* biomass and penicillin on ethanol and acetate and therefore no comparison with previous determined values is possible. As can be seen from Table VII.3 the mycelium and penicillin yields on ethanol are slightly higher than on glucose while the maintenance coefficient is slightly lower. For acetate as the carbon source mycelium and product yields are significantly lower, while the maintenance is significantly higher.

4.5 Maximum theoretical penicillin yield

It can be seen from table VII.5 that the estimated high amount of extra energy associated with penicillin production (K_P = 73 mol ATP/mol penicillin) has a large impact on the theoretical maximum yield. Comparison with previously calculated values (0.43 - 0.60 mol penicillin per mol of glucose) shows that our calculated maximum yield of penicillin on glucose is more than a factor of 2 lower. This is a direct consequence of the high extra energy costs (K_P) estimated in this study, which can be verified by substitution of K_P = 0 in the expression for Y_{SP}^{max} for glucose as the carbon source (Table VII.8). This results in a maximum theoretical penicillin yield of 0.39 mol/mol which is comparable to the values published by Jørgensen et al., 1995[5] (Table VII.5).

Table VII.5 Calculated values of the maximum theoretical yield of penicillin on glucose (Y_{SP}^{max}).

Reference	Y_{SP}^{max} (mol/mol)
This work	0.18
Jørgensen et al., 1995[5]	0.43 - 0.50
Hersbach et al., 1984[11]	0.47
Cooney and Acevedo, 1977[10]	0.60

4.6 Maximum yield and maintenance parameters on oxygen

In addition also the maximum yield parameters and maintenance coefficients with respect to oxygen have been calculated for the three different substrates, the results are shown in table VII.6.

Table VII.6 Calculated yield and maintenance parameters of penicillin and biomass on oxygen with their 95% confidence intervals.

C-source	Y_{OX}^{max} (Cmol/mol)	Y_{OP}^{max} (mol/mol)	Y_{OP}^{max} (mol/Cmol/h)
Glucose	2.15 ± 0.143	0.039 ± 0.008	0.0088 ± 0.0032
Ethanol	0.97 ± 0.04	0.028 ± 0.005	0.0106 ± 0.0038
Acetate	0.77 ± 0.03	0.024 ± 0.004	0.0117 ± 0.0042

It can be seen from this table that the maximum yields of biomass and penicillin on oxygen for growth on the C-2 substrates ethanol and acetate are much lower than for growth on glucose. This is a direct consequence of the higher energy demand for biomass assimilation from these compounds due to the need for gluconeogenesis. The low values of the maximum biomass and product yields on oxygen for both ethanol and acetate have direct implications for a possible use as carbon source in a penicillin production process. Because oxygen transfer is often a bottleneck in industrial fermentations (e.g. penicillin-G production) both C-2 compounds seem no attractive alternative industrial substrates.

4.7 Validation of estimated additional energy needs for penicillin production

If the estimated extra energy needs for penicillin production of 73 mol ATP per mol of penicillin indeed represents a realistic figure, changes in the specific penicillin production rate should have a significant effect on the operational biomass yield and thus on the mycelium concentration in carbon limited chemostat culture. Moreover, from the relation between these two rates the validity of the estimated value of K_P can be verified. This was done in two ways 1) by performing glucose limited steady state chemostat experiments without the side chain precursor PAA resulting in the absence of Pen-G production and only formation of small amounts of the by-products 6-APA, 8-HPA, IPN and OPC and 2) studying the relation between the pseudo steady state biomass and Pen-G concentration in a degenerating culture where the penicillin-G production rate gradually declines to low values. It has been observed that at relatively low dilution rates (below D = 0.015 h^{-1} for glucose as the carbon source) the capability of our *P. chrysogenum* strain to produce Pen-G is progressively lost (Figure. VII.10), which is a well known phenomenon in high producing strains (vanGulik et al., 2000[17] and references therein).

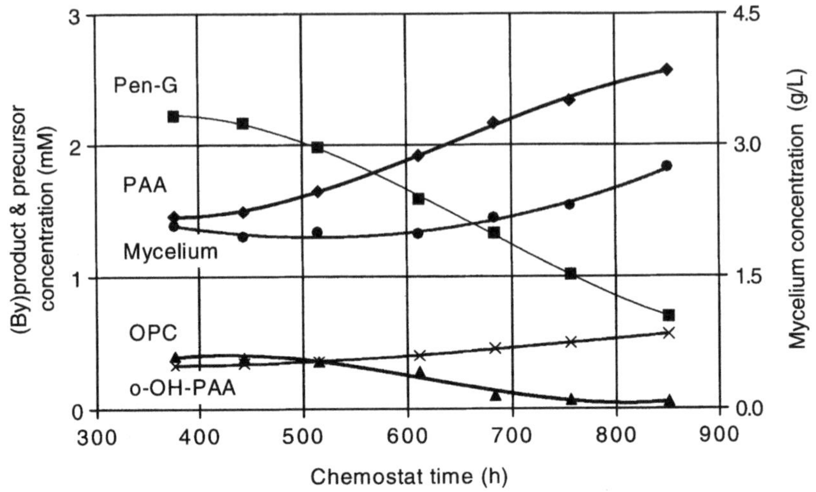

Fig. VII.10 Measured broth concentrations of mycelium, penicillin-G (Pen-G), phenylacetic acid (PAA), 6-oxopiperide-2-carboxylic-acid (OPC) and ortho-hydroxyphenylacetic acid (o-OH-PAA) in a glucose limited chemostat at a dilution rate of 0.01 h^{-1}.

This loss of penicillin productivity is accompanied (Figure VII.10) by a significant increase in mycelium concentration. Also the concentration of residual phenylacetic acid (PAA) increases strongly because less is consumed for the production of penicillin-G. However, one should be aware of the fact that a too high extracellular concentration of PAA may have influence on cellular energetics due to protonophoric uncoupling of the plasma membrane potential. As it is known that PAA is taken up by passive diffusion (Hillenga et al., 1995[36]) it may act as an uncoupling agent as has been described for other weak acids, e.g. acetate, propionate and benzoate in *Saccharomyces cerevisiae* (Verduyn et al. 1990[61], 1992[14]) and phenoxyacetic acid in *P. chrysogenum* (Henriksen, 1998[4]). However, we found from a separate experiment that under otherwise identical conditions the residual PAA concentration could be increased to approximately twice the value usually reached in our steady state experiments without any measurable effect on the steady state concentrations of mycelium and penicillin-G (results not shown). This shows that in the concentration range applied PAA had no significant uncoupling effect in our experiments.

In both cases, i.e. by omission of PAA and in a degenerating culture, a significant reduction (> 80%) in β-lactam production was accomplished. The measured mycelium concentrations during reduced β-lactam production are compared with the measured mycelium concentrations for the control situation (i.e. chemostat cultures with PAA and before the onset of degeneration). Using the stoichiometric model with the estimated ATP-stoichiometry parameters (Table VII.2) the mycelium concentrations were calculated for the high and low β-lactam producing conditions.

The measured and calculated mycelium concentrations are shown in Table VII.7 It can be inferred from these results that the mycelium concentrations under both high and low β-lactam producing conditions are predicted quite satisfactory. This supports the validity of the estimated extra energy needs for penicillin production in *P. chrysogenum*. At present it is not clear which process or processes, associated with β-lactam production, are responsible for the extra energy demand.

Table VII.7 Measured and calculated effect of the specific β-lactam production on the steady state biomass concentration in glucose limited chemostat cultures of *P. chrysogenum*.

	Measured biomass concentration (g/L)		Calculated biomass concentration (g/L)	
	High production	Low production	High production	Low production
Chemostat at $\mu = 0.01\ h^{-1}$	1.95	2.74	1.98	2.60
Chemostat at $\mu = 0.01\ h^{-1}$	2.77	3.27	2.63	3.28
Chemostat at $\mu = 0.01\ h^{-1}$	3.25	3.69	3.31	3.65

4.8 Expressions for the maximum yield and maintenance coefficients as a function of the ATP stoichiometry parameters

Expressions were derived for the parameters of the linear equation of substrate consumption (Eq. 1a) $Y_{SX}^{\max}$, $Y_{SP}^{\max}$ and m_S as a function of the ATP-stoichiometry parameters (, K_X, K_P and m_{ATP}. This was accomplished by solving the set of linear equations representing the mass balances for all compounds present in the metabolic network model for the specific growth rate, μ, and the specific rate of penicillin production, q_{Pen}, as input variables and the ATP-stoichiometry parameters δ, K_X, K_P and m_S as parameters. The result is a set of linear equations expressing all intracellular fluxes and net conversion rates as a function of μ and q_{Pen}. As an example the linear equation expressing the specific substrate uptake rate, q_S, as a function of μ and q_{Pen} for ethanol as the carbon source is derived as:

$$q_S = \frac{0.732\delta + 0.357K_X + 0.82}{\delta - 0.179}\cdot\mu + \frac{7.71\delta + 0.357K_P + 9.29}{\delta - 0.179}\cdot q_{Pen} + \left(\frac{5-2\delta}{13\delta - 2.32} + 0.154\right)\cdot m_{ATP} \tag{5}$$

Table VII.8 Derived relations for the calculation of the maximum biomass and penicillin yields and maintenance coefficients on substrate and oxygen from the estimated ATP-stoichiometry parameters.

Growth on glucose

$$Y_{SX}^{max} = \frac{\delta + 0.283}{1.07\delta + 0.566K_X + 1.02} \qquad Y_{OX}^{max} = \frac{\delta + 0.283}{0.0256\delta + 0.566K_X + 0.727}$$

$$Y_{SP}^{max} = \frac{\delta + 0.283}{11.2\delta + 0.566K_P + 12.3} \qquad Y_{OP}^{max} = \frac{\delta + 0.283}{1.72\delta + 0.566K_P + 9.58}$$

$$m_S = \left(\frac{5 - 2\delta}{9.83\delta + 2.78} + 0.203\right) \cdot m_{ATP} \qquad m_O = \left(\frac{5 - 2\delta}{9.83\delta + 2.78} + 0.203\right) \cdot m_{ATP}$$

Growth on ethanol

$$Y_{SX}^{max} = \frac{\delta - 0.179}{0.732\delta + 0.357K_X + 0.82} \qquad Y_{OX}^{max} = \frac{\delta - 0.179}{0.0471\delta + 0.536K_X + 1.42}$$

$$Y_{SP}^{max} = \frac{\delta - 0.179}{7.71\delta + 0.357K_P + 9.29} \qquad Y_{OP}^{max} = \frac{\delta - 0.179}{2.07\delta + 0.536K_P + 15.6}$$

$$m_S = \left(\frac{5 - 2\delta}{13\delta - 2.32} + 0.154\right) \cdot m_{ATP} \qquad m_O = \left(\frac{5 - 2\delta}{8.67\delta - 1.55} + 0.231\right) \cdot m_{ATP}$$

Growth on acetate

$$Y_{SX}^{max} = \frac{\delta - 0.278}{1.14\delta + 0.556K_X + 1.37} \qquad Y_{OX}^{max} = \frac{\delta - 0.278}{0.0833\delta + 0.556K_X + 1.66}$$

$$Y_{SP}^{max} = \frac{\delta - 0.278}{11.9\delta + 0.556K_P + 16.8} \qquad Y_{OP}^{max} = \frac{\delta - 0.278}{2.36\delta + 0.556K_P + 19.4}$$

$$m_S = \left(\frac{5 - 2\delta}{8\delta - 2.22} + 0.25\right) \cdot m_{ATP} \qquad m_O = \left(\frac{5 - 2\delta}{8\delta - 2.22} + 0.25\right) \cdot m_{ATP}$$

Clearly the first term on the right hand side of Equation 5 can be recognized as substrate uptake for growth, $\frac{1}{Y_{SX}^{max}} \cdot \mu$, the second term represents substrate uptake for penicillin production $\frac{1}{Y_{SP}^{max}} \cdot q_{Pen}$, where the last term represents

substrate uptake for maintenance. Similar equations were derived for glucose and acetate as carbon sources. The obtained expressions for Y^{max}_{SX}, Y^{max}_{SP} and m_s for the three carbon sources are shown in Table VII.8.
An analogous approach was used to derive expressions for the parameters of the linear equation for oxygen consumption Y^{max}_{OX}, Y^{max}_{OP} and m_o. The results are also shown in Table VII.8.

§5 Application of metabolic flux analysis for the identification of metabolic bottlenecks in the biosynthesis of penicillin-G

5.1 Identification of principal nodes for Penicillin production

A priori analysis of the metabolic network model for growth and product formation of *P. chrysogenum* was carried out in order to identify the subset of metabolic nodes of which the flux partitioning strongly depends on the rate of product formation. These nodes are referred to as principal nodes (Stephanopoulos and Vallino, 1999[7]).

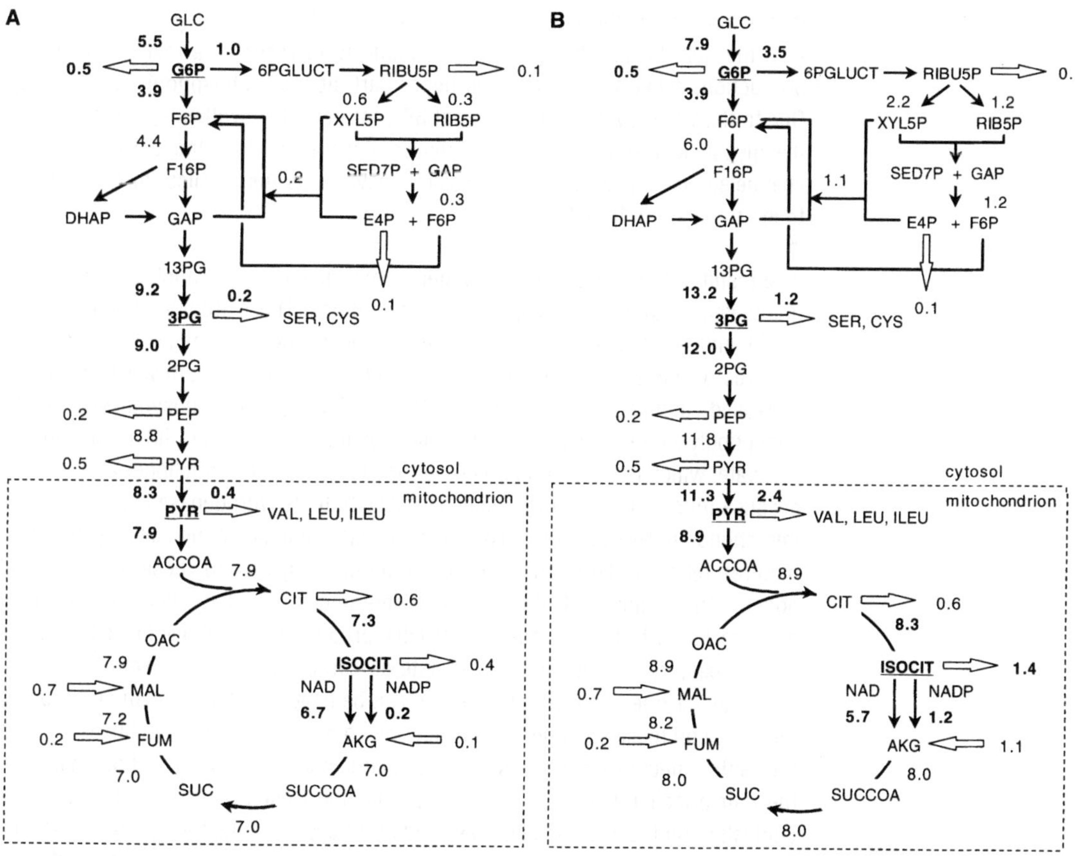

Fig. VII.11 A,B Calculated fluxes through the central metabolic pathways of *P. chrysogenum* from the stoichiometric model. The model contains two degrees of freedom, namely the specific growth rate and the specific penicillin production rate. Fluxes were calculated for a growth rate of 0.01 h^{-1}; A without production of penicillin; B with production of penicillin at a hypothetical rate of 1 mmol/Cmol biomass/h. Branch points showing significant changes in flux partitioning are underlined.

The analysis was carried out by calculating the flux distribution through the complete metabolic network as a function of the production rates of biomass and penicillin-G. For this a priori analysis literature values for the ATP stoichiometry parameters (P/O-ratio and maintenance energy requirements) have been used (Jørgensen et al.; 1995[5]). Fluxes through the metabolic network were calculated for a growth rate of 0.01 h^{-1} without production of penicillin (Figure VII.11 A) and with production of penicillin (Figure VII.11 B). For the second case the specific penicillin production rate was set to a hypothetical value of 1 mmol/Cmol/h which is twice the maximum value observed by Jørgensen et al (1995[5]) during fed-batch cultivation of *P. chrysogenum*. Although this seems to be a high value it can be calculated that the yield of penicillin on glucose under these conditions is still only 30% of the published theoretical maximum of 0.43 mol/mol glucose (Jørgensen et al., 1995[5]). It can be inferred from a comparison of the calculated flux patterns for both conditions that penicillin production leads to an increase of the (biomass specific) glucose uptake rate and, accordingly, to increased specific fluxes through the central metabolic pathways. It can be seen from Figure VII.11 that all additional glucose consumed is channeled through the pentose phosphate pathway, thus increasing the flux through this pathway significantly.
Also the flux through the glycolysis is found to increase as a result of penicillin production, especially through the upper part, above 3-phosphoglycerate (3PG). Due to the withdrawal, via cysteine, of 3PG for penicillin synthesis, the increase in the glycolytic flux below 3PG is smaller. Because part of the pyruvate flux is channeled, via valine, towards penicillin biosynthesis the smallest flux increase is found to occur in the TCA-cycle.

As a result of these changing flux patterns through central metabolism four branch points show significant changes in flux partitioning. These are underlined in Figure VII.11 and are located at i) glucose-6-phosphate, because penicillin production requires an increased flux through the pentose phosphate pathway to meet the increased demands for cytosolic NADPH for cysteine biosynthesis, ii) 3-phosphoglycerate, which is the carbon precursor of cysteine, iii) mitochondrial pyruvate, which is the carbon precursor of valine, and iv) mitochondrial isocitrate, because mitochondrial NADP dependent isocitrate dehydrogenase is the source of mitochondrial NADPH for valine biosynthesis. Significant penicillin production requires relatively large changes in flux partitioning around these four principal nodes. The rigidity of these nodes determines whether they are potential bottlenecks for further increase of penicillin production. Metabolic flux calculations for glucose, xylose, ethanol and acetate as carbon sources and ammonium and nitrate as nitrogen sources revealed, as can be expected, that growth and product formation on these substrates leads not only to dramatic changes in fluxes through the central metabolic pathways but also in large changes in flux partitioning around the four principal nodes for penicillin production (results not shown). Chemostat cultivation on these substrates was used as an experimental tool to manipulate the fluxes through the central metabolic pathways of *P. chrysogenum* and thus investigate the rigidity of these principal nodes for increased penicillin production.

5.2 Glucose, ethanol and acetate limited chemostat cultures

Carbon limited chemostat cultures of *Penicillium chrysogenum* DS12975 have been carried out on glucose, ethanol and acetate at dilution rates between 0.005 and 0.12 h^{-1}. Stable steady states with respect to mycelium concentration, respiration rate, penicillin-G production and byproduct formation could be obtained in glucose limited chemostat culture for dilution rates of 0.015 h^{-1} and higher.

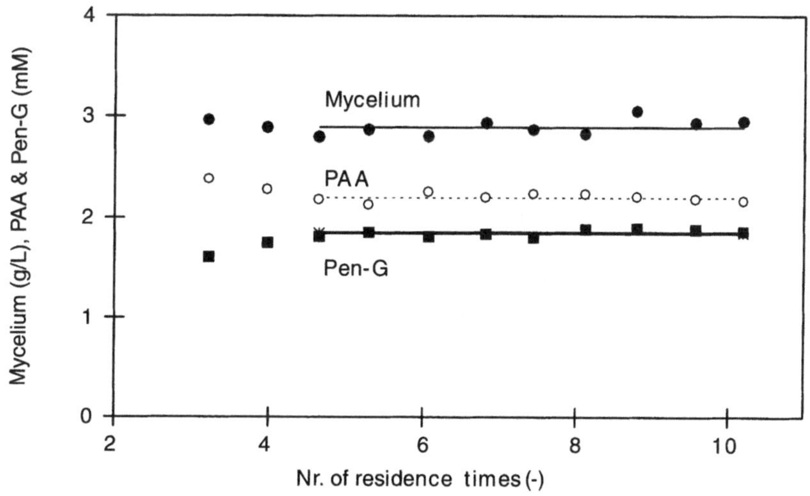

Fig. VII.12 Example of data from a glucose limited chemostat culture operated at a dilution rate of 0.029 h^{-1}.

An example of data from a steady state is shown in FigureVII.12. However, at dilution rates of 0.01 h^{-1} and below the cultures were observed to degenerate, i.e. gradually loose their capacity to produce penicillin-G before a stable steady state could be reached. This was accompanied by an increasing mycelium concentration. Loss of penicillin production in glucose limited chemostat cultures of *P. chrysogenum* is a known phenomenon (Righelato, 1976[62]; Christensen et al., 1995[63]). In both reported cases the loss of penicillin production was caused by the partial replacement of the parent strain by low or non-producing mutants/variants.It was found from our experiments that degeneration in ethanol and acetate limited cultures was more pronounced and occurred already at dilution rates of 0.03 h^{-1} and below.

5.3 Balancing of steady state data

From each steady state a set of 14 specific conversion rates was calculated (see materials and methods). The consistency of each set was checked from the carbon, nitrogen and degree of reduction balances. It was found that the recoveries were in all cases 95% or higher. Subsequently the quality of the data was checked through black box balancing and statistical detection of suspect measurements according to the algorithm of Wang and Stephanopoulos (1983[64]). Only a few measurements were found to be erroneous. Erroneous measurements were not used in the subsequent metabolic balancing of the data. This was not a problem because with the measured set of 14 specific conversion rates the metabolic system was 4 times overdetermined. From a chi-square test performed for each data set it appeared that in all cases the fit of the experimental data to the stoichiometric model was satisfactory (confidence level for errors in data and/or model less than 75%). The result of the metabolic flux balancing is a set of balanced conversion rates, a set of calculated conversion rates which were not or could not be measured, e.g. the production/consumption rates of water, protons, sulfate and phosphate, and the complete set of intracellular fluxes for each steady state. An example of a set of measured and balanced fluxes is shown in Table VII.9.

Table VII.9 Example of a set of biomass specific conversion rates, obtained from a glucose limited chemostat culture of *P. chrysogenum* operated at a dilution rate of 0.0302 h^{-1}, before and after metabolic flux balancing. Production rates of biomass, excreted peptides and polysaccharides are expressed in mCmol/Cmol/h, all other rates are expressed in mmol/Cmol/h.

Compound	Measured flux	Balanced flux
Glucose	-13.3	-13.1
Biomass	30.2	30.2
O_2	-36.2	-35.8
CO_2	37.5	38.1
NH_4^+	-7.14	-6.79
PAA	-0.691	-0.628
Pen-G + PIO	0.562	0.584
OPC	0.174	0.178
IPN	0	0
6-APA	0.0344	0.0351
8-HPA	0.0257	0.0263
o-OH-PAA	0.0427	0.0448
Excr. Peptides	1.12	1.15
Excr. Polysacch.	2.61	2.59
SO_4^{2-}	n.d.	-0.702
HPO_4^{2-}	n.d.	-0.240
H_2O	n.d.	55.2
H^+	n.d.	5.31

5.4 Penicillin production as a function of the growth rate

5.4.1 Glucose limited chemostat cultures

In Figure VII.13 the observed relation between the biomass specific production of penicillin-G and the specific growth rate is plotted for growth on glucose, ethanol and acetate.

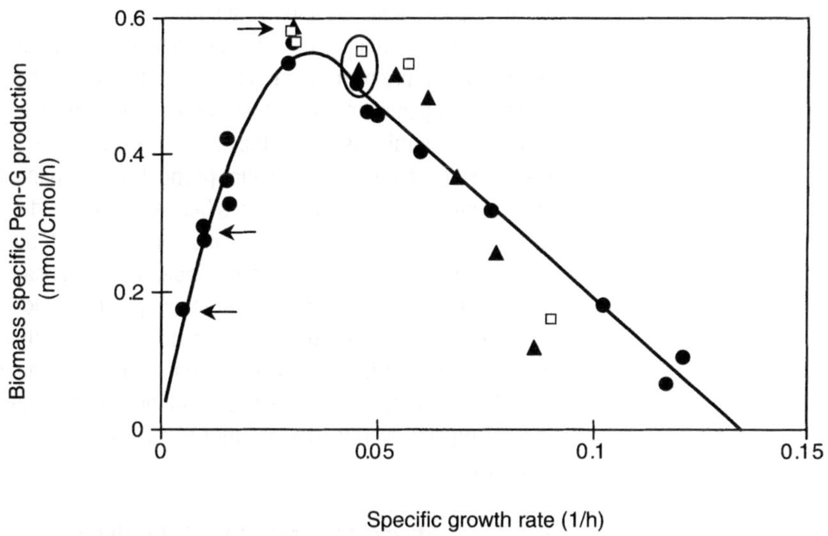

Fig. VII.13 Measured specific production of penicillin-G in glucose (●), ethanol (Δ) and acetate (◇) limited cultures of *Penicillium chrysogenum* DS12975. Points indicated with arrows represent no stable steady states but observed maximum specific penicillin production rates during the experiments. Solid line: q_{Pen-m} relation for the glucose limited chemostats.

In glucose limited chemostats penicillin production was observed to increase strongly at increasing dilution rate (which is equal to the growth rate in steady state chemostat culture) to a maximum of 0.56 mmol Pen-G / Cmol biomass / h. at a growth rate of 0.03 h^{-1}. When the dilution rate was further increased specific Pen-G production declined in a linear fashion.

To our knowledge this is the first time that the relation between specific penicillin production rate and growth rate for such a broad range of specific growth rates has been investigated. Previously published investigations on q_{Pen-m} relations have always been restricted to much smaller m ranges. Our observation of a positive correlation between penicillin production in *P. chrysogenum* and growth at relatively low growth rates ($\mu < 0.03$ h-1) is in good agreement with the findings reported in literature. Kluge et al. (1992[65]) found that q_{Pen} increased with the growth rate in glucose limited fed-batch cultures at extremely low growth rates ($0 < \mu < 0.004$ h-1). Wittler and Schugerl (1985[66]) observed a linear increase in specific production of penicillin-V in carbon limited fed-batch cultures for $0.003 < \mu < 0.018$. Ryu and

Hospodka (1980[67]) found that in carbon limited chemostat cultures carried out at dilution rates between 0.004 and 0.026 h^{-1} q_{Pen} increased with dilution rate until a dilution rate of 0.015 h^{-1} and thereafter remained constant.
With respect to higher growth rates the literature is less consistent. In an early paper of Pirt and Callow (1960[68]) it is reported that the relation between q_{Pen} and the dilution rate in glucose limited chemostat culture follows an optimum curve for dilution rates between 0.046 and 0.080 h^{-1}, with the maximum located at D = 0.051 h^{-1}. In a later paper Pirt and Righelato (1967[69]) reported for a different strain of *P. chrysogenum* that q_{Pen} was independent of the dilution rate in glucose limited chemostat culture over the range 0.014 to 0.086 h^{-1}. It should be realized, however, that specific penicillin production in these early strains was a factor 10 to 15 lower than in the strain used for the current work. More recently Christensen et al. (1995[63]) measured the specific production of penicillin-V in glucose limited chemostat culture for dilution rates between 0.023 and 0.078 h^{-1}. They used a strain of *P. chrysogenum* of which q_{Pen} was comparable to the strain used in the present work. Unfortunately their strain appeared genetically highly unstable. When cultivated in glucose limited chemostats specific penicillin production increased to a certain value and declined thereafter to a significantly lower steady state value. It was reported that the specific penicillin production was independent of the dilution rate but that specific Pen-V production during steady state increased with increasing dilution rate to a maximum for D = 0.050 h^{-1} and remained constant up to D = 0.078 h^{-1}.

5.4.2 Ethanol and acetate limited cultures

Due to the fact that degeneration of Pen-G production for growth on ethanol or acetate already occurred at dilution rates ≤ 0.03 h^{-1}, the lower part of the $q_{Pen} - \mu$ curve could not be explored for these two substrates. In case of a less pronounced loss of penicillin productivity the specific production was calculated as a function of time from the dynamic mass balance for penicillin-G, which for a constant volume system can be written as:

$$\frac{dC_{pen}}{dt} = q_{pen} \cdot C_X - D \cdot C_{pen} \qquad (6)$$

This was done for the glucose limited chemostats operated at dilution rates of 0.005 and 0.01 h^{-1} and for the ethanol and acetate limited chemostats operated at a dilution rate of 0.03 h^{-1}. The calculated maxima of the q_{Pen} against time curves are plotted in Figure VII.13. These points are indicated by small arrows.
A surprising result is that there are no large differences in the specific productivity of penicillin-G for the three different substrates for growth rates between 0.03 and 0.075 h^{-1}. This might indicate that the influence of primary metabolism on penicillin production is less important than was anticipated.

5.5 Changes in flux partitioning around principal nodes

The calculated fluxes around the four principal nodes for steady state glucose, ethanol and acetate limited chemostats performed at the same dilution rate of 0.045 h^{-1} are shown in Figure VII.14 A to D. At this dilution rate specific Pen-G production was almost identical for the three substrates (see Figure VII.13).

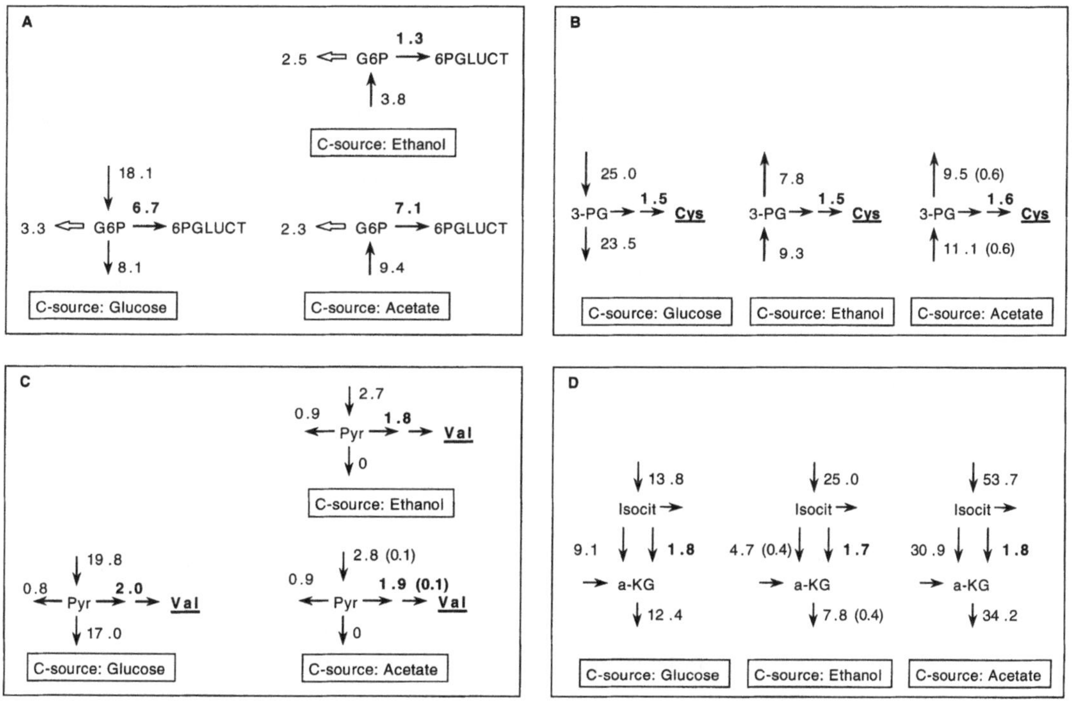

Fig. VII.14 A-D Calculated fluxes around the principal metabolic nodes for penicillin production for glucose, ethanol and acetate limited chemostat growth at a dilution rate of 0.045 h^{-1} A) glucose-6-phosphate node B) 3-phosphoglycerate node C) mitochondrial pyruvate node D) mitochondrial isocitrate node. The standard deviations of the calculated fluxes generally varied between 0.7 and 2.5 %. For fluxes with standard deviations higher than 2.5 % the corresponding standard deviations are indicated between brackets.

Figure VII.14A shows the different flux patterns around the glucose-6-phosphate branch point for growth on glucose, ethanol and acetate. In case glucose is the carbon source the glucose-6-phosphate produced by hexokinase is partitioned between glycolysis, pentose phosphate pathway and anabolism. For C-2 carbon sources like ethanol and acetate glucose-6-phosphate is generated by gluconeogenesis. The flux is now partitioned between two main pathways; the PP-pathway and anabolism. For growth on ethanol it is assumed that cytosolic NADPH is generated mainly through NADP dependent acetaldehyde dehydrogenase and that the PP-pathway has only an anabolic function. This leads to a significantly lower flux of glucose-6-phosphate entering the PP-pathway than for growth on acetate and glucose.

The fluxes around the 3-phosphoglycerate node are shown in Figure VII.14B. Because growth on the C-2 carbon sources ethanol and acetate requires gluconeogenesis, this leads to a reversal of the flux. Nevertheless the flux of 3-phosphoglycerate via cysteine to penicillin is approximately the same in all cases.

The flux partitioning around the mitochondrial pyruvate node is shown in Figure VII.14C. For growth on C-2 carbon sources, carbon enters the TCA cycle as AcetylCoA and therefore the fluxes around mitochondrial pyruvate are largely reduced when the cells are grown on ethanol or acetate instead of glucose as the sole carbon source. This has apparently no influence on the flux of pyruvate via valine to penicillin.

In Figure VII.14D the fluxes around isocitrate dehydrogenase are shown. Also in this case large differences occur, the flux for growth on acetate being the largest because of the relatively large energetic cost of biomass and product formation from acetate.

From these observations it can be concluded that in spite of the fact that growth on the carbon substrates glucose, ethanol and acetate introduced large changes in flux partitioning around principal nodes for penicillin production in primary metabolism, the observed changes in the specific penicillin production rate were marginal. This indicates that the four principal nodes are highly flexible and are unlikely to form potential bottlenecks for (a further increase of) penicillin production. We like to stress that although some, well founded, assumptions had to be made with respect to the biochemistry of *P. chrysogenum* (e.g. anaplerotic routes, pathway for cysteine biosynthesis) the influence on the flux results as displayed in Figure VII.14 was found to be limited and therefore of no influence on the conclusion drawn above.

5.6 Supply of cytosolic NADPH

An aspect, which was further examined, is the supply of cytosolic NADPH. Although the glucose-6-phosphate branch point appeared to be flexible, the pentose phosphate pathway is not the only source of cytosolic NADPH. Possible additional sources are cytosolic NADP dependent isocitrate dehydrogenase (Loftus et al., 1994[70]), NADP dependent mannitol dehydrogenase (Hult et al., 1980[71]) and in case of growth on ethanol, the NADP dependent acetaldehyde dehydrogenase (Postma et al., 1989[72]). Furthermore additional NADPH consumption might occur if the cells contain an NADPH oxidase (Djavadi et al., 1980[73]; Bruinenberg et al., 1985[74]). The relative contribution of these additional NADPH producing and consuming processes is not known. Therefore further experiments have been carried out in order to investigate if the supply of NADPH could be a potential bottleneck. In order to investigate this, the NADPH needs for primary metabolism were stepwise increased by using combinations of two carbon sources, glucose and xylose, and two nitrogen sources, ammonium and nitrate. The metabolism of xylose and nitrate requires additional consumption of cytosolic NADPH. In the first step of the catabolism of xylose it is reduced to xylitol, using NADPH as the electron donor. Xylitol is subsequently oxidized to xylulose using NAD as electron acceptor. Finally xylulose is phosphorylated to xylulose-5-phosphate which is catabolized further via the pentose phosphate pathway. This implies that when the cells are grown on xylose as the sole carbon source for each mol of xylose consumed one mol of NADPH is converted into NADH. When nitrate is used as nitrogen source it has to be reduced to ammonium through the action of nitrate and nitrite reductases. The cofactor for nitrate reductase in fungi is NADPH. For nitrite reductase both NADH and NADPH can serve as cofactors in *Candida utilis* and *Neurospora crassa* (Prodouz et al., 1981[75]; Sengupta et al., 1997[76]). Therefore the impact of nitrate reduction on the demand of cytosolic NADPH is not exactly known. In the worst case 4 mol of NADPH is required for the reduction of one mol of nitrate, in the best case only one mol of NADPH and 3 mol of NADH are required.

Table VII.10 Manipulation of the total demand for cytosolic NADPH through growth on different combinations of two C- and N- sources and the effect on penicillin production.

	Total NADPH demand (-)	Measured Pen-G production (-)
Glucose & NH_3	1	1
Xylose & NH_3	2	0.79
Glucose & NO_3	2.8	0.64
Xylose & NO_3	3.8	0.54

Steady state chemostats have been conducted on the four possible combinations of both carbon and nitrogen sources. All experiments were performed at the same dilution rate of 0.03 h^{-1}. In all cases stable steady states were reached. The results with respect to the worst case calculation of the NADPH demand and the measured penicillin production are shown in Table VII.10. It can be seen from this table that an increasing demand for cytosolic NADPH leads to a drastic reduction in the production of penicillin. It can be inferred from these results that the supply of cytosolic NADPH for penicillin production, which is mainly needed for sulfate reduction, may be a potential metabolic bottleneck. Hence, the role of primary metabolism in the cofactor supply/regeneration seems more critical than the supply of C-compounds.

§6 Application of ^{13}C-NMR for the estimation of intracellular fluxes

6.1 Problems in steady state metabolic flux analysis

Determining the steady state fluxes in a metabolic network is an important step in metabolic engineering. A good knowledge of the reaction stoichiometries and steady state flux map of a metabolic network form the foundation on which a kinetic metabolic model can be built. Such a kinetic model is needed to perform an *in silico* analysis of the network. This analysis serves the purpose of identifying the bottlenecks of the metabolic network, which are the targets of choice for genetically improving a micro-organism.

Intracellular fluxes cannot be directly measured. Steady state fluxes can be calculated by solving a set of linear equations consisting of both the mass balances of intracellular metabolites and the measured net consumption or net production rates of a number of membrane-crossing compounds. The calculation of steady state fluxes is based on the assumption that no net accumulation of intracellular metabolites occurs. For the metabolic network shown in Figure.VII.15, the calculation of the intracellular fluxes is as follows:

$$\begin{pmatrix} \underline{\underline{S}} \\ \underline{\underline{M}} \end{pmatrix} \cdot \underline{v} = \begin{pmatrix} \underline{0} \\ \underline{\varphi} \end{pmatrix} \Leftrightarrow \begin{matrix} mass\ balance\ I \\ measured\ consumption\ S \\ measured\ production\ X \\ measured\ production\ P_1 \\ measured\ production\ P_2 \end{matrix} \begin{pmatrix} -1 & 1 & 1 & 1 & 1 \\ \hline -1 & 0 & 0 & 0 & 0 \\ 0 & 0 & 0 & 0 & 1 \\ 0 & 1 & 0 & 0 & 0 \\ 0 & 0 & 1 & 1 & 0 \end{pmatrix} \cdot \begin{pmatrix} v_1 \\ v_2 \\ v_3 \\ v_4 \\ v_5 \end{pmatrix} = \begin{pmatrix} \underline{0} \\ \varphi_s \\ \varphi_x \\ \varphi_{p1} \\ \varphi_{p2} \end{pmatrix} \quad (7)$$

$$rank\begin{pmatrix} \underline{\underline{S}} \\ \underline{\underline{M}} \end{pmatrix} = 4$$

The answer to the question whether all intracellular fluxes can be calculated from the measured extracellular rates, depends on the number of extracellular fluxes that can be measured and on the structure of the intracellular network. Some structural features of metabolic networks prohibit determination of all intracellular fluxes, irrespective of the extracellular measurements that are performed. One such feature is the occurrence of parallel pathways in the cell, *i.e.* when a substrate or intermediate can be turned into a product by means of various different routes. In such cases mass balances only yield the sum of the fluxes through the parallel pathway branches but never the separate fluxes.

Consider for example the schematic network in Figure.VII.15, where measuring the consumption rate of substrate S and the production rates of X, P_1 and P_2 does not suffice to calculate the separate fluxes v_3 and v_4. Another way of saying this, is stating that the set of flux constraints is underdetermined, since Eq. 7 shows that the rank of the combined stoichiometric (S) and measurement (M) matrices is lower than the number of fluxes (v). A real-life example of a parallel pathway is found in lysine biosynthesis (see e.g. [Sonntag et al., 1993[77]]).

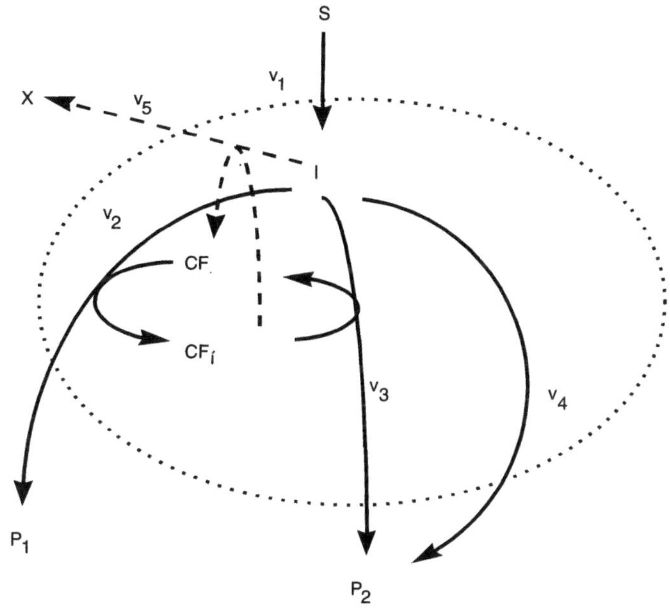

Fig. VII.15 A metabolic network in which substrate S is converted to intermediate I that is converted to biomass X and products P_1 and P_2. CF is a cofactor; the prime indicates its activated state.

Another inobservability problem in mass balancing stems from bi-directional fluxes [Wiechert et al.,1997[78]]. Mass balancing only yields net fluxes between the metabolite pools, but it does not give the information needed to conclude what forward and backward fluxes constitute that net flux. From a cell regulatory point of view, it is interesting to know whether reactions in the cell are uni or bidirectional. Reversibility assumptions can be made on the basis of thermodynamic calculations. However, thermodynamic data that are determined *in vitro* are often found to be invalid *in vivo*, due to the complex properties of the viscous, protein rich cytosol of the cell.

6.2 Solutions to overcome inobservability of fluxes

In order to determine fluxes that cannot be found using mass balances, one needs additional information about the reactions in the cell. Enzyme assays are one of these additional sources of information. By determining which enzymes are expressed in the metabolic system under investigation, one knows which pathways are active. Enzyme assays, however, do not give information about the sizes of the fluxes through these pathways unless the exact enzyme kinetics and intracellular metabolite concentrations are known. At present, the validity of in vitro enzyme kinetics under *in vivo* circumstances is under debate and methods for accurate measurements of intracellular metabolites are under development. Therefore, determining all intracellular fluxes from enzyme and metabolite measurements is not feasible yet.

An alternative approach to solve the problem of parallel pathways involving different cofactors, is to include mass balances of metabolic cofactors such as ATP, NAD(P)H and Coenzyme-A in the set of linear flux constraints. (Note that some researchers state that pathways employing different cofactors may not be called 'parallel'.) A hypothetical example of how a cofactor balance can make the fluxes through two parallel pathways observable, is shown in Figure.VII.15. In this figure the fluxes through the parallel reactions v_3 and v_4 leading from intermediate I to product P_2 are different with respect to their cofactors. If the activated cofactor is assumed to be ATP, one can imagine that reaction v_4 is efficient and does not involve the hydrolysis of ATP, whereas reaction v_3 does. If the cofactor balance is taken into account, it can be seen that the rate at which product P_1 is produced, determines the rate at which ATP is generated intracellularly. This rate fixes the fluxes requiring ATP (i.e. v_3 and v_5). Now, all intracellular fluxes can be determined, since the set of flux constraints is determined. Eq.8 shows that the rank of the combined stoichiometric and measurement matrices equals the number of fluxes.

$$\begin{array}{r} \textit{mass balance I} \\ \textit{mass balance CF} \\ \textit{measured consumption S} \\ \textit{measured production X} \\ \textit{measured production } P_1 \\ \textit{measured production } P_2 \end{array} \begin{pmatrix} -1 & 1 & 1 & 1 & 1 \\ 0 & -1 & 1 & 0 & 1 \\ \hline -1 & 0 & 0 & 0 & 0 \\ 0 & 0 & 0 & 0 & 1 \\ 0 & 1 & 0 & 0 & 0 \\ 0 & 0 & 1 & 1 & 0 \end{pmatrix} \cdot \begin{pmatrix} v_1 \\ v_2 \\ v_3 \\ v_4 \\ v_5 \end{pmatrix} = \begin{pmatrix} 0 \\ \underline{0} \\ \varphi_s \\ \varphi_x \\ \varphi_{p1} \\ \varphi_{p2} \end{pmatrix} \qquad (8)$$

$$rank\left(\frac{\underline{\underline{S}}}{\underline{\underline{M}}}\right) = 5$$

An example of the above situation is encountered in the primary carbon metabolism. The glycolytic and pentose phosphate pathways may be considered as parallel pathways converting glucose 6-phosphate to pyruvate. Whereas the pentose phosphate branch of this parallel pathway yields ATP, NADH and NADPH, the glycolytic branch yields more ATP and NADH, but no NADPH. In contrast to NADH, NADPH is known to be involved in the biosynthesis of biomass and (in the case of *Penicillium chrysogenum*) penicillin. Therefore, equating the consumption of NADPH in biosythetic reactions to its production in the oxidative steps of the pentose phosphate is one way of determining how much glucose is consumed via the pentose phosphate pathway.

Although cofactor balances have been shown to offer a solution to some flux analyses in parallel pathways, they have some serious disadvantages. Firstly, cofactor balances do not enable determination of the separate forward and backward fluxes in bidirectional reaction steps. Besides, cofactor mass balances rely on estimated reaction stoichiometries and on controversial assumptions (van Gulik et al.,1995[12]; Marx et al.,1996[79],1999[80]; Schmidt et al., 1998[81]; Wiechert et al.,1997[78]. Examples of the latter are a growth-independent ATP yield of the respiratory system (P/O-ratio), fixed growth and non-growth associated maintenance requirements of the cell, the presence of ATP-wasting futile cycles, NAD(P)H specificities of various enzymes and the presence or absence of transhydrogenases that interconvert NADH and NADPH. Furthermore, cofactors are involved in many reactions in the cell to such extent that it is hardly possible to take all of them into account. As a consequence, cofactor mass balances are most probably far from complete and may lead to erroneous flux estimates when included in the mass balances.

6.3 Using ^{13}C-NMR for metabolic flux analysis

The ^{13}C-labelling technique is a method that has the potential to resolve fluxes flowing through a complicated metabolic network. This technique makes use of the fact that the pathways in the cell split and form carbon-carbon bonds in well-known ways. The method comprises to feeding a ^{13}C-labeled compound to the metabolic network that is investigated and measuring the amounts of labeled carbon atoms at various carbon positions of the metabolic intermediates. These measurements either rely on the different magnetic momentum of ^{12}C- and ^{13}C-atoms (Nuclear Magnetic Resonance spectroscopy, NMR) or on the different atomic mass of the two isotopes (Mass Spectrometry, MS).
Metabolic fluxes can be calculated from the ^{13}C-labeling data that are obtained from NMR or MS [Schmidt et al.,1997[82]; Szyperski, 1998[83]]. Without going into details regarding the method, the general approach of flux analysis based on labeling experiments is outlined in Figure.VII.16. As the figure shows, fluxes cannot be directly determined from the measured labeling data. Due to non-linearities in the model that relates the ^{13}C-distributions of metabolites to the steady state metabolic fluxes, the fluxes have to be found using an iterative optimization.

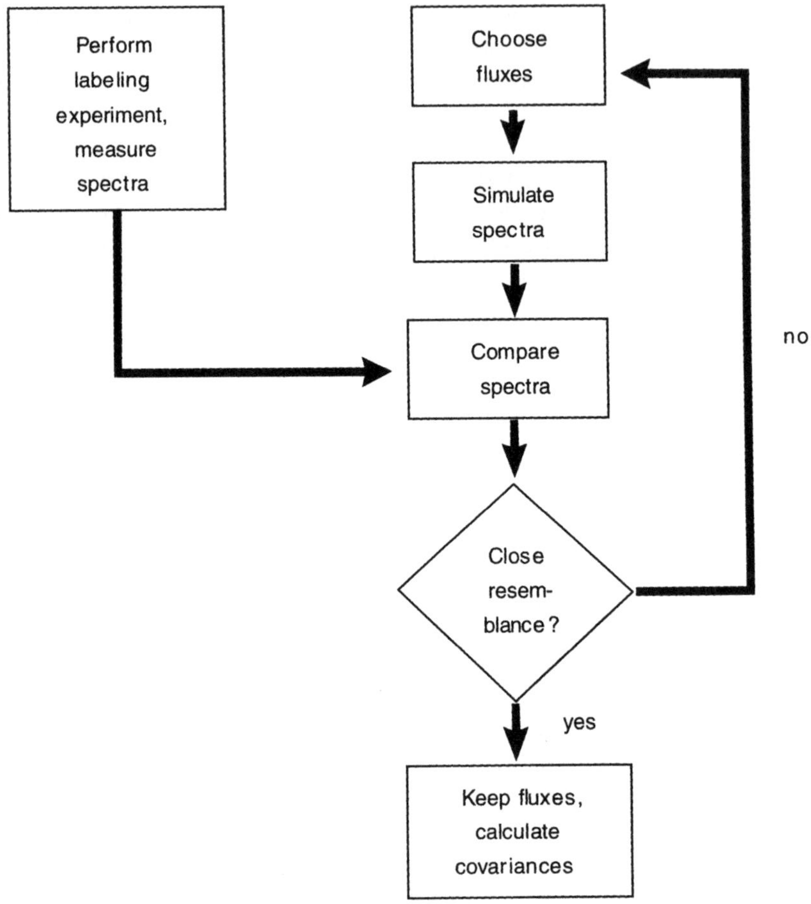

Fig. VII.16 The overall algorithm used for determining a set of fluxes from measured NMR spectra.

Returning to the case of the parallel glycolytic and pentose phosphate pathway, one can show that both branches cause a different rearrangement of the carbon atoms that enter the cell in the form of glucose. This fact forms the basis of our research that has as an aim to determine the fluxes through the glycolysis and the pentose pathway in *P. chrysogenum* from NMR measurements of the ^{13}C-label distributions of intermediates of both pathways. It is interesting to see whether these labeling studies result in different flux estimations than those that were previously determined using NADPH-balances (van Gulik et al..2000[17]). This may shed new light on the question whether the availability of the cofactor NADPH is a potential bottleneck in the production of penicillin.

6.4 Results of ^{13}C-labeling studies

Results that have been obtained so far in this research include a preliminary estimation of steady state fluxes in cultures of *P. chrysogenum* growing on defined medium (see [van Gulik et al., 2000[17]]) containing either NO_3^- or NH_4^+ as a nitrogen source and containing 10% uniformly labeled [$^{13}C_6$]-glucose. The analyses indicate that fluxes into the pentose phosphate pathway may be higher than previously determined. However, so far the procedure shown in Figure.VII.16 has not yet yielded satisfactory fits between measured and simulated ^{13}C-labeling data. Therefore, we have first been focusing our research on some methodological aspects of the ^{13}C-labeling technique.

Problems we have addressed, include the sensitivity of the outcomes of ^{13}C-labeling studies to modeling errors [van Winden et al.,2000a[84]]. Two potential errors in metabolic model are the oversimplification of the non-oxidative part of the pentose phosphate pathway and the ignorance of the effects of so-called 'channeling', i.e. the direct enzyme-enzyme transfer of intermediates. Besides, we have studied which fluxes in a metabolic network are theoretically identifiable from a given set of ^{13}C-labeling data [van Winden et al.,2000b[85]]. Finally, we have developed spectral analysis software in order to improve the quality of the measurement data that are extracted from the NMR spectra [van Winden et al.,2000c[86]]. Besides accurate labeling data, this spectral analysis software yields covariances of the measurement data that are indispensable for the determination of the reliability of the fluxes that are found using the 13C-labeling technique.

Based on the obtained methodological insights we plan to make a final estimation of the fluxes in the central carbon metabolism of *P. chrysogenum* in the near future.

§7 A first attempt to model the regulation of penicillin-G production in *Penicillium chrysogenum*

7.1 Control networks and product formation

It must be realized that the complete reaction network of the cell is controlled through a multi-layered control network. This control network consists of the genome, containing the blueprint of the complete cellular machinery (including the control structures), the transcriptome, i.e. the messenger RNA molecules which code for the proteins to be synthesized and the proteome, i.e. all metabolically active proteins. The control network responds to signals from the reaction network (metabolite levels) and controls the enzyme levels in the reaction network.

It is clear that a better understanding of the regulation of product formation on the level of the control network of enzyme expression is of key importance for a rational improvement of product formation through reprogramming of microbial metabolism. Identification of (a set of) regulatory genes that controls expression of all catalytic components of a pathway is highly relevant for industrial application. This approach may circumvent some of the intrinsic limitations that occur when metabolic engineering is targeted at the structural genes encoding these catalytic components (e.g. the tendency of flux control to shift to other enzymes when an enzyme with a high degree of control is over-expressed).

In this paragraph a simple model for the regulation of the penicillin biosynthesis pathway is derived, based on a few assumptions and on experimental observations. It is shown that with this simple model not only a satisfactory description can be obtained of the observed relation between specific growth rate and specific penicillin production rate under steady state conditions. In addition the model provides also a qualitative description of the dynamics of the specific penicillin production rate during the transition between feast conditions (i.e. unlimited batch growth) and glucose limited growth.

7.2 Simple dynamic model for penicillin production

The model is based on the following assumptions:

1) The specific penicillin production rate is proportional to the intracellular concentration of one, rate limiting, enzyme in the biosynthesis pathway (Theilgaard and Nielsen, 1999[87])
2) The transcription of the gene which codes for this enzyme is repressed by glucose (Martin et al., 1999[88])
3) The decay rate of this enzyme is first order in the intracellular enzyme concentration.

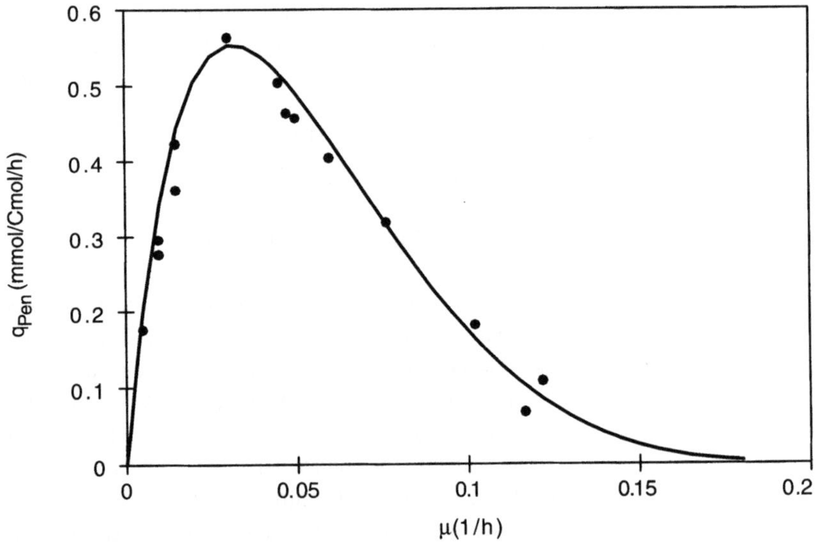

Fig. VII.17 Model description of the relation between specific growth rate and specific penicillin production under steady state conditions (●) measurements, (—) model description.

A detailed mathematical derivation of the dynamic model is given elsewhere (vanGulik et al., in preparation[89]). The model contains only a moderate number of parameters, namely five. It was found that the model is rather insensitive for the values of two of these parameters and therefore these were maintained at a constant value while optimizing the three remaining parameters. With the assumption that the specific penicillin production rate is proportional to the intracellular concentration of the rate limiting enzyme E, it was investigated whether the observed relation between the specific penicillin production and the specific growth rate could be described with this model. A good fit was obtained for the steady state situation as can be seen from Figure VII.17.

The dynamic response of the model, that is the description of the increase of the specific penicillin production rate during the transition from the batch phase (where no penicillin production takes place) to the steady state chemostat phase is shown in Figure VII.18. It can be inferred from this figure that the model predicts that the rate of increase of the specific penicillin production rate is strongly dependent on the dilution rate of the chemostat (which is equal to the biomass specific growth rate under steady state conditions).

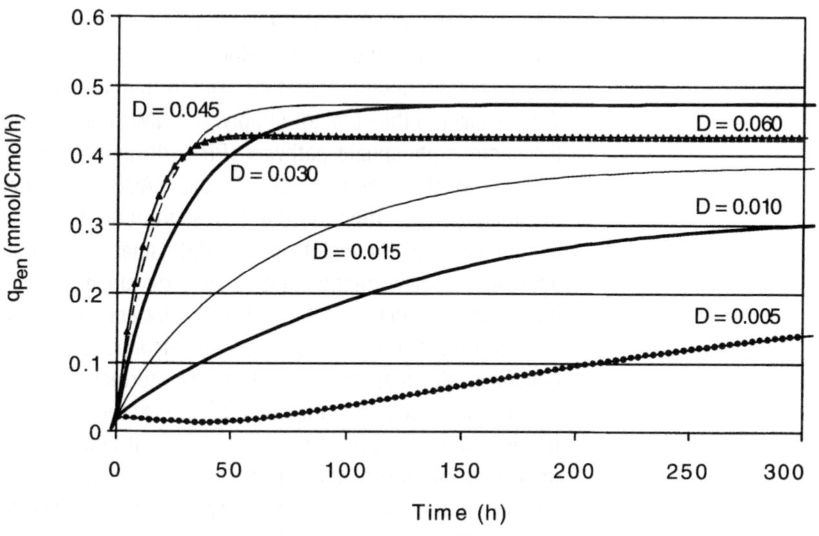

Fig. VII.18 Model prediction of the dynamics of specific penicillin production during the transition between the batch phase and the steady state chemostat phase at different dilution rates.

At increasing dilution rate the increase of the specific penicillin production occurs more rapid and vice versa. Experiments during which the transition between batch and chemostat phase was monitored at different dilution rates have shown the same phenomenon (experimental results not shown). It appeared that the model provided a fairly good qualitative description of the dynamics of specific penicillin production during the transition between batch and chemostat phase.

§8 Conclusions and outlook

Chemostat experiments with the penicillin producing filamentous fungus *Penicillium chrysogenum* performed under well defined conditions have led to a better understanding of the possible locations of bottlenecks for further increase of the productivity. For this work it was of great value that a high producing industrial strain was available. The main conclusions which can be drawn from this work are that, at least for the strain used, no bottlenecks are expected to be located in primary carbon metabolism. However, this was not true with respect to the supply of cofactors. It was found that especially the supply of reducing equivalents in the form of NADPH appears to be critical, since induction of the increase of the metabolic NADPH demand resulted in a decrease of the specific penicillin production rate.
An in dept analysis of the energetics of penicillin-G production lead to a rather unexpected finding, namely that the in-vivo ATP costs of penicillin production are much higher than expected from the well known biochemistry of the biosynthesis pathway. This led to a drastic reduction of the maximum theoretical yield of penicillin on the carbon source. Both the cofactor supply and the in-vivo energetics of penicillin production therefore need further attention in future research.
With respect to the cofactor supply, C^{13}-NMR measurements of the fluxes through the pentose phosphate pathway (PP-pathway), which is considered as the main pathway for the generation of cytosolic NADPH, have been carried out. Besides a preliminary estimation of the fluxes through the PP-pathway under different conditions, a better insight in the difficulties and pitfalls of this technique was obtained and new methodologies were developed. An important notion was that, due to the fact that also for C^{13}-NMR flux estimations stoichiometric models are used, errors in these models lead to erroneous flux estimations. It is therefore indispensable that such models are correct. This might require additional biochemical research to unravel the correct stoichiometry of the metabolic network under study. Although, from an economical point of view, *P. chrysogenum* is an important micro-organism, relevant information concerning primary metabolism and the connection between primary metabolic pathways and product biosynthesis routes is still lacking. As in other eukaryotic cells metabolic compartmentation is an important phenomenon in the interpretation and modification of cellular metabolism. Especially microbodies are important, because part of the biosynthesis pathway of penicillin is located in these organelles. Another important aspect is the excretion of penicillin. From the fact that penicillin excretion occurs against a concentration gradient (inside low, outside high) it can be inferred that this is an active process. The exact mechanism, however, is still not known. All these aspects should be addressed in further physiological research on *P. chrysogenum*.
Future research on metabolic engineering of β-lactam antibiotic production should not only focus on the stoïchiometry and energetics but also strongly on the in-vivo kinetics of primary and secondary metabolism. Although the biosynthesis pathways of β-lactam antibiotics are usually relatively short, they are always interconnected at various points with the primary metabolic pathways, e.g. for the supply of carbon precursors, various cofactors and metabolic energy. Furthermore, intracellular

transport steps play an important role because these pathways are divided over several compartments of the cell. This means that recombinant-DNA based improvement of product formation and/or introduction of novel pathways driven by intuition alone is not feasible. A detailed insight in the in-vivo kinetics of the biosynthesis pathway(s) and interconnected primary routes, in the form of a mathematical model, is therefore indispensable. It has recently been shown that such models may contain approximate kinetic equations (e.g. a linear / logarithmic approximation) leading to a large reduction of the model complexity and the advantage that analytical solutions are possible (Heijnen, 2000[90]). Nevertheless, such models are well suited to predict which changes have to be made to a certain metabolic network to achieve a certain goal (e.g. increased product formation), (Visser and Heijnen, paper in preparation[91]).

§9 References

1 Nielsen J, Jørgensen HS. 1995. Metabolic control analysis of the penicillin biosynthetic pathway in a high-yielding strain of *Penicillium chrysogenum.* Biotechnol. Prog. **11**: 299-305.
2 Pissara P, Nielsen J, Bazin MJ. 1996. Pathway kinetics and metabolic control analysis of a high-yielding strain of *Penicillium chrysogenum* during fed-batch cultivations. Biotechnol. Bioeng. **51**: 168-176.
3 Theilgaard HA, Nielsen J. 1999. Metabolic control analysis of the penicillin biosynthetic pathway: the influence of the LLD-ACV : bisACV ratio on the flux control. Antonie Van Leeuwenhoek **75**: 145-154.
4 Henriksen CM, Nielsen J, Villadsen J. 1998. Cyclization of alpha-aminoadipic acid into the delta-lactam 6-oxo-piperidine-2-carboxylic acid by *Penicillium chrysogenum.* J. Antibiot. **51**: 99-106.
5 Jørgensen HS, Nielsen J, Villadsen J, Møllgaard H. 1995. Metabolic flux distributions in *Penicillium chrysogenum* during fed-batch cultivations. Biotechnol. Bioeng. **46**: 117-131.
6 Henriksen CM, Christensen LH, Nielsen J, Villadsen J. 1996. Growth energetics and metabolic fluxes in continuous cultures of *Penicillium chrysogenum.* J. Biotechnol. **45**: 149-164.
7 Stephanopoulos G, Vallino JJ. 1991. Network rigidity and metabolic engineering in metabolite overproduction. Science **252**: 1675-1681.
8 Vallino JJ, Stephanopoulos G. 1994a. Carbon Flux Distribution at the Pyruvate Branch Point in *Corynebacterium glutamicum* during Lysine Overproduction. Biotechnol. Prog. **10**: 320-326.
9 Vallino JJ, Stephanopoulos G. 1994b. Carbon Flux Distribution at the Glucose 6-phosphate Branch Point in *Corynebacterium glutamicum* during Lysine Overproduction. Biotechnol. Prog. **10**: 327-334.
10 Cooney CL, Acevedo F, 1977. Theoretical conversion yields for penicillin synthesis. Biotechnol.Bioeng. **19**:1449-1462.
11 Hersbach GJM, Van der Beck CP, Van Dijk PWM, 1984. The penicillins: properties, biosynthesis and fermentation. Biotechnol.Ind.Antibiot. **22**:45-140.

12 vanGulik WM, Heijnen JJ, 1995. A metabolic network stoichiometry analysis of microbial growth and product formation. Biotechnol.Bioeng. **48**:681-698.

13 Vanrolleghem PA, deJong-Gubbels P, vanGulik WM, Pronk JT, vanDijken JP, Heijnen JJ, 1996. Validation of a metabolic network for *Saccharomyces cerevisiae* using mixed substrate studies. Biotechnol.Prog. **12**:434-448.

14 Verduyn C, Stouthamer, AH, Scheffers, WA, vanDijken, JP (1991) A theoretical evaluation of growth yields of yeasts. Antonie van Leeuwenhoek **59**:49-63.

15 Verduyn C, Postma E, Scheffers WA, Dijken JPv, 1992. Effect of benzoic acid on metabolic fluxes in yeasts: A continuous-culture study on the regulation of respiration and alcoholic fermentation. Yeast **8**:501-517.

16 Nielsen J, 1995. Physiological engineering aspects of *Penicillium chrysogenum.* Habilitationsschrift, Technical University of Denmark, Lyngby, Denmark, 1995.

17 van Gulik, de Laat, W.T.A.M., Vinke, J.L., Heijnen, J.J. (2000) Application of Metabolic Flux Analysis for the Identification of Metabolic Bottlenecks in the Biosynthesis of Penicillin-G. Biotech. Bioeng., **68**, **6**, 602-618.

18 Osmani SA, Scrutton C, 1983. The sub-cellular localisation of pyruvate carboxylase and of some other enzymes in *Aspergillus nidulans.* Eur.J.Biochem. **133**:551-560.

19 Osmani SA, Scrutton MC, 1985. The sub-cellular localisation and regulatory properties of pyruvate carboxylase from *Rhizopus arrhizus.* Eur.J.Biochem. **147**:119-128.

20 Pronk JT, Steensma HY, vanDijken JP, 1996. Pyruvate metabolism in *Saccharomyces cerevisiae.* Yeast **12**:1607-1633.

21 Bercovitz A, Peleg Y, Battat E, Rokem JS, Goldberg I, 1990. Localization of pyruvate carboxylase in organic acid-producing *Aspergillus strains.* Appl.Environ.Microbiol. **56**:1594-1597.

22 Bruinenberg PM, vanDijken JP, Scheffers WA, 1983. An enzymic analysis of NADPH production and consumption in *Candida utilis.* J.Gen.Microbiol. **129**:965-971.

23 Treichler H-J, Liersch M, Nüesch J, Döbeli H. 1979. Role of Sulfur Metabolism in Cephalosporin C and Penicillin Biosynthesis In: OK Sebek and AL Laskin (eds) Genetics of Industrial Microorganisms. Washington DC: American Society for Microbiology pp 97-104.

24 Dobeli H, Nuesch J. 1980. Regulatory properties of O-acetyl-L-serine sulfhydrylase of *Cephalosporium acremonium*: evidence of an isoenzyme and it's importance in Cephalosporin C biosynthesis. Antimicrob. Agents Chemother. **18**: 111-117.

25 Østergaard S, Theilgaard HA, Nielsen J. 1998. Identification and purification of O-acetyl-L-serine sulphhydrylase in *Penicillium chrysogenum.* Appl. Microbiol. Biotechnol. **50**: 663-668.

26 Bender DA, 1985. Amino acid metabolism. John Wiley & Sons, New York.

27 Smith JE, Berry DR, 1976. The filamentous fungi, volume 2: Biosynthesis and metabolism. Edward Arnold publishers Ltd., London.

28 Lendenfeld T, Ghali D, Wolschek M, Kubicek-Pranz EM, Kubicek CP. 1993. Subcellular compartmentation of penicillin biosynthesis in *Penicillium chrysogenum.* J. Biol. Chem. **268**: 665-671.

29 Müller WH, VanderKrift TP, Krouwer AJ, Wosten HA, VanderVoort LH, Smaal EB, Verkleij AJ. 1991. Localization of the pathway of the penicillin biosynthesis in *Penicillium chrysogenum.* EMBO. J. **10**(**2**): 489-495.

30 Serrano R. 1988. Structure and function of proton translocating ATPase in plasma membranes of plants and fungi Biochim. Biophys. Acta **947**: 1-28.

31 Boles E, Hollenberg CP. 1997. The molecular genetics of hexose transport in yeasts. FEMS Microbiol. Rev. **21**: 85-111.

32 Does AL, Bisson LF. 1989. Characterization of xylose uptake in the yeasts *Pichia heedii* and *Pichia stipitis.* Appl. Environ. Microbiol. **55**: 159-164.

33 Kilian SG, VanUden N. 1988. Transport of xylose and glucose in the xylose-fermenting yeast *Pichia stipitis.* Appl. Microbiol. Biotechnol. **27**: 545-548.

34 Cassio F, Leao C, VanUden N. 1987. Transport of lactate and other short-chain monocarboxylates in the yeast *Saccharomyces cerevisiae.* Applied and Environmental Microbiology **53-3**: 509-513.

35 Leao C, VanUden N. 1986. Transport of lactate and other short-chain monocarboxylates in the yeast *Candida utilis.* Appl. Microbiol. Biotechnol. **23**: 389-393.

36 Hillenga DJ, Versantvoort HJ, VanderMolen S, Driessen AJ, Konings WN. 1995. *Penicillium chrysogenum* takes up the penicillin G precursor Phenylacetic acid by passive diffusion. Appl. Environ. Microbiol. **61**(**7**): 2589-2595.

37 Hackette SL, Skye GE, Burton C, Segel IH. 1970. Characterization of an ammonium transport system in filamentous fungi with methylammonium-^{14}C as the substrate. J. Biol. Chem. **245**: 4241-4250.

38 Roon RJ, Levy JS, Larimore F. 1977. Negative interactions between amino acid and methylamine/ammonia transport systems of *Saccharomyces cerevisiae.* The Journal of Biological Chemistry 252-**11**: 3599-3604.

39 Eddy AA. 1982. Mechanisms of solute transport in selected eukaryotic microorganisms. Adv. Microb. Physiol. **23**: 1-78.

40 Blatt MR, Maurousset L, Meharg AA. 1997. High-affinity NO_{3-}-H^{+} cotransport in the fungus *Neurospora* - induction and control by pH and membrane voltage. J. Membr. Biol. **160**: 59-76.

41 Walker JE. 1992. The mitochondrial transporter family. Current opinion in structural biology **2**: 519-526.

42 Palmieri F. 1994. Mitochondrial carrier proteins. FEBS Lett. **346**: 48-54.

43 Nicolay K, Veenhuis M, Douma AC, Harder W. 1987. A 31P NMR study of the internal pH of yeast peroxisomes. Arch. Microbiol. **147**: 37-41.

44 Douma AC, Veenhuis M, Sulter GJ, Harder W. 1987. A proton translocating adenosine triphosphatase is associated with the peroxisomal membrane of yeasts. Arch. Microbiol. **147**: 42-47.

45 Elgersma Y, Vanroermund CWT, Wanders RJA, Tabak, HF. 1995. Peroxisomal and mitochondrial carnithine acetyltransferases of *Saccharomyces cerevisiae* are encoded by a single gene. EMBO. J. **14**: 3472-3479.

46 Roels, JA 1983. Energetics and kinetics in biotechnology. Elsevier Biomedical, Amsterdam.

47 Andre B, 1995. An overview of membrane transport proteins in *Saccharomyces cerevisiae.* Yeast **11**:1575-1611.
48 Hinnebusch AG, Liebman SW, 1991. Protein synthesis and translational control in *Saccharomyces cerevisiae.* In: Broach JR, Pringle JR, Jones EW (eds.), The molecular biology of the yeast *Saccharomyces.* Cold Spring Harbor Laboratory Press, pp. 627-735.
49 Rigoulet M, Leverve X, Fontaine E, Ouhabi R, Guerin B, 1998. Quantitative analysis of some mechanisms affecting the yield of oxidative phosphorylation: Dependence upon both fluxes and forces. Mol.Cell.Biochem. **184**:35-52.
50 Kallow W, von Döhren H, Kleinkauf H, 1998. Penicillin biosynthesis - energy requirement for tripeptide precursor formation by delta-(l-alpha-aminoadipyl)-l-cysteinyl-d-valine synthetase from *Acremonium chrysogenum.* Biochemistry **37**:5947-5952.
51 Senior AE, 1988. ATP synthesis by oxidative phosphorylation. Physiol.Rev. **68** (1):177-231.
52 Zubay, G. 1988. Biochemistry, 2. ed.., Macmillan Publishing Company, New York.
53 vanGulik, WM, Antoniewicz, MR, deLaat, WTAM, Vinke, JL, Heijnen, JJ. (2001) Energetics of growth and penicillin production in a high producing strain of *Penicillium chrysogenum.* Biotechnol. Bioeng. **72**:185-193.
54 Light PA, Garland PB, 1971. A comparison of mitochondria from *Torulopsis utilis* grown in continuous culture with glycerol , iron, ammonium, magnesium or phosphate as the growth limiting nutrient. Biochem.J. **124**:123-134.
55 Aiking H, Sterkenburg A, Tempest DW, 1977. Influence of specific growth limitation and dilution rate on the phosphorylation efficiency and cytochrome content of mitochondria of *Candida utilis* NCYC 321. Arch.Microbiol. **113**:65-72.
56 Onishi, T., 1973. Mechanisms of electron transport and energy conversion in the site 1 region of the respiratory chain. Biochim. Biophys. Acta **301**:105-128.
57 Bonnet JABAF, deKok, HE, Roels, JA, 1980. The growth of *Saccharomyces cerevisiae* CBS 426 on mixtures of glucose and ethanol: a model. Antonie van Leeuwenhoek **46**:565-576.
58 Ferguson SJ, 1986. The ups and downs of P/O ratios (and the question of non-integral coupling stoichiometries for oxidative phosphorylation and related processes). TIBS **11**:351-353.
59 Henriksen CM, Nielsen J, Villadsen J, 1998. Modelling of the protonophoric uncoupling by phenoxyacetic acid of the plasma membrane potential of *Penicillium chrysogenum.* Biotechnol.Bioeng. **60**:761-766.
60 Mason HRS, Righelato RC, 1976. Energetics of fungal growth: the effect of growth-limiting substrate on respiration of *Penicillium chrysogenum.* J.Appl.Chem.Biotechnol. **26**:145-152.
61 Verduyn C, Postma E, Scheffers WA, Dijken JPv, 1990. Energetics of *Saccharomyces cerevisiae* in anaerobic glucose- limited chemostat cultures. J.Gen.Microbiol. **136**:405-412.

62 Righelato RC, Trinci AP, Pirt SJ, Peat A, 1968. The influence of maintenance energy and growth rate on the metabolic activity, morphology and conidation of *Penicillium chrysogenum.* J.Gen.Microbiol. **50**:399-412.

63 Christensen LH, Henriksen CM, Nielsen J, Villadsen J, Egel-Mitani M, 1995. Continuous cultivation of *Penicillium chrysogenum.* Growth on glucose and penicillin production. J.Biotechnol. **42**:95-107.

64 Wang NS, Stephanopoulos G. 1983. Application of macroscopic balances to the identification of gross measurement errors. Biotechnol. Bioeng. **25**: 2177-2208.

65 Kluge M, Siegmund D, Diekmann H, Thoma M. 1992. A model for penicillin production with and without temperature shift after the growth phase. Appl. Microbiol. Biotechnol. **36**: 446-451.

66 Wittler R, Schügerl K. 1985. Interrelation between penicillin productivity and growth rate. Appl. Microbiol. Biotechnol. **21**: 348-355.

67 Ryu DDY, Hospodka J. 1980. Quantitative physiology of *Penicillium chrysogenum* in penicillin fermentations. Biotechnol. Bioeng. **22**: 289-298.

68 Pirt SJ, Callow DS. 1960. Studies of the Growth of *Penicillium chrysogenum* in Continuous Flow Culture with Reference to Penicillin Production. J. Appl. Bacteriol. **23**: 87-98.

69 Pirt SJ, Righelato, RC. 1967. Effect of growth rate on the synthesis of penicillin by *Penicillium chrysogenum* in batch and chemostat cultures. Appl. Microbiol. **15**: 1284-1290.

70 Loftus TM, Hall LV, Anderson SL, McAlister-Henn L, 1994. Isolation, characterization, and disruption of the yeast gene encoding cytosolic NADP-specific isocitrate dehydrogenase. Biochemistry **33**:9661-9667.

71 Hult K, Veide A, Gatenbeck S, 1980. The distribution of the NADPH regenerating mannitol cycle among fungal species. Arch.Microbiol. **128**:253-255.

72 Postma E, Verduyn C, Scheffers WA, vanDijken JP (1989) Enzymic analysis of the Crabtree effect in glucose-limited chemostat cultures of *Saccharomyces cerevisiae.* Appl. Env. Microbiol. **53**:468-477.

73 Djavadi FHS, Moradi M, Djavadi-Ohaniance L, 1980. Direct oxidation of NADPH by submitochondrial particles from *Saccharomyces cerevisiae.* Eur.J.Biochem. **107**:501-504.

74 Bruinenberg PM, vanDijken JP, Kuenen JG, Scheffers WA, 1985. Oxidation of NADH and NADPH by mitochondria from the yeast *Candida utilis.* J.Gen.Microbiol. **131**:1043-1051.

75 Prodouz KN, Garret RH. 1981. *Neurospora crassa* NAD(P)H-nitrite-reductase. J. Biol. Chem. **256**: 9711-9717.

76 Sengupta S, Shaila MS, Rao GR. 1997. In vitro and in vivo regulation of assimilatory nitrite reductase from *Candida utilis.* Arch. Microbiol. **168**: 215-224.

77 Sonntag, K., Eggeling, L., De Graaf, A.A., Sahm, H. (1993) Flux Partitioning in the Split Pathway of Lysine Synthesis in *Corynebacterium glutamicum.* Eur. J. Biochem., **213**, 1325-1331.

78 Wiechert, W., de Graaf, A.A. (1997) Bidirectional reaction steps in metabolic networks: I. Modeling and Simulation of Carbon Isotope Labeling Experiments. Biotech. Bioeng., **55**,**1**, 101-117.

79 Marx, A., de Graaf, A.A., Wiechert, W., Eggeling, L., Sahm, H. (1996) Determination of the Fluxes in the Central Metabolism of *Corynebacterium glutamicum* by Nuclear Magnetic Resonance Spectroscopy Combined With Metabolic Balancing. Biotech. Bioeng., **49**, **2**, 111-129.

80 Marx, A., Eikmans, B.J., Sahm, H., de Graaf, A.A., Eggeling, L. (1999) Response of the Central Metabolism in *Corynebacterium glutamicum* to the use of an NADH-Dependent Glutamate Dehydrogenase. *Metab. Eng.*, **1**, **1**, 35-48.

81 Schmidt, K., Marx, A., de Graaf, A.A., Wiechert, W., Sahm, H.,Nielsen, J., Villadsen, J. (1998) ^{13}C Tracer Experiments and Metabolite Balancing for Metabolic Flux Analysis: Comparing Two Approaches. B*iotech. Bioeng.*, **58**, **2&3**, 254-257

82 Schmidt, K, Carlsen, M., Nielsen, J., Villadsen, J. (1997) Modelling isotopomer distributions in biochemical networks using isotopomer mapping matrices. *Biotech. Bioeng.*, **55**,**6**, 831-840.

83 Szyperski, T. (1998) 13C-NMR, MS and Metabolic Flux Balancing in Biotechnology Research. Quart. Rev. Biophys., **31**, **1**, 41-106.

84 van Winden, W.A., Verheijen, P.J.T., Heijnen, J.J. (2000a) Possible Pitfalls of Flux Calculations Based on ^{13}C-Labeling, Accepted for publication in Metab. Eng., October 2000

85 van Winden, W.A., Heijnen, J.J., Verheijen, P.J.T., Grievink, J. (2000b) A Priori Analysis of Metabolic Flux Identifiability from ^{13}C-Labeling Data, Submitted to Biotech. Bioeng., August 2000

86 van Winden, W.A., Schipper, D., Verheijen, P.J.T., Heijnen, J.J. (2000c) Innovations in the Generation and Analysis of 2D [^{13}C,^{1}H] COSY Spectra for Flux Analysis Purposes, Manuscript in preparation

87 Theilgaard HA, Nielsen J, 1999. Metabolic control analysis of the penicillin biosynthetic pathway: the influence of the LLD-ACV : bisACV ratio on the flux control. Antonie Van Leeuwenhoek International Journal Of General And Molecular Microbiology **75**:145-154.

88 Martin JE, Casqueiro J, Kosalkova K, Marcos AT, Gutierrez S, 1999. Penicillin and cephalosporin biosynthesis: Mechanism of carbon catabolite regulation of penicillin production. Antonie Van Leeuwenhoek International Journal Of General And Molecular Microbiology **75**:21-31

89 vanGulik, WM, Noorman, H, Vinke, JL, Heijnen, JJ (2001) Kinetics of penicillin-G production in *Penicillium chrysogenum.* Paper in preparation.

90 Heijnen, JJ. (2000) Unified kinetic and MCA based models in metabolic engineering. Paper presented at Metabolic Engineering III, Colorado Springs, USA.

91 Visser, D, Heijnen, JJ. (2001) Dynamic simulation and metabolic re-design of a branched pathway using LinLog kinetics. Paper in preparation.

MIX
Papier aus verantwortungsvollen Quellen
Paper from responsible sources
FSC® C105338

If you have any concerns about our products,
you can contact us on
ProductSafety@springernature.com

In case Publisher is established outside the EU,
the EU authorized representative is:
Springer Nature Customer Service Center GmbH
Europaplatz 3, 69115 Heidelberg, Germany

Printed by Libri Plureos GmbH
in Hamburg, Germany